BLOOD OXYGENATION

BLOOD OXYGENATION

Proceedings of the International Symposium on Blood Oxygenation, held at the University of Cincinnati, December 1-3, 1969

Edited by Daniel Hershey

Professor of Chemical Engineering
Department of Chemical and Nuclear Engineering
University of Cincinnati, Cincinnati, Ohio

PLENUM PRESS • NEW YORK – LONDON • 1970

Library of Congress Catalog Card Number 74-122626
ISBN 978-1-4684-1859-0 ISBN 978-1-4684-1857-6 (eBook)
DOI 10.1007/978-1-4684-1857-6

Softcover reprint of the hardcover 1st edition 1970

A Division of Plenum Publishing Corporation
227 West 17th Street, New York, N.Y. 10011

United Kingdom edition published by Plenum Press, London
A Division of Plenum Publishing Company, Ltd.
Donington House, 30 Norfolk Street, London W.C. 2, England

INTRODUCTION

Under the broad heading of blood oxygenation there may be specific areas of study, such as the kinetics of the oxygen-hemoglobin reaction, diffusion of gases through the red cell, blood preservation, blood chemistry, oxygen electrode design and the design and evaluation of artificial blood oxygenators. Blood oxygenation is of interest to many disciplines including physicians, chemists, physicists, biologists, physiologists and engineers.

The International Symposium on Blood Oxygenation was organized in order to bring together the people working in the various areas of blood oxygenation. This multidiscipline meeting was held at the University of Cincinnati on December 1, 2 and 3 of 1969. It was jointly sponsored by the U. S. Army Medical Research and Development Command and the University of Cincinnati. Participants came from Australia, England, Israel, Italy, Japan and the United States. There were 122 persons registered for the Symposium.

From the nature of the discussion during the meeting, it seemed apparent that the participants were benefiting from the contacts with colleagues in other disciplines. The result was a significant contribution to the present fund of knowledge of blood oxygenation and an enhancement of the future work.

The papers presented at the Symposium are contained in BLOOD OXYGENATION. This book should be a valuable reference source for those working in the field. It can also serve as a convenient starting point for those about to embark on work in the area of blood oxygenation. With a little imagination, an instructor may also be able to adapt the material to serve as a textbook.

Daniel Hershey

LIST OF CONTRIBUTORS

Eraldo Antonini
Institutes of Biological Chemistry of the Universities of Rome, Italy

H. I. Bicher
Department of Anatomy, Medical University of South Carolina, Charleston, South Carolina, USA

Kenneth B. Bischoff
University of Maryland, College Park, Maryland, USA

G. Blauer
The Hebrew University, Jerusalem, Israel

D. F. Bruley
Department of Chemical Engineering, Clemson University, Clemson, South Carolina, USA

M. J. Buckley
Massachusetts General Hospital, Boston, Massachusetts, USA

Jerry M. Calkins
Harry Diamond Laboratory, Washington, D. C., USA

R. Ben Dawson
United States Army Medical Research Laboratory, Fort Knox, Kentucky, USA

R. P. deFilippi
Abcor Incorporated, Cambridge, Massachusetts, USA

Leopold Dintenfass
Sydney Hospital, Department of Medicine, University of Sidney, Sydney, Australia

William J. Dorson, Jr.
Chemical Engineering Faculty, Arizona State University, Tempe, Arizona, USA

Marvin Fleischman
University of Louisville, Louisville, Kentucky, USA

Robert Franco
Department of Chemical and Nuclear Engineering, University of Cincinnati, Cincinnati, Ohio, USA

J. L. Gainer
Department of Chemical Engineering, University of Virginia, Charlottesville, Virginia, USA

Eric J. Guilbeau
Department of Chemical Engineering, Louisiana Polytechnic Institute, Ruston, Louisiana, USA

K. R. Hall
Department of Chemical Engineering, University of Virginia, Charlottesville, Virginia, USA

G. W. Harris
Abcor Incorporated, Cambridge, Massachusetts, USA

Daniel Hershey
Department of Chemical and Nuclear Engineering, University of Cincinnati, Cincinnati, Ohio, USA

Robert A. B. Holland
School of Physiology, University of New South Wales, Kensington, New South Wales, Australia

Wolfgang H. Koch
Department of Chemical and Nuclear Engineering, University of Cincinnati, Cincinnati, Ohio, USA

Theodor Kolobow
National Heart and Lung Institute, National Institutes of Health, Bethesda, Maryland, USA

M. H. Knisely
Department of Anatomy, Medical University of South Carolina, Charleston, South Carolina, USA

Clifford E. Lanham
Harry Diamond Laboratory, Washington, D. C., USA

Ronald C. Menke
Department of Chemical and Nuclear Engineering, University of Cincinnati, Cincinnati, Ohio, USA

Masaji Mochizuki
Research Institute of Applied Electricity, Hokkaido University, Sapporo, Japan

R. M. Navari
Department of Chemical Engineering, University of Virginia, Charlottesville, Virginia, USA

J. H. Porter
Abcor, Incorporated, Cambridge, Massachusetts, USA

Thomas M. Regan
University of Maryland, College Park, Maryland, USA

D. D. Reneau
Department of Chemical Engineering, Louisiana Polytechnic Institute, Ruston, Louisiana, USA

Robert L. Sigman
National Heart and Lung Institute, National Institutes of Health, Bethesda, Maryland, USA

John A. Sirs
Department of Physics, St. Mary's Hospital Medical School, University of London, London, England

Edmund E. Spaeth
Department of Chemical Engineering, Washington University, St. Louis, Missouri, USA

F. C. Tompkins
Abcor, Incorporated, Cambridge, Massachusetts, USA

Fernando Villarroel
Harry Diamond Laboratory, Washington, D. C., USA

Warren M. Zapol
National Heart and Lung Institute, National Institutes of Health, Bethesda, Maryland, USA

CONTENTS

FACTORS DETERMINING THE VELOCITY OF GAS UPTAKE BY INTRACELLULAR HEMOGLOBIN

Robert A. B. Holland, Assoc. Prof. of Physiology

School of Physiology, University of New South Wales

Kensington, New South Wales, Australia

This paper will consider some of the factors which affect the rate at which red cells can take up oxygen and carbon monoxide. A considerable part of it will be devoted to carbon monoxide and there are several reasons for this. Carbon monoxide combines with hemoglobin at the same site as does oxygen, both ligands attaching to the iron atom of the porphyrin group. The slow rate at which CO dissociates from carboxyhemoglobin causes it to have an affinity for hemoglobin approximately two hundred times as great as the affinity of oxygen. This enables CO in small concentrations to displace oxygen from hemoglobin thus reducing the oxygen carrying capacity of the blood, and is largely responsible for the toxicity of carbon monoxide. The tight binding of CO to hemoglobin has made it of great interest to physiologists and it has been widely used experimentally both as a means of producing hypoxemia and for measuring lung diffusing capacity. Another useful property of CO is that it combines with hemoglobin less rapidly than does oxygen. This enables one to measure some of the CO reaction rates more easily and accurately than the corresponding oxygen reaction rates. From these measurements we can draw general conclusions on the factors affecting gas uptake by the cells.

The original methods for measuring the rapid reactions of hemoglobin were published in 1923 by Hartridge and Roughton and much of our present knowledge is due to the work of Roughton and co-workers. The subject of gas uptake rates by red cells has been reviewed in the last ten years by Roughton (29) and by Forster (7) and so the present paper will concentrate on more recent work and draw attention to certain unsolved problems.

It is divided into sections as follows:

SYMBOLS AND ABBREVIATIONS

The symbols used in the text and equations are standard (8, 29) but for convenience are restated here.

The velocity constants for the four successive stages of oxygenation of the hemoglobin tetramer are given by the symbols k'_1, k'_2, k'_3, k'_4, and the constants for the deoxygenation of the tetramer by k_4, k_3, k_2, and k_1. Throughout the use of the prime (') indicates an association velocity constant and its absence indicates a dissociation velocity constant. For the reactions of CO with hemoglobin ℓ is used instead of k. In cases where the overall rate of gas combination with reduced hemoglobin is measured, the symbol k' or ℓ' is used without a subscript.

To describe the rate at which CO replaces oxygen from the fully saturated tetramer, Roughton (1945) introduced the symbol m' defined by the equation

$$\frac{d\,[COHb]}{dt} = m' \frac{[CO][O_2Hb]}{[O_2]} \tag{1}$$

The value of m' is not constant but varies with the ratio $[CO]/[O_2]$ and thus is referred to as a pseudoconstant rather than as a constant. Its relation to the more fundamental constants is expressed by the equation

$$m' = \frac{k_4 \ell_4}{4 k'_4 \left\{ 1 + \frac{\ell'_4 [CO]}{k'_4 [O_2]} \right\}} \tag{2}$$

Roughton, Forster, and Cander (27) introduced the term m'_∞ to describe the rate of the replacement reaction when the ratio $[CO]/[O_2]$ is negligible, for example when lung diffusing capacity is measured with low concentrations of CO. From equation (2) it follows that

$$m'_\infty = \frac{k_4 \ell'_4}{4 k'_4} \tag{3}$$

Symbols such as ℓ', k'_4 without the subscript "c" refer to reactions of hemoglobin in solutions. In describing the velocities of uptake of oxygen and CO by reduced red cells the symbols k'_c and ℓ'_c are used and m'_c is used for the rate of CO replacing O_2 in cells. The definition of k'_c is

$$\frac{d[O_2Hb]}{dt} = k'_c [O_2][Hb] \tag{4}$$

$[O_2]$ being the concentration of oxygen in the fluid immediately surrounding the red cells, and the reverse reaction again being neglected. Similar equations apply for ℓ'_c and m'_c.

The use of expressions for the apparent velocity constants for gas uptake by cells has certain limitations. The rate must be measured as soon after the reactants are mixed as possible. This is because the equations of Roughton describing the combined processes of diffusion and chemical reaction into the cells are not valid except at zero time (29). The reverse reaction must also be neglected in order to solve the equations; the error introduced by neglecting it is minimized by measuring the rate at zero time and, in the case of CO, by the inherent slowness of the dissociation reaction.

All the combination velocity constants referred to in this paper are second order and are expressed in units of "liters per millimole per second" which is equivalent to "per millimolar per sec" abbreviated as $mM^{-1}\ sec^{-1}$. The dissociation velocity constants, and m', m'_∞ and m'_c are first order and are expressed in units of sec^{-1}.

MODEL OF THE RED CELL

The factors affecting red cell reaction rates have usually been analyzed using one of Roughton's models of the red cell. It has not so far been possible to consider the red cell as a biconcave disk (its form in mammals) but equations are available for the uptake rate of gases by a sphere of equivalent volume (26), by a membrane-less layer of hemoglobin (25) and by a layer of hemoglobin bounded by a membrane whose permeability to gases can take different values. Of these, the last mentioned, described by Nicolson and Roughton (22) is the most useful.

In this model the cell is regarded as being a sheet of infinite extent, and of thickness equal to the average half-thickness of the red cell. The gas is regarded as entering through the membrane from one side only. The thickness of the layer is expressed by the symbol b (in cm) and the concentration of hemoglobin is taken as being that in the red cell, normally approximately 20 mM (hemoglobin millimolarity throughout is stated with reference to concentration of the monomer, this being the unit which combines with one molecule of gas). At 37°C the diffusion coefficient of oxygen in this concentration of hemoglobin is 7.1×10^{-6} cm^2 sec^{-1} and for CO the value is 7.6×10^{-6} cm^2 sec^{-1} (15, 19, 29). The diffusion coefficient (D) is taken as having a temperature coefficient of 2.5% per degree centigrade.

The permeability of the red cell membrane to the gas is expressed as λ where λ is the ratio of the permeability of the membrane to the permeability of the cell interior. It can be expressed in terms of the diffusion coefficients as follows

$$\lambda = \frac{D_2}{b_2} \Big/ \frac{D}{b} = \frac{D_2\, b}{D\, b_2} \tag{5}$$

where D_2 is the diffusion coefficient of the gas in the red cell membrane, b_2 is the thickness of the red cell membrane, D and b are as defined above. A high value for λ indicates high membrane permeability.

The equations of Nicolson and Roughton predict two slowing factors which cause the rate at which reduced cells take up the gas to be less than the rate at which hemoglobin solution combines with the gas. To express these factors a term ω is introduced, ω being defined by the equation

$$\omega = b\sqrt{\frac{k'[Hb]}{D}} \tag{6}$$

where [Hb] is the initial concentration of reduced hemoglobin inside the red cells. For the CO reaction ℓ' is substituted for k' in equation 6.

The first slowing factor is that attributable to the process of simultaneous diffusion and chemical reaction in the substance of the red cell. It predicts that the cell reaction will be slowed, when compared to the solution reaction, by a factor of $(\tanh \omega)/\omega$ so that in a hemoblogin layer without a membrane

$$\frac{k_c'}{k'} = \frac{\tanh \omega}{\omega} \quad (7)$$

The second slowing is caused by a barrier to gas diffusion at the cell membrane and slows the reaction by a further factor of $1/[1 + (\omega/\lambda) \tanh \omega]$. Thus the combined equation relating cell and solution rates is

$$\frac{k_c'}{k'} = \frac{\tanh \omega}{\omega} \cdot \frac{1}{1 + \frac{\omega}{\lambda} \tanh \omega} \quad (8)$$

Similar equations describe the slowing of the reaction with carbon monoxide.

The use of these equations to analyze some recent data is described in the subsequent sections.

THE IMPORTANCE OF RED CELL SIZE

In applying equation 8 to the analysis of the reaction rate a value has to be inserted for b, the mean half-thickness of the red cell. Mammalian erythrocytes are normally non-nucleated biconcave circular discs, except for those of camel and llama which are ovoid rather than circular. Thus there is no obvious value to use for half-thickness. Roughton (25) took the value of 0.7 microns in the case of the sheep; this was a compromise between half-thicknesses at the thinnest and thickest parts and made some allowance for gas entering at the disc edge. This value has been used for half-thickness of the sheep cell in other publications (22, 31) and also for the half thickness of the human red cell (6). However, the mean corpuscular volume (MCV) of the erythrocytes in these two species differs considerably. Sheep MCV is 33 cubic microns and human MCV is 90 cu. microns. Of the common mammals the goat has the smallest MCV, 20 cu. microns, while high values are found in lower forms, particularly in amphibia (1).

It was therefore of interest to investigate whether the size of the cells affected the velocity of gas uptake. In the case of oxygen uptake by reduced cells there was a considerable effect. Holland and Forster (11) found k'_c for the goat to be more than double the value found for k'_c in the dog and in the rabbit (MCV of each approximately 60 cu. microns) and in general found a fair

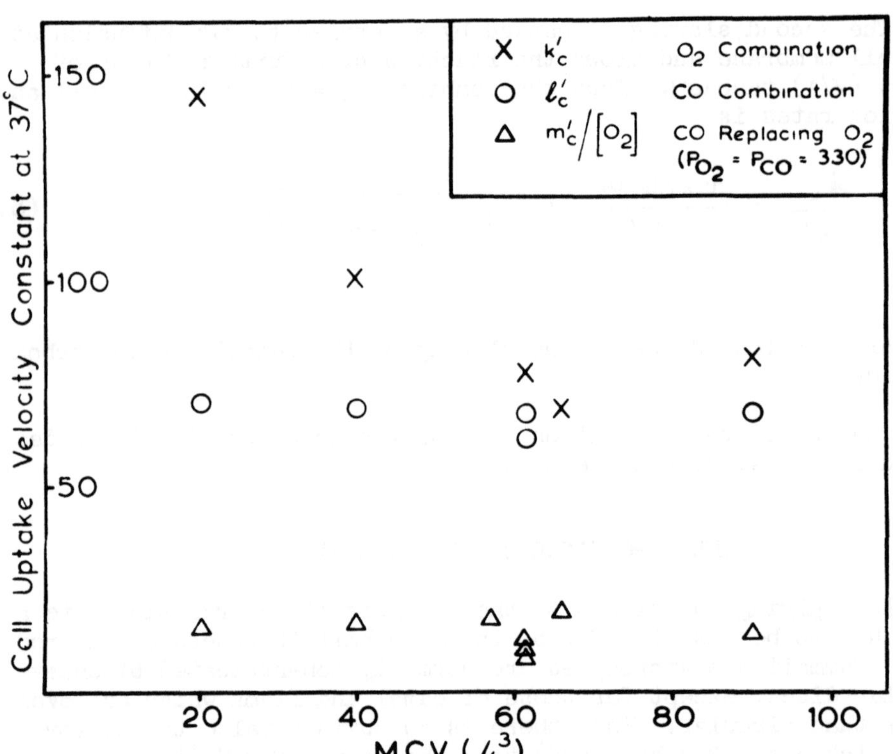

Fig. 1 The effect of cell size on the cell uptake rate constant or pseudoconstant at 37°C. The effect of cell size is greatest for the most rapid reactions. The points for the replacement reaction pseudoconstant σ' cell ($= m'_{cell}/[O_2]$) are from Holland (14) and for the uptake constants for the CO and O_2 reactions the points are from references 10 and 11 respectively. The species represented in the figure are, from left to right; goat, horse, cat, rabbit, dog and adult man. Mean values are shown except in the rabbit where there was considerable intraspecies variation.

relationship between smallness of red cell and rapidity of oxygen uptake (Fig 1). The value of k'_c was also determined for the bullfrog whose erythrocytes are nucleated, ovoid, and biconvex with a value of 680 cu. microns for MCV. Bullfrog k'_c is not shown in Fig 1 but was 19 mM^{-1} sec^{-1}, which is about one-eighth of the value in the goat and one-quarter of the value in man.

We had to consider two possible explanations other than cell size to account for the interspecies differences in oxygen uptake rate. The first was the possibility that the species with smaller cells also had hemoglobin molecules which reacted faster with oxygen. However, this reaction rate in solution was approximately the same in all species where we measured it. Also from equation 8, for a reaction as rapid as this, the cell rate is rather insensitive to changes in the solution rate. Therefore, despite the difficulty in accurately measuring the rate of a reaction as rapid as the combination of oxygen with reduced hemoglobin solution, we felt we had excluded differences in the hemoglobin molecule as a cause of the observed differences in k'_c.

The second possible explanation was that the cell membrane permeability (λ) might be higher in the species with the smaller cells. There is no experimental way of excluding this; in equation 8 the value of b and λ are interdependent, and unless one is known with accuracy, it is impossible to know the other. However, it seems inherently unlikely that membrane permeability would vary inversely with cell size. We therefore consider that the variations found in k'_c represent a cell size effect although the other factors mentioned may well play a part any may be responsible for the irregularities in the graph of k'_c against MCV.

We are not able to say what value should be taken for cell half-thickness. Ponder (23) gives values for the greatest and least thicknesses of the erythrocytes of man, rabbit, and sheep, showing the sheep (MCV = 33 cu microns) to have thicker cells than the rabbit (MCV = 61 cu microns). Holland and Forster (11) found k'_c to be higher in the sheep (137 mM^{-1} sec^{-1}) than in the rabbit (77 mM^{-1} sec^{-1}). Clearly the value taken for b in equation 8 can be at best an approximation giving some expression to the importance of cell size.

When one measures the relative rates of slower reactions one finds less cell size effect (Fig 1). The reaction between CO and reduced hemoglobin solution at 37°C is about one-eighth as rapid as the oxygen reaction. The cell reaction rates are not slowed in proportion but the cell size effect was found to be much less (10). When the much slower two stage reaction of CO replacing oxygen from oxyhemoglobin was studied (14), size effects were found to be minimal (Fig 1, 2).

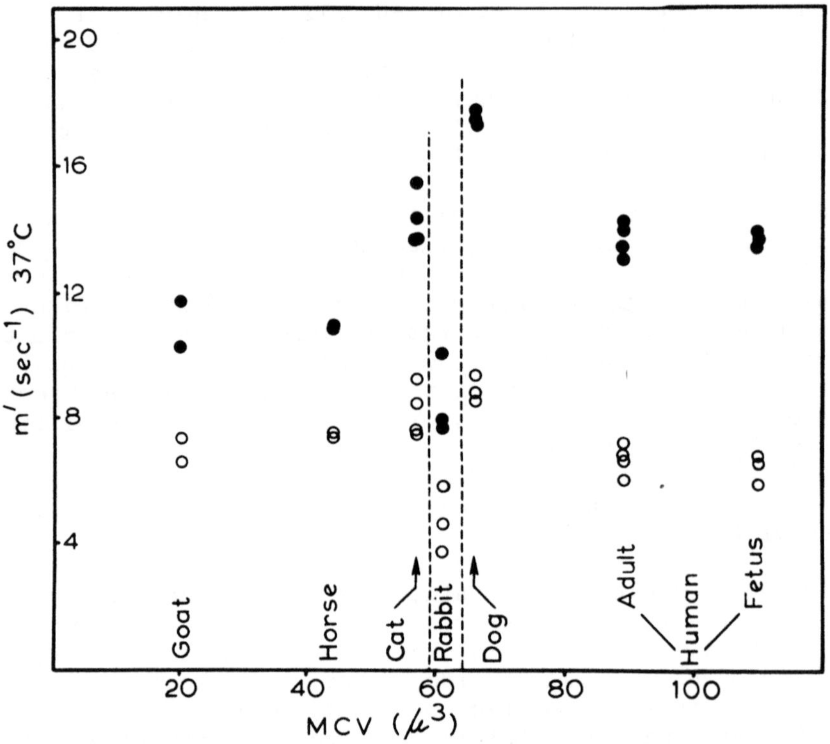

Fig. 2 Pseudoconstants for cell reaction (open circles) and solution reaction (closed circles) at 37°C plotted against mean corpuscular volume for different mammalian red cells. For each point the ratio $[O_2]/[CO]$ immediately after mixing in the apparatus was 1.32. Most intraspecies variation was shown by the rabbit.

Thus, for very rapid reactions, cell size is a factor in determining cell gas uptake rate. Its importance has not been measured for reactions faster than the reaction between O_2 and reduced hemoglobin (two faster reactions are: (I) $Hb + NO \rightarrow HbNO$; and (II) $Hb_4X_3 + O_2 \rightarrow Hb_4X_3O_2$, where Hb_4X_3 is any three-parts saturated tetramer) but it is a reasonable extrapolation of our findings to infer that cell size will be of importance in these reactions, too.

THE RED CELL MEMBRANE AS A BARRIER TO GAS DIFFUSION

There are at present conflicting views as to whether the red cell membrane presents an important barrier to diffusion of gases into the cells. Much of the evidence has been discussed by Forster (7). Attention was drawn in the previous section of the present paper to the interdependence of b and λ when Roughton's equations are applied to the cell reaction rates. The effect of the value taken for effective cell half-thickness on the value found for λ is shown in Table 1. When the reaction of either CO or O_2 with intracellular reduced hemoglobin is measured by the stopped-flow apparatus, and a value of 0.7 microns taken for human cell half-thickness, the value of λ is found to be about 1.5. If half-thickness is taken as 1.2 microns, λ is found to be either 8.0 or 4.9. Since 2.3 microns is stated by Ponder to be the maximum thickness of the human red cell (23), these would appear to be the maximum values for λ. All values shown for λ in the first two columns represent a considerable barrier to gas uptake. This is most easily demonstrated with respect to the oxygen reaction. For the values shown in table I, $\omega = 5.9$ when b = 0.7 microns and ω is higher for the other values of b. Thus $\tanh \omega$ may be taken as unity. Equation 7 reduces to

$$\frac{k_c'}{k'} = \frac{1}{\omega} \qquad (9)$$

for the case where the membrane resistance is zero, and equation 8 for the case where there is membrane resistance reduces to

$$\frac{k_c'}{k'} = \frac{1}{\omega} \cdot \frac{1}{1 + \frac{\omega}{\lambda}} \qquad (10)$$

If b is taken as 0.7 microns, $\omega = 5.9$, and equation 9 predicts a slowing (by a factor of 5.9) to a value for k'_c of 425 compared to the observed value of 81 mM^{-1} sec^{-1}. The further slowing to the observed value (by a factor of 5.2) is caused by the membrane's having finite resistance. In the case where b is taken as 1.2 microns, $\omega = 10$, and a slowing by a factor of 10 is predicted by equation 9; a further slowing by a factor of 3 is due to membrane factors.

Similar results are found if λ and membrane slowing are calculated for the results of Forster et al. (6) on human blood. Their measurements were made with the continuous flow apparatus using optical methods of following the change from Hb to COHb or to O_2Hb. Holland (10, 12) and Holland and Forster (11) used a modification

TABLE I

Change in value found for λ as value taken for cell half-thickness is varied. Effect of method used for k'_c measurement is also shown.

b (microns)	CO + Reduced Hb Stopped-flow Optical detection	O_2 + Reduced Hb Stopped-flow Optical detection	O_2 + Reduced Hb Continuous Flow P_{O_2} measurement
	$\ell' = 363$	$k' = 2500$	$k' = 2500$
	$\ell'_c = 67.4$	$k'_c = 81$	$k'_c = 165$
0.7	1.5	1.4	3.7
0.8	2.1	1.9	5.4
0.9	3.0	2.4	7.5
1.0	4.2	3.1	10.8
1.1	5.8	3.9	14.3
1.2	8.0	4.9	19.7

Calculations are for human cells at 37°C with initial [Hb] = 20 mM. D in the red cell was taken as 7.6×10^{-6} cm^2 sec^{-1} for CO and as 7.1×10^{-6} cm^2 sec^{-1} for O_2. For the CO + Hb reaction, the results of Holland (10) for ℓ' and ℓ'_c were used. For the O_2 + Hb reactions, k' was taken as 2.5×10^3 mM^{-1} sec^{-1}, this being consistent with Gibson et al. (8) and with Holland and Forster (11). The value of 81 for k'_c is from Holland and Forster (11) and the value of 165 for k'_c is from Staub, Bishop and Forster (32). All values of reaction constants are in units of mM^{-1} sec^{-1}.

of the stopped flow apparatus of Sirs and Roughton (30) which also follows the reaction optically. These papers reported significant membrane barriers to gas diffusion in the human adult and fetus, and in the goat, horse, rabbit, and dog. Sirs (31) also showed a significant barrier in the sheep. Thus, the membrane has appeared to decrease the rate of gas exchange when gas uptake rates have been measured by optical methods.

A value of λ as high as 8.0 still implies a decreased diffusion coefficient of gas in the membrane. The thickness of the cell membrane is given by Davson (5) as approximately 50 Å. If this value is used in equation 5 for the case where cell half-thickness is taken as 1.2 microns, the diffusion coefficient of CO in the membrane is found to be only one-thirtieth of its diffusion coefficient in the cell interior.

If the results in the first two columns are examined using Roughton's equation for gas uptake by a membrane-bounded sphere (29) we still find a significant slowing due to the cell membrane. As a sphere has the lowest possible surface/volume ratio, the membrane resistance found cannot be merely an artefact caused by possible defects in the sheet model of the cell.

A lower membrane resistance is suggested by measurements of k'_c with the oxygen electrode. Staub, Bishop, and Forster (32) measured k'_c using the rapid continuous flow apparatus with an oxygen electrode located at varying distances downstream, thus "seeing" the reaction at varying times after mixing of the reactants. As shown in Table 1 they found k'_c for reduced cells to be much higher than had been found by optical methods using continuous flow (6) or stopped flow (11). An implication of this was that Staub et al. found higher values for λ. Lawson (18) has pointed out another instance where use of the oxygen electrode in the continuous flow apparatus gives higher figures for reaction rates than are obtained with optical methods. Nakamura and Staub (21) used this system to measure the velocity of dissociation of oxygen from oxyhemoglobin in red cells. They found rates approximately three times as great as found by Lawson (18) and by Lawson, Holland, and Forster (17) and Lawson suggested that the presence of a stagnant layer of fluid on the surface of the electrode was responsible for falsely high apparent reaction rates.

Kreuzer and Yahr (16) compared the rates of diffusion of oxygen into thin films of: (a) packed red cells, and (b) concentrated hemoglobin solution. They found that the rates were approximately the same and inferred that the cell membrane offered little resistance to gas passage. They also suggested that in the experiments of Forster, et al. (6) many of the red cells may have been crenated giving falsely low values for ℓ'_c and k'_c and hence for λ. However in the experiments where the present author has measured ℓ'_c or k'_c (10, 11, 12) the cells were examined for crenation and only in the dog was it visible significantly. We cannot at present explain why the value of λ differs when determined by these different methods.

There are several ways in which the interpretation of the kinetic data may be questioned. We may examine these objections to see if they might invalidate the finding of resistance to gas diffusion at the cell membrane.

(i) The model may take insufficient account of the shape of the cell. This possibility does not alter the general finding of membrane resistance to gas uptake. As outlined above, significant membrane resistance is found even when the model taken has minimum surface/volume ratio (sphere) or has a thickness clearly greater than the mean half-thickness of the red cell. These findings apply to all species studied.

(ii) The model does not allow for possible diffusion by hemoglobin molecules. The effect of this on the reaction rate has recently been studied theoretically by Moll (20) who found that it did not affect the early stage of the reaction. It is in this early part that we measure k'_c. Also any allowance for acceleration of gas uptake by hemoglobin diffusion would cause an increase in the predicted value of k'_c and require us to postulate even greater resistance at the cell membrane.

(iii) The reaction rates measured in dilute hemoglobin solution may not indicate the rate at which the hemoglobin molecules react inside the red cells. Roughton (28) has summarized the evidence for regarding the reactions in solution and in the cells as being essentially similar. However, more recently the finding by Chanutin and Curnish (4) and by Benesch and Benesch (3) that 2, 3-diphosphoglycerate (DPG) alters the oxygen equilibrium of red cells has reopened the question. DPG is found in high concentration in red cells of many species and causes a shift to the right of the O_2Hb equilibrium curve. It is likely that the binding of DPG to hemoglobin is the reason why the equilibrium curve of whole blood lies to the right of that for hemoglobin solution. The effect of DPG on the reaction rates has not been measured but the kinetic explanation of these equilibrium findings could be that DPG causes an increase in velocity of the dissociation reaction, a decrease in velocity of the combination reaction, or both. If the effect of DPG were to cause a slowing of the combination of hemoglobin with ligands, this could account for some of the slowing of k'_c and ℓ'_c relative to k' and ℓ' and less resistance to gas uptake would then appear to be located in the membrane.

There are, however, two reasons for discounting the possibility that the observed difference between cell and solution reaction rates is largely due to DPG. The first is that considerable cell membrane resistance to gas passage has been found in the cells of the sheep (11, 22, 30) and goat (11, 12). These species have very little DPG or other organic phosphate in their red cells (24). Secondly, Bauer et al. (2) found that DPG has less effect on the O_2Hb equilibrium curve in fetal blood than in maternal blood. Despite the differing DPG effects, Holland (12) found λ for fetal blood to be the same as he found λ for adult blood (see Table 1).

The red cell membrane thus has been shown to decrease the rate of gas uptake significantly whenever its effects have been measured using Roughton's equations. The use of the color change appears the most valid way to follow the reactions and on present evidence it appears that the membrane does present a genuine barrier to gas entry.

THE EFFECT OF SOLUTION REACTION RATE ON CELL RATE

The rate of gas uptake by cells is always found to be lower than the rate of combination of the gas with hemoglobin solution. The slowing rate for reactions of different velocities is shown by Roughton (29) and the general finding is that the faster the reaction in solution the more it is found to be slowed when the cell rate is measured. This is consistent with predictions from eqs 7 and 8 because a high solution rate raises the value of ω. The value of tanh ω does not rise proportionately with ω and for ω values above about 2.5 tanh ω is very close to its limiting value of unity. The results of the present author are essentially confirmatory of Roughton's findings. Of the results shown in Fig. 1, the slowing effect of the cell became less as the reaction rate slowed. The ratio of cell rate to solution rate was approximately 1:30 for the combination of O_2 and reduced Hb, approximately 1:5 for the combination of CO and reduced Hb, and about 1:2 for the reaction where CO replaced O_2 from oxyhemoglobin under conditions where the tensions of the gases were the same.

If one uses eq 7 to describe the cell uptake rate, thus ignoring any resistance at the cell membrane, one predicts that the cell reaction rate will continue to rise as the solution rate is increased. Eq 7 rearranges to

$$k_c' = \frac{1}{b}\sqrt{\frac{k'D}{[Hb]}} \tanh \omega \tag{11}$$

which predicts that the cell rate will be proportional at least to the square root of the solution rate even when the rate is so rapid than tanh $\omega = 1$. This does not adequately explain the results in Table 1. In the goat, the reaction rate in the cells continues to rise as the solution rate becomes faster. However, in the larger human cells k'_c is very little larger than ℓ'_c despite the fact that the ratio $k' : \ell'$ is 7 : 1.

A rearrangement of equation 8 may be made to show that the model which allows for the cell membrane resistance is consistent with the low value of human k'_c. Equation 8 rearranges to

$$k_c = \frac{1}{b\sqrt{\frac{[Hb]}{D}}\left\{\frac{1}{\sqrt{k'}\tanh\omega} + \frac{b}{\lambda}\sqrt{\frac{[Hb]}{D}}\right\}} \qquad (12)$$

In eq. 12, the first term inside the brackets in the denominator is the only term in which k' appears. As k' increases this term decreases and the predicted value for k'_c increases. However, if this term is relatively small compared with the second term inside the brackets, as it is when k' is large, further increase in k' can have little effect on k'_c. In the goat b/λ is smaller than in man (10, 11). Thus the term in k' is relatively more important in the goat and therefore the solution rate can rise higher in the goat before reaching the level where further increases in solution rate are ineffective in raising the cell rate.

Further interaction of cell and solution rates may be seen when the temperature is changed but as temperature change also affects the diffusion coefficient, the effect of temperature is considered in a separate section.

THE EFFECT OF TEMPERATURE ON CELL REACTION RATE

The effect of an increase in temperature is to cause an increase in the rate at which gas is taken up into the cells. From equation 10, this could be caused by an increase in solution rate, in diffusion coefficient of the gas, or in cell membrane permeability.

The most suitable reaction for studying the effect of temperature is the uptake of CO by reduced cells. The reaction of oxygen with reduced hemoglobin is not temperature sensitive (9) and the replacement of oxygen by CO is a more complex process which involves the diffusion of two gases through the cell.

The velocity constants for the reaction of reduced hemoglobin with CO show normal temperature effects when plotted against the reciprocal of Absolute temperature (Fig. 3). From this plot the energy of activation is obtained by the Arrhenius equation:

$$\frac{d(\ln l')}{d(10^3/T)} = -\frac{E}{R} \qquad (13)$$

where E is the energy of activation and R is the gas constant, 1.987 calories $mole^{-1}$ $degree^{-1}$. As the abscissa in Fig. 3 is in units of $10^3/T$, E is in units of kcal $mole^{-1}$. There is some interspecies variation but in all cases the energy of activation is in the region of 9 kcal $mole^{-1}$.

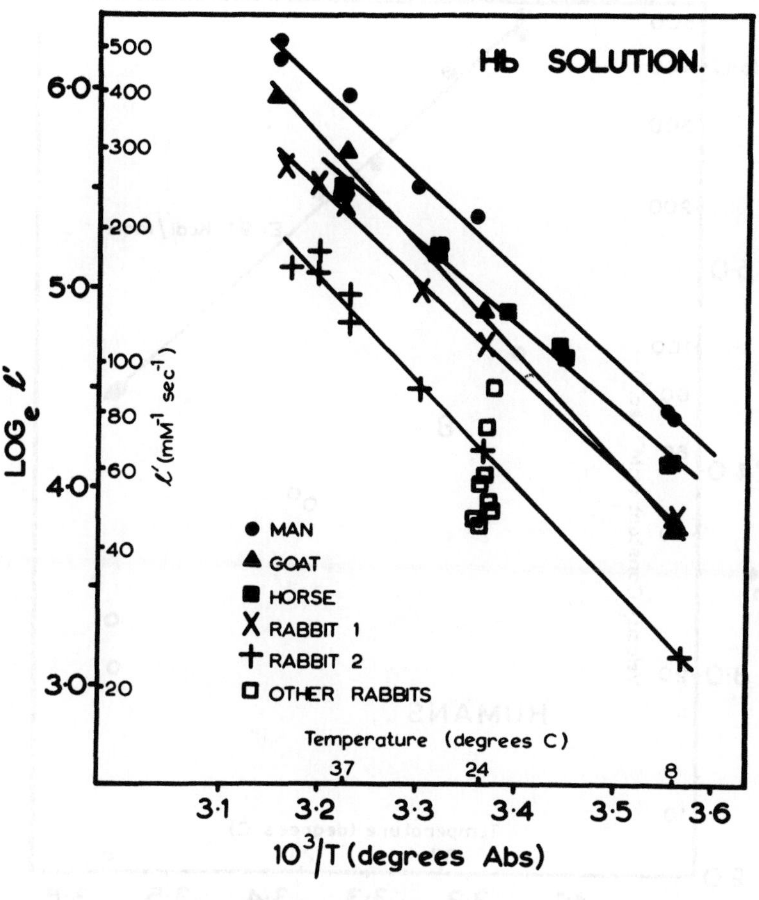

Figure 3. Arrhenius plots of the velocity constants for the reaction rate of CO with reduced hemoglobin solution in four species. The rabbit was the only species studied to show intraspecies variation. Each open rectangle represents a different rabbit studied at room temperature only.

The temperature coefficient of gas diffusion through concentrated hemoglobin solutions has not been measured directly but Roughton applies a factor of 2.5% per centigrade (29). The energy of activation of many diffusion processes is given by Taylor as 3.5 kcal $mole^{-1}$ which corresponds to Roughton's factor in the region

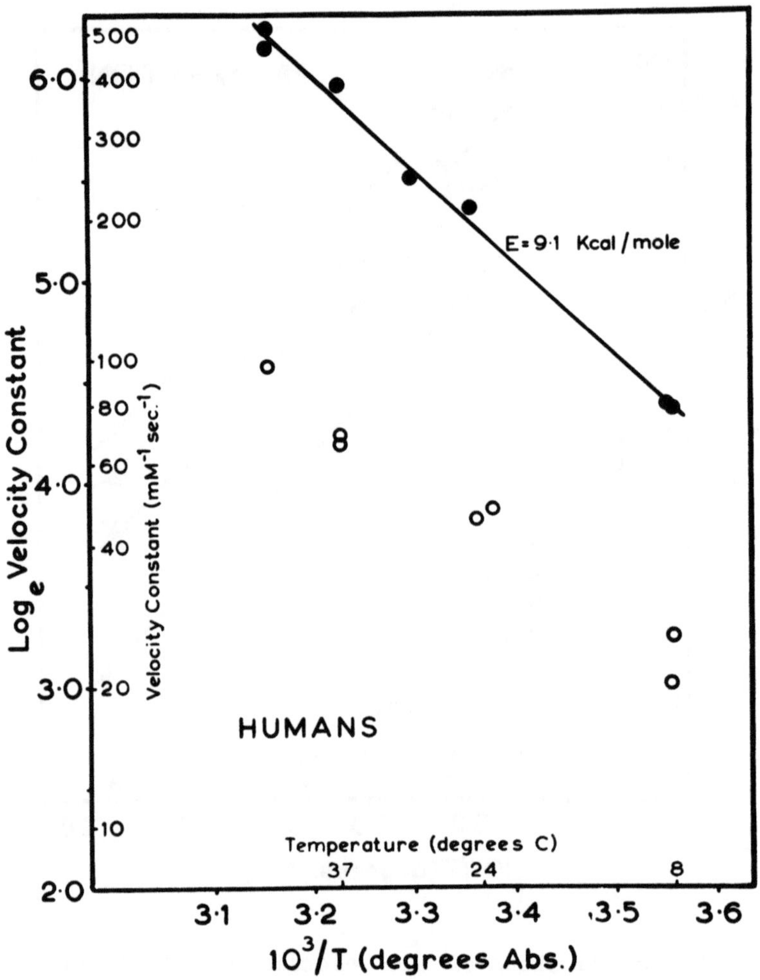

Figure 4. Arrhenius plot of the velocity constants for the uptake of CO by reduced human hemoglobin in solution (closed circles) and in the red cells (open circles).

of body temperature. In the absence of direct measurements, the effect of temperature on the gas diffusion is taken in this paper as corresponding to the energy of activation given by Taylor.

The comparison of temperature effect on cell and solution reaction rates is shown for man in Fig. 4 taken from reference 10. The cell reaction is less affected by temperature than the solution reaction. The scatter of points is greater in the case of the reaction but an apparent energy of activation was calculated for the cell rate and was 6.8 kcal $mole^{-1}$ compared with 9.1 kcal $mole^{-1}$ for the solution reaction. This fitted well with the value which could be predicted from the equations of Roughton. Holland (10) by differentiation of equation 7 derived the equation

$$\frac{d(\ln l'_c)}{d(1/T)} = \frac{1}{2}\left[\left\{\frac{E_{Hb}}{R} + \frac{E_{DIFF}}{R}\right\} + \frac{w}{\sinh w + \cosh w} \cdot \left\{\frac{E_{Hb}}{R} - \frac{E_{DIFF}}{R}\right\}\right] \quad (14)$$

which is equivalent to

$$\frac{d(\ln l'_c)}{d(1/T)} = -\frac{1}{2}\left[\left\{\frac{E_{Hb}}{R} + \frac{E_{DIFF}}{R}\right\} + \frac{2w}{\sinh 2w}\left\{\frac{E_{Hb}}{R} - \frac{E_{DIFF}}{R}\right\}\right] \quad (15)$$

E_{Hb} is the activation energy of the combination of CO with reduced Hb and E_{DIFF} is the activation energy of the diffusion process (taken as 3.5 kcal $mole^{-1}$). In the limiting cases of equation 15, as $w \to 0$, $(2w/\sinh 2w) \to 1$ and $d(\ln l'_c)/d(1/T) \to -E/R$ which is the equation for hemoglobin solution. This is to be expected intuitively because w becomes large when the features which make cell suspensions differ from hemoglobin solution are accentuated. These features are a large cell size (b), a high concentration of reduced hemoglobin inside the cell ([Hb]) and a high ratio of reaction rate to gas diffusion coefficient in the cell (k'/D). As w increases, $2w/\sinh 2w$ approaches zero. For $w = 2$, $2w/\sinh 2w < 0.008$; and for $w = 2.5$, $2w/\sinh 2w < 0.04$. Thus as w increases

$$\frac{d(\ln l'_c)}{d(1/T)} \longrightarrow -\frac{1}{2}\left\{\frac{E_{Hb}}{R} + \frac{E_{DIFF}}{R}\right\} \quad (16)$$

the apparent activation of the cell reaction then being the mean of the activation energies of the reaction and diffusion processes. For human reduced cells taking up CO at 37°C, w = 2.5, its exact value depending on the value taken for b. The value found for E_{cells} was 6.8 kcal mole^{-1} and the value prediced by eq. 16 was 6.4 kcal mole^{-1}, in good agreement with the observed figure.

Equations 14 - 16 take no account of resistance at the cell membrane. If the membrane resistance does not change with temperature the predicted value for E_{cells} will be less. In view of the uncertainties in cell size and membrane permeability, a semiquantitative treatment is all that is possible.

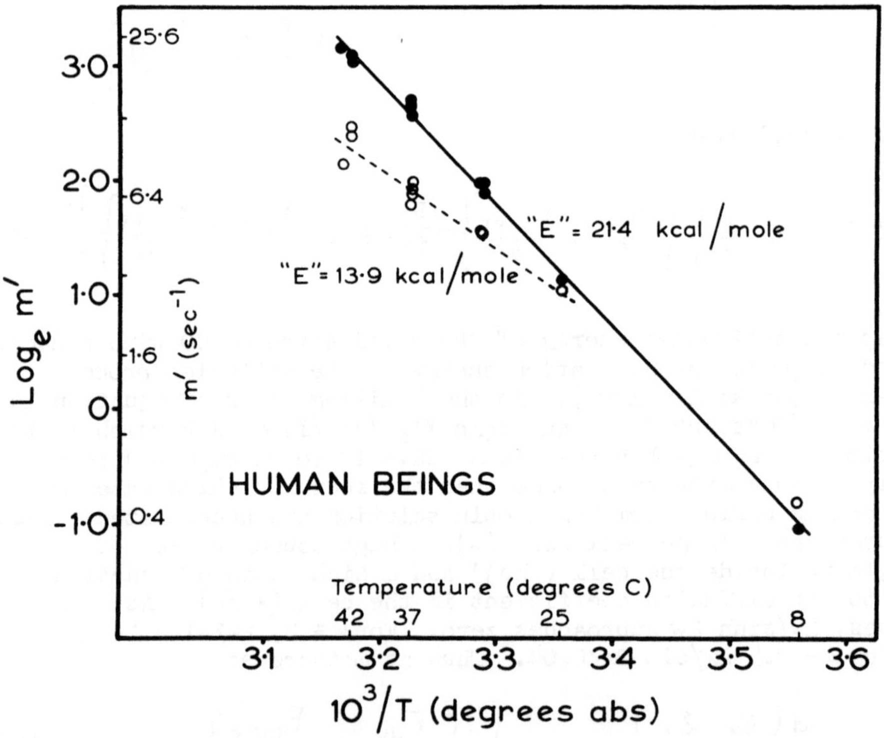

Fig. 5 The effect of temperature on the rate at which carbon monoxide replaced oxygen from fully saturated hemoglobin in solution (closed circles) and in the red cells (open circles). For each point the ratio $[O_2]/[CO]$ immediately after mixing in the apparatus was 1.32.

If we examine the results of Holland (10) for ℓ'_c in other species, we find that the temperature effect on the cell rate was less than on the solution rate in horses, goats, and rabbits. In all cases the magnitude of the effect above 20°C was approximately as predicted from eq. 16 if allowance is made for the simplifications necessary in arriving at this equation. In all these three species the temperature effect on cell rate was much greater at low temperatures, being almost as great as the effect on the solution rate. This finding was in accordance with the predictions from eq. 15 for temperatures where the solution reaction is slow and hence ω is small.

Holland (14) also measured the effect of temperature on the rate of the replacement reaction, $CO + O_2Hb \rightarrow COHb + O_2$, in the red cells of man (Fig. 5) and the horse. This reaction is more complex at high CO concentrations because the diffusion of two gases is going on at the same time as the chemical reaction. However for each species the temperature effect was close to that predicted by the equations given above and both species showed a greater temperature effect at low temperatures.

PHYSIOLOGICAL IMPORTANCE OF HEMOGLOBIN REACTION RATES

The reaction in which CO displaces oxygen from oxyhemoglobin has been one of the most useful studied. This reaction is not of importance in the functioning of the organism but it is the reaction which occurs when the diffusing capacity of the lung is measured. The velocity of the replacement reaction varies with the oxygen tension and Roughton and Forster (28) have used this property to measure the diffusing capacity of the alveolar-capillary membrane (D_M) and the pulmonary capillary blood volume (Vc). It is not within the scope of this paper to describe their method in detail but it requires a knowledge of the rate of the replacement reaction when the oxygen tensions are above 200 mm Hg and the CO tension is very low. This rate cannot be measured directly but can be derived from measurements made of m'c, m' and m'_∞. It is usually abbreviated as θ and expressed in units of ml CO taken up per minute into 1 ml of blood of fixed hemoglobin concentration, per mm Hg CO tension difference between the outside and inside of the red cell. However the only figures available for θ were those of Roughton, Forster, and Cander (27) who gave θ for human red cells at 37°C. To measure Vc with validity requires knowledge of θ for the appropriate species and temperature. Holland (14) has given additional values of θ to enable Vc to be measured under different conditions. Also Holland pointed out that in determining Vc, one need not know the absolute value of θ but merely how it is varied with P_{O_2}. Under the conditions of diffusing capacity measurements $1/\theta$ is a linear function of P_{O_2}. Vc can be calculated from lung diffusing capacity at two levels of oxygen (provided P_{O_2} is above about 150 mm Hg) and a knowledge of the slope of $1/\theta$ on P_{O_2}. Furthermore, the slope of $1/\theta$ on P_{O_2} can be found without measuring any cell reaction rates; it is a function of m'_∞ in hemoglobin solution. Values of m'_∞ are available with greater accuracy than the cell reaction rates (13) and more accurate measurement of Vc on exercise and in species other than man may now be made.

Roughton and Forster (28) showed that in the uptake of low concentrations of CO at sea level, a significant part of the resistance to uptake was in the combination of the gas with intracellular hemoglobin. However measurements of other reaction rates of hemoglobin has not shown any where the slowness of the reaction limited any function of the whole animal. It is possible that all the hemoglobin reactions occur rapidly enough to leave a substantial reserve but measurements of reaction kinetics under a variety of physiological stresses will be needed before one can finally reach this conclusion.

BIBLIOGRAPHY

1. Altman, P. L., Gibson, J. F. and Wang, C. C. Handbook of Respiration. Philadelphia, W. B. Saunders Co., 1958.

2. Bauer, C., Ludwig, M., Ludwig, I. and Bartels, H. Factors governing the oxygen affinity of human adult and fetal blood. Respir. Physiol. 7:271-277, 1969.

3. Benesch, R. and Benesch, R. E. The effect of organic phosphates from the human erythrocyte on the allosteric properties of hemoglobin. Biochem. Biophys. Res. Commun. 26:162-167, 1967.

4. Chanutin, A. and Curnish, R. R. Effect of inorganic phosphates on the oxygen equilibrium of human erythrocytes. Arch. Biochem. Biophys. 121:96-102, 1967.

5. Davson, H. Growth of the concept of the paucimolecular membrane. Circulation 26:1022-1037, 1962.

6. Forster, R. E. Rate of gas uptake by red cells. Handbook

of Physiology, section 3. Respiration Vol. I (W. O. Fenn and H. Rahn, eds.) Washington, D.C., The American Physiological Society, 827-837, 1964.

7. Forster, R. E., Roughton, F. J. W., Kreuzer, F., and Briscoe, W. A. Photocolorimetric determination of rate of uptake of CO and O_2 by reduced human red cell suspensions at 37°C. J. Appl. Physiol. 11:260-268, 1957.

8. Gibson, Q. H., Kreuzer, F., Meda, E. and Roughton, F. J. W. The kinetics of human haemoglobin in solution and in the red cell at 37°C. J. Physiol. 129:65-89, 1955.

9. Gibson, Q. H. The kinetics of reactions between haemoglobin and gases. Progr. Biophysics and Biophysic. Chem. 9:2-53, 1959.

10. Holland, R. A. B. Cell and solution velocity constants for the reaction CO + Hb → COHb at different temperatures in mammals with different red cell sizes. J. Gen. Physiol. 49:199-220, 1965.

11. Holland, R. A. B. and Forster, R. E. The effect of size of red cells on the kinetics of their oxygen uptake. J. Gen. Physiol. 49:727-742, 1966.

12. Holland, R. A. B. Kinetics of combination of O_2 and CO with human hemoglobin F in cells and solution. Respir. Physiol. 3:307-317, 1967.

13. Holland, R. A. B. Rate of O_2 dissociation from O_2Hb and relative combination rate of CO and O_2 in mammals at 37°C. Respir. Physiol. 7:30-42, 1969.

14. Holland, R. A. B. Rate at which CO replaces O_2 from O_2Hb in red cells of different species. Respir. Physiol. 7:43-63, 1969.

15. Klug, A., Kreuzer, F. and Roughton, F. J. W. The diffusion of oxygen in concentrated hemoglobin solutions. Helv. Physiol. Acta. 14:121-128, 1956.

16. Kreuzer, F. and Yahr, W. Z. Influence of red cell membrane on diffusion of oxygen. J. Appl. Physiol. 15:1117-1122, 1960.

17. Lawson, W. H., Holland R. A. B. and Forster, R. E. Effect of temperature on deoxygenation rate of human red cells. J. Appl. Physiol. 20:912-918, 1965.

18. Lawson, W. H. Interrelation of pH, temperature, and oxygen on deoxygenation rate of red cells. J. Appl. Physiol. 21:905-914, 1966.

19. Longmuir, I. S. and Roughton, F. J. W. The diffusion coefficients of carbon monoxide and nitrogen in hemoglobin solutions. J. Physiol. 118:264-275, 1952.

20. Moll, W. The influence of hemoglobin diffusion on oxygen uptake and release by red cells. Respir. Physiol. 6:1-15, 1969.

21. Nakamura, T. and Staub, N. C. Synergism in the kinetic reactions of O_2 and CO_2 with human red blood cells. J. Physiol. (London), 173:161-177, 1964.

22. Nicolson, P. and Roughton, F. J. W. A theoretical study of the influence of diffusion and chemical reaction velocity on the rate of exchange of carbon monoxide and oxygen between the red blood corpuscle and the surrounding fluid. Proc. Roy. Soc. London, Series B, 138:241-264, 1951.

23. Ponder, E. Hemolysis and Related Phenomena. New York, Grune and Stratton, 1948.

24. Rapaport, S., and Guest, G. M. Distribution of acid-soluble phosphorus in the blood cells of various vertebrates. J. Biol. Chem. 138:269-282, 1941.

25. Roughton, F. J. W. Diffusion and chemical reaction velocity as joint factors in determining rate of uptake of oxygen and carbon monoxide by red blood cells, Proc. Roy. Soc. London, Series B, 111:1-36, 1932.

26. Roughton, F. J. W. Diffusion and chemical reaction velocity in cylindrical and spherical systems of physiological interest, Proc. Roy. Soc. London, Series B, 140:203-229, 1952.

27. Roughton, F. J. W., Forster, R. E. and Cander, L. Rate at which carbon monoxide replaces oxygen from combination with human hemoglobin in solution and in the red cell. J Appl. Physiol. 11:269-276, 1957.

28. Roughton, F. J. W. and Forster, R. E. Relative importance of diffusion and chemical reaction rates in determing rate of exchange of gases in the human lung, with special reference to true diffusing capacity of pulmonary membrane and volume of blood in the lung capillaries. J. Appl. Physiol. 11:290-302, 1957.

29. Roughton, F. J. W. Diffusion and simultaneous chemical reaction velocity in haemoglobin solutions and red cell suspension. Progr. in Biophysics and Biophysic. Chem. 9:56-104, 1959.

30. Sirs, J. A. and Roughton, F. J. W. Stopped-flow measurements of CO and O_2 uptake by hemoglobin in sheep erythrocytes. J. Appl. Physiol. 18:158-165, 1963.

31. Sirs, J. A. Uptake of O_2 and CO by hemoglobin in sheep erythrocytes at various temperatures. J. Appl. Physiol. 18:166-170, 1963.

32. Staub, N. C., Bishop, J. M. and Forster, R. E. Velocity of O_2 uptake by human red blood cells. J. Appl. Physiol. 16:511-516, 1961.

33. Taylor, H. S. The temperature variation of diffusion processes. J. Chem. Physics. 6:331-334, 1938.

Author's work referred to in this article was supported by the National Heart Foundation of Australia and by the Joint Coal Board of New South Wales.

OXYGENATION VELOCITY OF THE RED CELL AND PULMONARY DIFFUSING CAPACITY

Masaji Mochizuki

Research Institute of Applied Electricity, Hokkaido University, Sapporo, Japan

I. INTRODUCTION

From recent studies it would seem that one of the fundamental factors determining the oxygenation velocity of the red cell in the lung, namely, the diffusion barrier of oxygen across the capillary wall requires investigation. According to Miyamoto and Moll (1968) the red cell becomes deformed in the capillary forced by the kinetics of the blood flow and, moreover, the hematicrit value in the capillary is considerably lower than that in larger arteries. Staub and Schultz (1968) ascertained morphologically that the red cell flows via 2 to 3 alveoli in succession, but in regard to the duration of the contact time during which the red cell is in contact with alveolar gas, some questions still remain.

As a term to indicate the diffusion rate through this barrier, the diffusing capacity was introduced by Bohr (1909) 60 years ago. At that time the alveolar membrane was considered to be a unique factor which determined the oxygenation velocity. However, Roughton (1945) clarified that the reaction rate at which CO replaces O_2 from the oxygenated hemoglobin (O_2Hb) was actually lower than the expected rate and that it was significantly affected by oxygen tension P_{O_2}. Since his contribution the chemical reaction rate of the red cell with O_2 or CO has come to be considered as an important factor in the treatment of the pulmonary diffusing capacity. Mochizuki and Fukuoka (1958) said that the diffusing capacity for O_2 would be better referred to as the "reacting capacity", insofar as the oxygen saturation, S_{O_2}, is used in the calculation of D_{LO_2} instead of the actual P_{O_2} inside the red cell. They suggested that the diffusing capacity for O_2 should be closely correlated with hemoglobin saturation, S_{O_2}.

The correlation between the rate of chemical reaction of CO with hemoglobin and the CO diffusing capacity $D_{L_{CO}}$ was confirmed by Kruhoffer (1954), who showed the dependency of $D_{L_{CO}}$ on the alveolar gas P_{O_2}. However, the diffusion rate across the alveolar membrane D_M was not determined before the study of Forster and others (1957). They derived a simple relation between the reaction rate, the diffusing capacity and the diffusion rate across the membrane and determined the last quantity by using the measured values of the former two quantities. The reaction rate which was used for the calculation of D_M in their work was evaluated according to a diffusion equation derived by Nicolson and Roughton (1951) with special reference to the data of Roughton, Forster and Cander (1957). As to whether the above calculated reaction rate was feasible or not, questions were later raised on the following points: 1) The rate was measured at a comparatively high level of P_{CO} of about 100 mmHg, although $D_{L_{CO}}$ was measured at a low P_{CO} range of 1 to 3 mmHg. 2) The diffusion rate through the red cell membrane was not measured. 3) In vivo, in the capillary the red cell deforms from the normal biconcave structure, whereas in the observation tube it was normal.

In general, the oxygenation velocity of the hemoglobin depends on S_{O_2}: It decreases as the hemoglobin becomes saturated with O_2, as described by Roughton (1932) in the following manner:

$$\frac{d[O_2Hb]}{dt} = K' [O_2][Hb]$$

where $[O_2]$, $[Hb]$, $[O_2Hb]$ are the molar concentrations of oxygen, hemoglobin and oxyhemoglobin. The oxygenation velocity of the Hb molecule is very rapid as compared with the rate of CO with O_2Hb. Especially, when S_{O_2} is low, the reaction is of such rapidity that it is completed within 20 msec at P_{O_2} of about 100 mmHg, as shown by Gibson and others (1955) and Mochizuki et al.(1966). Apart from the measurement in a Hb solution, when the oxygenation of Hb is measured in a red cell suspension, the reaction rate is markedly suppressed by diffusion into and out of the red cell. According to Mochizuki (1966) it becomes almost constant in a S_{O_2} range of 0 to 50%, regardless of the S_{O_2}, though in an initial transition stage of the reaction, the rate usually decreased exponentially. In a range higher than 50%, however, the rate decreased gradually as the S_{O_2} increased. This finding appeared to support the experimental fact of Bartels and others (1955) in that the $D_{L_{CO}}$ increases at a hypoxic level of the alveolar P_{O_2}. However, an accurate evaluation of the diffusion rate through the capillary wall could not be attained, since it was shown by Koyama and Mochizuki (1967, 1968, 1969) that the reaction rate varied significantly according to the flow conditions in the observation tube.

In a theoretical study on the diffusion with chemical reactions in the red cell, Nicolson and Roughton (1951) introduced as one of the diffusion barriers a cell membrane surrounding the red cell. On the other hand, Mochizuki and Fukuoka (1958) took into account a barrier consisting of a stagnant boundary layer combined with the cell membrane, the diffusion property of which was characterized by a value "transfer coefficient". The difference between the characteristics of the former and the latter diffusion barriers consists of the dependency of the diffusion rate on the convection around the red cell. Thus, the experimental fact that the diffusion rate was increased with the flow rate seems to prove the existence of the stagnant boundary layer. On the other hand, it was suggested that the estimation of the oxygenation velocity in the lung would be too uncertain without knowing the flow conditions around the cell, especially in the lung capillary.

The flow velocity through the lung capillary may be as slow as 1 - 0.2 $mm \cdot sec^{-1}$, but the stagnant layer on the cell surface is largely disturbed by a flow through a 10 micron capillary. Thus, it may be said that the diffusion rate may rather be increased in the capillary as compared with a red cell suspension flowing through an observation tube, however rapid the flow may be. Thus, a necessity to reproduce the same flow conditions as that in the lung has arisen in in vitro experiments. Naturally, such requirement is almost impossible to be satisfied. Thus, it seems rather fruitful to make a direct measurement of the diffusion rate across the barrier which involves both the capillary wall as well as the stagnant layer of plasma in vivo. In the course of coming to the in vivo experiments we have attempted the following two studies. Namely, 1) The dependency of the oxygenation velocity on the flow rate in the observation tube was studied, in order to determine whether the maximum reaction rate could be obtained when the flow velocity was made as high as possible. 2) The reaction rate at which CO replaces O_2 from O_2Hb was measured in a laked Hb solution with a low P_{CO} of 3 to 6 mmHg.

II. DEPENDENCY OF THE OXYGENATION VELOCITY ON THE FLOW VELOCITY IN THE OBSERVATION TUBE

1) Photometric Method

In order to evaluate the oxygenation velocity, Roughton (1932) introduced the initial oxygenation rate K_c', which was obtained from a tangential line drawn on a curve showing the hemoglobin saturation (S_{O_2}) change against the reaction time at the beginning of the reaction. This represents the oxygenation velocity at the initial stage alone, and not that of a successive longer reaction time. Subsequently, Forster and others (1957) introduced another velocity constant θ by modifying the initial velocity rate K_c'. However, this too was limited to the oxygenation velocity at the initial stage. Hence, Mochizuki and Fukuoka (1958) proposed another constant Fc in order to express the oxygenation velocity not only at the initial stage, but also at the later stage of the reaction. This was tentatively called the velocity factor of the oxygenation of the red cell. Since the oxygenation process of the red cell in the pulmonary capillary has a long duration, the velocity factor was superior to other parameters representing the oxygenation velocity. In any event, three different types of velocity constants have hitherto been introduced. In addition, to further complicate matters, a considerable variation among the velocity constants measured by several workers is seen as shown in Table 1, where all K_c' and θ values have been converted to Fc by changing the respective units. As seen in the Table, the obtained values deviated from each other over a wide range of Fc from 0.01 to 0.11 $sec^{-1}mmHg^{-1}$. Such a wide variation makes it difficult to estimate the oxygenation velocity in the lung. Thus, it becomes necessary to determine whether some intrinsic factors by which the measured Fc values may be scattered exist either in the measuring system or in the red cell itself.

The relationship between the oxygenation velocity and the convection was already studied by Hartridge and Roughton (1927) and they reported that the velocity was not influenced by the convection. However, in order to treat the oxygenation velocity more accurately, a reinvestigation is required from a new point of view. The experimental procedure used was similar to that used by Hartridge and Roughton, in other words, the oxygenation velocity was measured under various velocities of flow passing through an observation tube in a rapid flow system.

Method

A rapid flow apparatus designed and constructed by Mochizuki (1966) was employed. Syringes of 100 ml were used instead of 20 ml, in such a way as to obtain flow velocities ranging from 300 to 74 $cm \cdot sec^{-1}$ through an observation tube. The red cell suspension was prepared by adding 7 ml of fresh heparinized blood to 2.8l of a buffer solution. The solution was composed of NaCl, 6.7 g; KCl, 0.26 g; KH_2PO_4, 3 mg; Na_2HPO_4, 0.86 g and $NaHCO_3$, 2.4 g. The pH was 7.4 at 40 mmHg P_{CO_2}. The red cell suspension was

Table 1

Oxygenation velocities obtained hitherto by several workers.

Author	Method	Reyn. No.	S_{O_2}	K_c'	Θ_{O_2}	Fc
Staub et al.(1961)	Rapid Fl. Pt	3500	0	164	2.8	(0.093)
	Rapid Fl.		25		2.0*	0.067*
Gibson et al.(1955)	Rapid Fl.		0	100 - 150		(0.057 - 0.085)
Forster et al.(1957)	Rapid Fl.	2330-2920	0	79 - 98	1.4-1.7	(0.047 - 0.057)
			35	84	1.4	(0.047)
Sirs (1963)	Stopped Fl.		0	50 - 60	0.83-1.0	(0.028 - 0.033)
Holland et al.(1966)	Stopped Fl.		0	70 - 85	1.2-1.45	(0.040 - 0.048)
Mochizuki et al.(1966)	Rapid Fl.	735	50 - 70			0.025
Koyama et al.(1967)	Stopped Fl. Pt		50 - 70			0.011
			0			(0.048)
Koyama et al.(1968)	Rapid Fl. Pt	5540	50 - 70			0.052
		1485	0			(0.095)
		6047	50 - 70			0.029
Koyama et al.(1969)	Rapid Fl.	3028	50 - 70			0.036
		1512	0			(0.079)
			50 - 70			0.030
			50 - 70			0.019

(Pt) indicates the values obtained by a polarographic method using a Pt electrode.
* is a value calculated from the figure shown in the literature.

previously deoxygenated by bubbling with a gas mixture of 5.70% CO_2 and 0.08% O_2. The buffer solution to be mixed with the suspension was equilibrated with a gas mixture of 5.76% CO_2 and 25.50% O_2. Both solutions were brought to a vigorous head-on collision in a mixing chamber connected with the observation tube by manipulating the two syringes. The oxygenation process was followed by a differential type colorimeter using two wave lengths of 599 and 572 mμ. The flow velocities through the observation tube were 294, 147 and 74 cm$\cdot$sec^{-1}. The entire apparatus was kept at 37°C.

In order to calibrate the output signal from the colorimeter with the S_{O_2}, it was necessary to measure the output current at least at two known levels of S_{O_2}. In the present study 100% and 0% S_{O_2} were used for the calibration as the known S_{O_2}. For obtaining a sample of 100% S_{O_2}, the deoxygenated red cell suspension was mixed with the buffer solution bubbled with pure O_2 and forced to flow down the observation tube, whose transmittance was measured. Since the oxygenation proceeds rapidly at a hyperoxic level, the required sample of 100% S_{O_2} was obtained soon after the mixing. The sample of 0% S_{O_2} was simply obtained by mixing the deoxygenated suspension with the buffer solution with O_2 free gas in a tonometer.

Results

The measurement was carried out on fresh blood samples drawn from five normal subjects. In most cases duplicate measurements were made; the differences between the two measured values were in general less than 5% in hemoglobin saturation, S_{O_2}. The S_{O_2}-time curves measured over a period of a few days showed a small diversity with an error of a few per cent in S_{O_2} in standard deviation. Hence, the day-by-day variation and individual errors seemed to be due to the same cause of error. In Fig. 1 the S_{O_2} change obtained at three different flow velocities are plotted against the reaction time. The plotted points represent the averages of four measured S_{O_2} values and the arrows at these points show the standard deviations. As clearly seen in this example the S_{O_2} increased most rapidly at the flow velocity of 294 cm$\cdot$sec^{-1} and most slowly at 74 cm$\cdot$sec^{-1} in each subject. This suggested that the oxygenation velocity was regularly affected by the flow velocity through the observation tube, though there were some variations in the mode of the S_{O_2} increase in the individual curves.

According to the method described by Mochizuki (1966) the Fc value was calculated from the smoothed S_{O_2}-time curve which was closest to the measured S_{O_2} points. The Fc showed large values at the initial stage of the reaction, but it decreased with the lapse

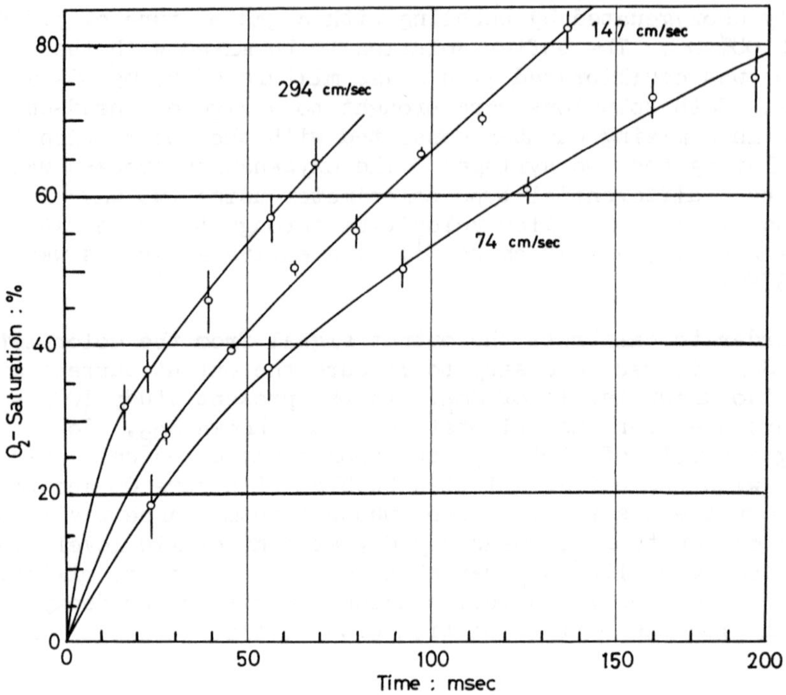

Fig. 1 The S_{O_2}-time curve obtained in the same subject at three different flow velocities of 294, 147 and 74 $cm \cdot sec^{-1}$.

of time as shown in Fig. 2. The curves in Fig. 2 were calculated from the S_{O_2}-time curves of Fig. 1. As may be surmised from the curves in Fig. 1, the Fc value was higher at higher velocities of the flow in the observation tube. The Fc at the initial stage showed values ranging from 0.065 to 0.093 $mmHg^{-1} \cdot sec^{-1}$ according to the flow velocity, and reduced in parallel with the reaction time to attain a constant value ranging from 0.02 to 0.035 $mmHg^{-1} \cdot sec^{-1}$ in 40 to 50 msec. The correlation between the Fc value and the flow velocity was further tested in another three subjects. The measured Fc values in the S_{O_2} range of 40 to 60% are summarized in Table 2, where the flow velocity is expressed by Reynolds number. The correlation between them was tested statistically using values described, and it was shown that the correlation was significant with the P-value, which was less than 0.02 in the T-test.

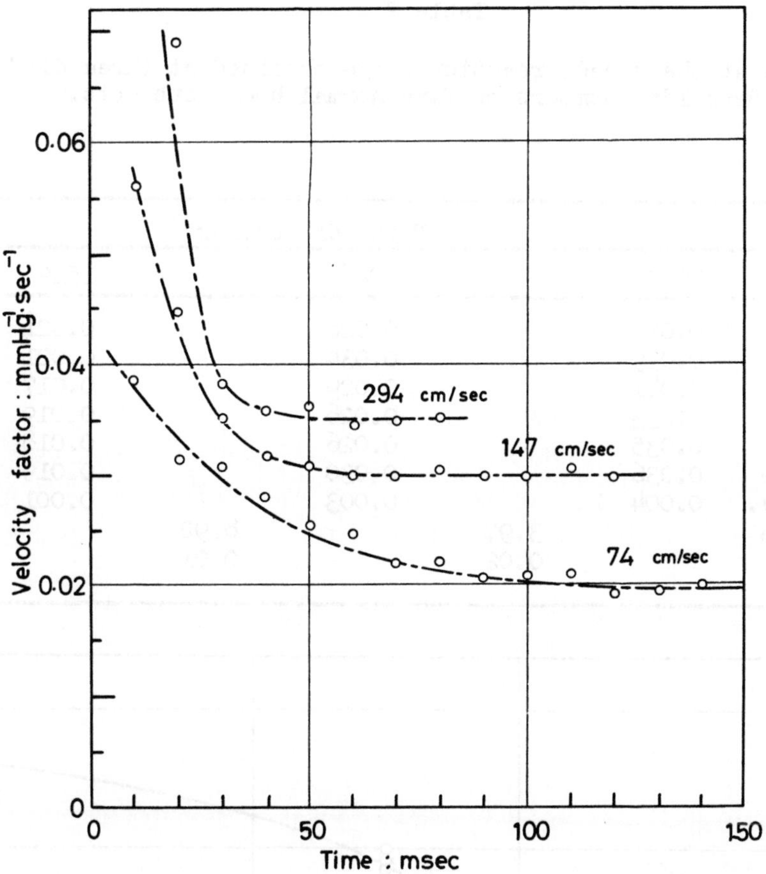

Fig. 2 Fc values on each 10 msec time interval plotted against the reaction time.

In Fig. 3 the Fc value is plotted against the Reynolds number. The plotted point at Reynolds number 0 was obtained previously by Koyama and Mochizuki (1967) by using a stopped flow method combined with a sputtered Pt electrode elaborated by Saito (1967). The points in Fig. 3 show that the value increased with the flow velocity linearly at a low range of the velocity, but at a higher range the rate of the increment gradually decreased.

Table 2

Fc values at the steady reaction stage obtained at three different Reynolds numbers on five normal human subjects.

	Reynolds number				
Subj.	6047		3028		1512
M. M.	0.032		0.028		0.021
Y. F.	0.039		0.034		0.021
Y. K.	0.041		0.029		0.017
M. S.	0.033		0.028		0.019
T. K.	0.035		0.026		0.018
Average	0.036		0.030		0.019
St. dev.	0.004		0.003		0.001
t-value		3.97		8.92	
p		0.02		0.01	

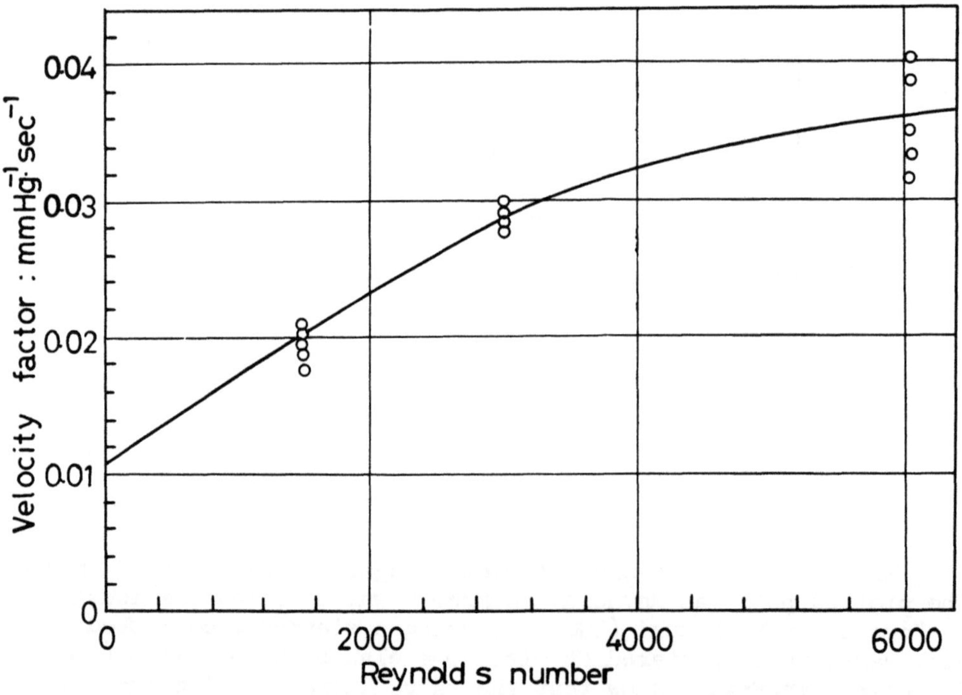

Fig. 3 Relationship between Reynolds number and obtained Fc values at an S_{O_2} range of 45 to 60%. Reynolds numbers were calculated from velocities of 294, 147 and 74 cm·sec^{-1}.

Discussion

The oxygenation velocity is very rapid in a laked Hb solution in a S_{O_2} range lower than 60%. In such a low range of S_{O_2} the diffusion equation can be expressed according to Mochizuki (1966) as follows:

$$P_a - P(x,t) = (P_a - P_o)\sum_{1}^{\infty} \frac{2 \sin\mu_m}{\mu_m \cdot S_m} \cos\frac{\mu_m \cdot x}{d} \exp(-\frac{t}{T_m}) \quad (1)$$

where P_a is the P_{O_2} in the bulk solution, $P(x,t)$ is the P_{O_2} in the red cell, P_o, the initial P_{O_2} in the red cell, d, half the thickness of a plate model of the red cell, μ_m's represent the eigen values characterizing the boundary conditions and the T_m's indicate the time constants. Among μ_m, T_m and S_m the following relations are involved;

$$T_m = \alpha' \cdot d^2 / \alpha \cdot D \cdot \mu_m^2 \quad (2)$$

$$S_m = 1 + G/(G^2 + \mu_m^2) \quad (3)$$

and

$$\mu_m \tan\mu_m = G = \eta d/\alpha D \quad (4)$$

where η is the transfer coefficient, D, the diffusion coefficient, α, the O_2 solubility and α', the gradient of the O_2 dissociation curve expressed by the same unit as that of α.

At the beginning of the diffusion process a number of the terms of Eq (1) should be summed up in order to evaluate the O_2 distribution in the red cell. However, at a later stage of the process, all terms except the first involving μ_1 becomes so small that Eq (1) turns to the following simple equation:

$$P_a - P(x,t) = (P_a - P_o) \frac{2 \sin\mu_1}{S_1} \cos\frac{\mu_1 x}{d} \exp \frac{t}{T_1} \quad (5)$$

When the diffusion process can be expressed by Eq (5), the diffusion at this stage may be referred to as "steady state diffusion," because the velocity factor Fc becomes a constant value, independent of the lapse of time. On the other hand, the period from the start to this steady state should be called the "transition period," since the velocity factor Fc of this period reduces rapidly with time. The duration of this period is appreciated by the next longest time constant to T_1, that is, T_2.

As described in Eq (2), T_2 is rewritten as follows:

$$T_2 = \alpha' d^2 / \alpha D \mu_2^2 \qquad (6)$$

Eq (6) shows that the time constant reduces as the gradient of the O_2 dissociation curve α' decreases and, because the eigen value increases along with the transfer coefficient as shown in Eq (4), T_2 is shortened as the transfer coefficient increases. The second eigen value can be evaluated from the first eigen value μ_1 which in turn is obtained from the velocity factor showing a steady state diffusion rate which is expressed as $Fc = \mu_1^2 \cdot \alpha \cdot D/d^2$. Thus, the time constant T_2 can be derived theoretically from the steady state diffusion rate obtained experimentally.

As shown in Table 2 and Fig. 3 the Fc value in the steady state increased almost linearly with the flow velocity in the observation tube. This experimental fact suggests that the flow velocity or the convection around the red cell surface has a close positive correlation with the transfer coefficient. In other words, when the transfer coefficient increases with the flow velocity, the eigen values μ_1 and μ_2 increase in a parallel manner; the increase of μ_1 appears as the increase of the Fc value in the steady state diffusion and the increase of μ_2 appears as the shortening of the initial transition period.

According to Mochizuki (1966) $\alpha D/d^2$ can be put as 0.023 $mmHg^{-1} \cdot sec^{-1}$. Therefore, in the following three cases where the steady state Fc values are 0.020, 0.029 and 0.035 $mmHg^{-1} \cdot sec^{-1}$ as shown in Table 3, the first eigen values μ_1's are calculated as 0.94, 1.13 and 1.24. Furthermore, referring to Fig. 4 showing the relationship between the eigen value and $G = \eta d/\alpha D$, the second eigen values μ_2's can roughly be evaluated from the above first eigen values as 3.2, 3.5 and 3.8. On the other hand, the gradient of the O_2 dissociation curve at an S_{O_2} range of 40 to 60% is about 3% S_{O_2} per 1 mmHg. This is converted to 0.01 $mmHg^{-1}$ if expressed in terms of O_2 concentration (ml) in 1 ml red cells per 1 mmHg. Thus, the time constants of the initial transition period are calculated as 43, 35 and 30 msec, as summarized in Table 3. These values agreed well with the experimental data shown in Fig. 2.

The very nature of the Fc which shows a rapid decrease at the initial stage of the oxygenation renders its determination fairly difficult, especially, it becomes very difficult at the initial stage of the reaction, since the measured value is strongly affected by errors in both the convection of fluid surrounding the red cell and the measured reaction time. As shown in Table 1, the velocity factors so far obtained by several workers show a considerable scattering over a wide range. This may be attributed to the above-

Table 3

The measured velocity factors at a steady diffusion state, the calculate eigen values, μ_1 and μ_2, and the time constant T_2 obtained from the above Fc values.

	No. 1	No. 2	No. 3
Velocity factor in steady state	0.020	0.029	0.035
μ_1	0.939	1.127	1.243
μ_2	3.20	3.55	3.80
$\mu_2$msec	43	35	30

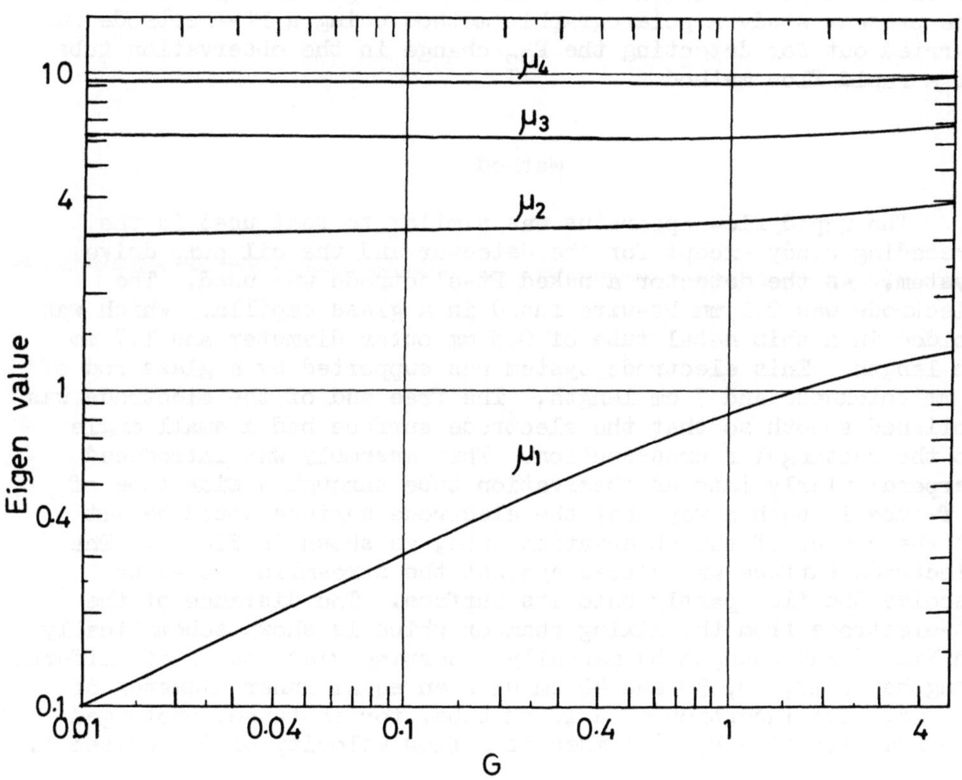

Fig. 4 Relation of the eigen values, μ_m, and the conditions of $G = \eta d/\alpha D$, where η is the transfer coefficient; d, the thickness of the model sheet; α, the O_2 solubility and D, the diffusion coefficient.

mentioned nature of the Fc itself. In this respect is is certainly desirable to measure the Fc value not only at the early stage but also in the later stage when the steady state diffusion is observed.

Roughton and others have stated in their numerous papers that the cell membrane would have a high resistance to the O_2 diffusion. However, if the cell membrane is the sole dominant resistive factor, the velocity factor would not be affected by the flow velocity in either the transition period or the steady state diffusion stage.

2) Polarographic Method

In the preceding study (1969) the Fc value of the red cell was measured by photometry and it was found to depend on the flow velocity employed. This finding seems to explain the unsettled problems concerning the characteristics of the resistive boundary layer around the red cell. The dependency, however, is suspected to be due to a technical error inherent to photometry. Thus, in the present study a polarographic method using a Pt-electrode was carried out for detecting the P_{O_2} change in the observation tube of a rapid flow method.

Method

The rapid flow apparatus was similar to that used in the preceding study except for the detector and the oil pump drive system. As the detector a naked Pt-electrode was used. The electrode was 0.1 mm Pt-wire fused in a glass capillary which was molded in a thin metal tube of 0.5 mm outer diameter and 1.7 mm in length. This electrode system was supported by a glass rod of 3 mm thickness and 7 cm length. The free end of the electrode was polished smooth so that the electrode surface had a small angle to the rectangular crosssection. This assembly was introduced perpendicularly into an observation tube through a side tube of a T-tube in such a way that the electrode surface would be set at the center of the observation tube, as shown in Fig. 5. The electrode surface was placed against the streamline so as to receive the flow partly onto its surface. The distance of the Pt-electrode from the mixing chamber which is shown schematically in Fig. 5 was changed by manually inserting glass tubes of different lengths of 10, 20, 30 and 40 cm with an equal inner diameter of 3.6 mm. The insertion of a 10 cm tube, for instance, prolonged the reaction time by 36.1 msec at a flow velocity of 277 $cm \cdot sec^{-1}$.

One set of measurements was composed of five actual measurements at five reaction times and two calibrations under a definite flow velocity. The red cell suspension was made by

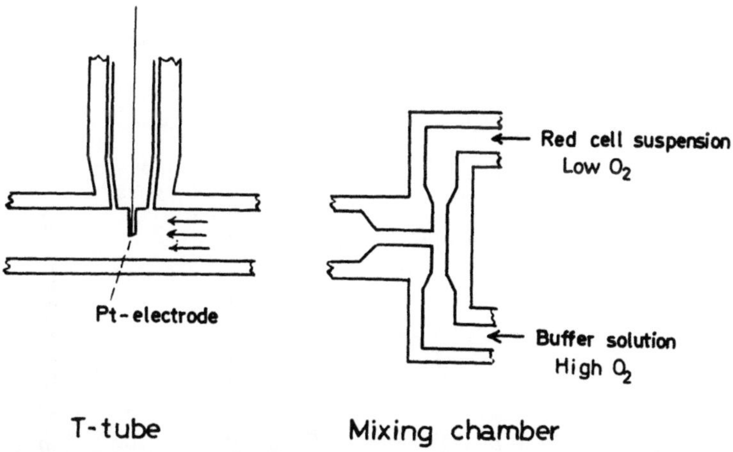

Fig. 5 Schematic illustration of the mixing chamber, T-tube and projecting Pt-electrode.

adding 20 ml blood to 500 ml saline buffer solution to attain a 25 fold dilution. This suspension was equilibrated with a gas mixture of 0.08% O_2 and 5.76% CO_2 and brought into head-on collision with the buffer solution equilibrated with a gas mixture of 28.4% O_2, 5.64% CO_2 or 32.4% O_2 and 5.62% CO_2. At the very moment of the collision the red cell began to take up oxygen, thus, P_{O_2} in solution began to fall. The decrease of P_{O_2} was followed by the Pt-electrode along the observation tube at points 3, 10, 20, 30 and 40 cm apart from the mixing part. For the purpose of the calibration a red cell suspension of 50 fold dilution was prepared. This was divided into two parts and equilibriated with gas mixtures of 0.08% O_2, 5.76% CO_2 and 12.45% O_2, 6.04% CO_2 or 8.82% O_2 and 5.69% CO_2 separately. In this way twelve sets of measurements were made at various flow velocities produced by controlling the pumping rate of the oil pump.

Results

The O_2 reduction current through the Pt-electrode increased with the increment of the flow velocity through the observation tube, but at a constant flow velocity the current was proportional to P_{O_2} in the suspension alone. The current, therefore, was calibrated at the same flow velocity as that of the actual measurements.

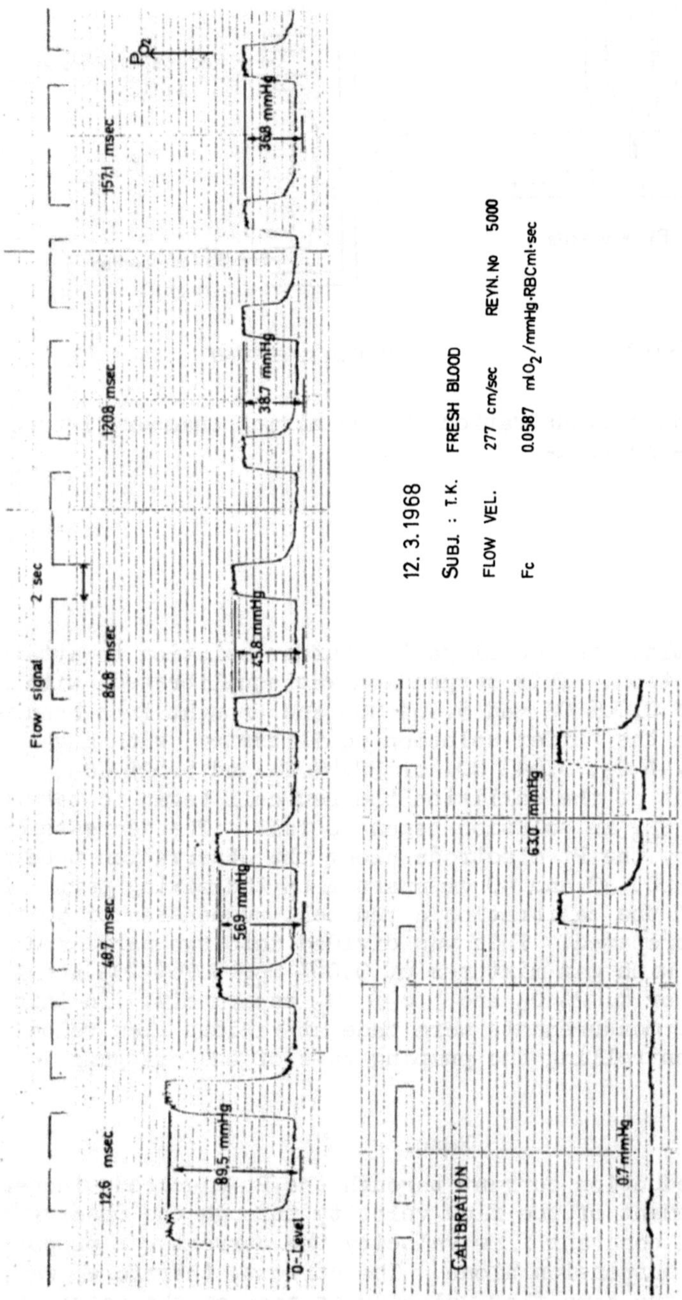

Fig. 6 An example of the actual registration of polarographic currents.

Fig. 6 shows an actual record obtained in the red cell suspension at a flow velocity of 277 $cm \cdot sec^{-1}$. At first the electrode was placed nearest the mixing chamber. At the moment of the onset of the flow the current through the electrodes increased rapidly and attained a plateau which was maintained so long as the fluid is made to flow at a constant flow rate. When the flow stopped, the current returned to the original value. The Pt-electrode was then moved further apart from the mixing chamber, namely, the reaction time was prolonged. In accordance with the reaction time the current at the plateau decreased, showing the progress of the O_2 saturation of the red cell. The last four curves were made for the calibration at P_{O_2}'s of 0.7 and 63.0 mmHg. By referring to the height of the plateaus, the current deflections were calibrated by P_{O_2}.

Plotting the calibrated P_{O_2} versus the reaction time, a P_{O_2}-time curve was constructed as shown in Fig. 7. Because the hemoglobin saturation(S_{O_2}) change almost stopped at 160 msec, the equilibrium point was determined by extrapolating this curve. Since at the equilibrium point the P_{O_2} in the red cell should be equal to that of the outer medium, the final S_{O_2} was obtained by referring to the O_2 dissociation curve and this P_{O_2} at the equilibrium point. Moreover, from the differences between P_{O_2} and S_{O_2} at the initial and final stages an accurate dilution ratio of the red cell was calculated, and then the S_{O_2}-time curve was evaluated by the use of the change in P_{O_2}-measured. The time course of the back pressure (Pc) in the red cell could also be obtained from the S_{O_2} values with reference to the O_2 dissociation curve.

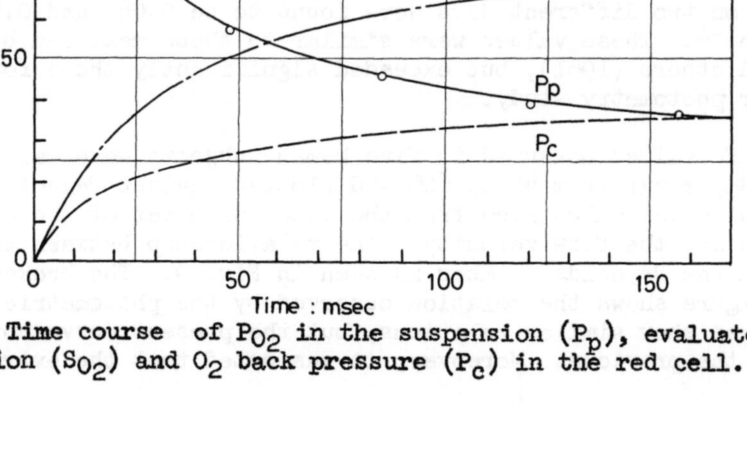

Fig. 7 Time course of P_{O_2} in the suspension (P_p), evaluated O_2-saturation (S_{O_2}) and O_2 back pressure (P_c) in the red cell.

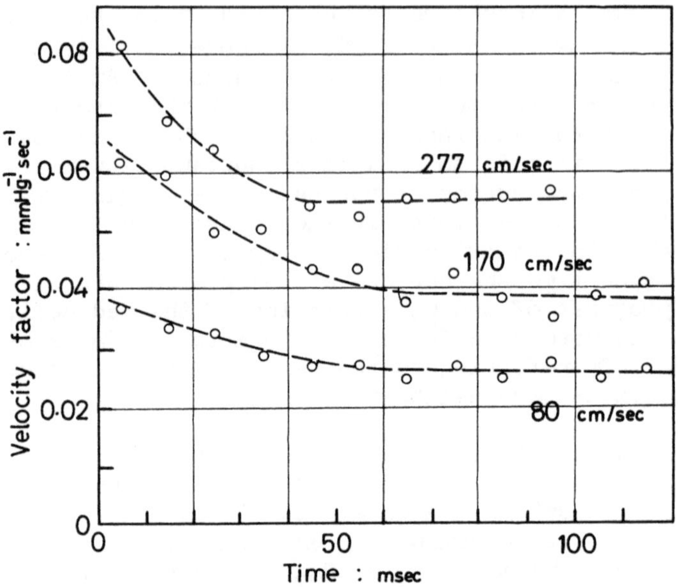

Fig. 8 Time course of sequential Fc values for 10 msec interval at three different flow velocities.

The Pc and S_{O_2} values thus estimated are shown in Fig. 7, and Fc values were calculated therefrom.

A set of the Fc values as a function of time was obtained from these curves are shown in Fig. 8, where the Fc was measured at three different flow velocities of 277, 170 and 80 cm·sec^{-1}. At the initial stage of the reaction the Fc value showed a high value and fell to a constant value soon after. The initial Fc values obtained on two different days were found to be 0.082 and 0.095 mmHg^{-1}·sec^{-1}. These values were similar to those measured by Staub and others (1961), but exceeded significantly those found by us in our photometry study.

The Fc values measured in five human subjects were calculated for the S_{O_2} range from 40 to 60% and plotted against Reynold s number which was calculated from the inner diameter of the observation tube and the flow velocity. The relationship between the Fc value and the Reynolds number is seen in Fig. 9. The broken line in the figure shows the relation observed by the photometric method. Both curves show similar relations, but the present curve notably exceeded the previous. Moreover, it was noted that the extrapolated

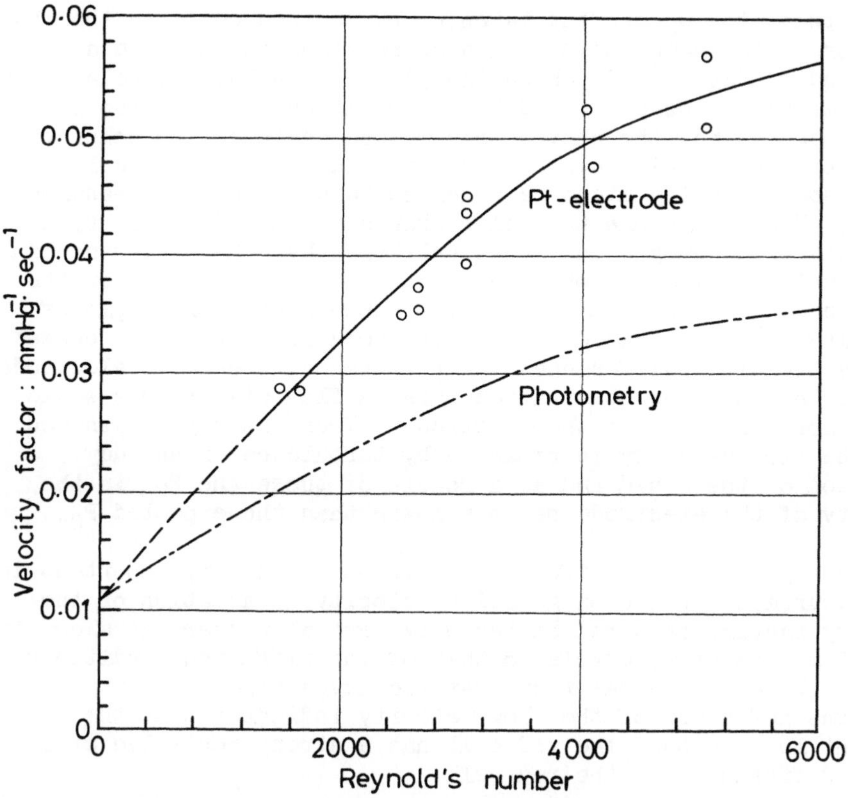

Fig. 9 Fc values plotted against Reynolds number.

point at the zero flow fell on about 0.01 $mmHg^{-1} \cdot sec^{-1}$ which was obtained by a stopped flow method.

Discussion

Through both the preceding and the present studies it was clarified that the oxygenation velocity was increased significantly with the flow velocity. In addition, a systematic difference was observed between the two values measured by the polarographic and the photometric methods. The former finding may certainly be explained as an increment of diffusibility through the boundary layer around the red cell. But the difference between the Fc values measured by the above two methods seems difficult to be explained by the same reason.

Between the O_2 content taken up by the red cell and the change in color of the hemoglobin, a linear relation should hold as described by Sirs and Roughton (1963). Naturally, the polarographic method measures the dissolved O_2 before it enters into the red cell, on the other hand, the photometric method measures the color change caused by the O_2 which has already entered into the red cell. Hence, some time lag might be observed between the two measured values. However, since the color change is caused solely by O_2 combining with hemoglobin, it is unlikely that the difference in the Fc values which was observed even at the later stage of the reaction can be ascribed to the difference in the principle of measuring methods. Rather it may be probable that it was caused by some technical differences. For instance, when a Pt-electrode is inserted into the observation tube, a flow disturbance would occur, more or less, at the electrode. Thus, it may be assumed that the flow velocity is retarded by turbulence or an eddy formation of the fluid and as a result of these the P_{O_2} in the vicinity of the electrode becomes lower than the expected P_{O_2}.

In any event the diffusion conditions seem very important in the measurement of the oxygenation velocity. Variation of the velocity factors measured hitherto by several workers as shown in Table 1 may be attributable to that of the diffusion conditions. Anyhow, the experimental fact that the oxygenation velocity is under the influence of the flow velocity indicates that the stagnant layer around the red cell has a strong resistive effect on the diffusion into the red cell.

III. A KINETIC STUDY ON THE OXYGEN REPLACEMENT REACTION BY CARBON MONOXIDE FROM OXYGENATED HEMOGLOBIN IN HUMAN HEMOGLOBIN SOLUTION

Originally, the terms of $D_{L_{O_2}}$ and $D_{L_{CO}}$ were derived as indicators of the velocities at which O_2 and CO diffused across the alveolar membrane into the capillary blood. However, as far as the reactions of these gases with hemoglobin are concerned, the diffusing capacities may reasonably be considered as indicators of the overall reaction velocities at which O_2 or CO reacts with hemoglobin over the entire pulmonary capillaries, where both the diffusion and the chemical reaction participate in the velocities at the same time. Provided that a diffusion barrier around a single red cell had a diffusion constant Fd/M in the pulmonary capillary for a test gas, where M is the number of the red cells in 1 ml red cells and provided that the reaction of the gas with Hb has a velocity factor Fc/M, then the overall reaction velocity of the red cells in the capillaries F_L/M may easily be derived as follows (1968):

$$F_L = \frac{Fc \cdot Fd}{Fc + Fd} \quad (7)$$

Furthermore, the pulmonary diffusing capacity may be given as

$$D_L = F_L \cdot Q \cdot Ht \cdot tc \quad (8)$$

where Q is the pulmonary blood flow, Ht, the hemtocrit value and tc, the contact time of the red cell with the alveolar air. Consequently, if two sets of measurements of Q·Ht and D_L are available, the values of Fd and tc may be obtained.

The Fc value in Eq (7), ideally, would be a velocity factor with a transfer coefficient of infinite greatness. When measured in vitro, however, it is influenced by the diffusion conditions around the red cell and such an extreme Fc value can not be obtained. Therefore, it seems rather advantageous or practical to calculate its value theoretically by using the velocity factor obtained in a laked Hb solution as Roughton and Forster (1957) actually did.

In the measurement of $D_{L_{CO}}$, the alveolar CO tension $P_{A_{CO}}$ usually ranged from 1 to 3 mmHg, but in the reaction rate of CO and O_2Hb in vitro, the P_{CO} in the solution used was as high as or exceeded 15 mmHg. Thus, as to whether the reaction rate obtained at such a high P_{CO} range is still applicable to the analysis of $D_{L_{CO}}$, the question still remains to be proved. According to Douglas, Haldane and Haldane (1912) CO saturation of Hb depends greatly on the P_{O_2} in the solution: At a low P_{CO} range the dependency on P_{O_2} is highly pronounced as shown by the following relation:

$$\frac{\%\ COHb}{\%\ O_2Hb} = 210\ \frac{P_{CO}}{P_{O_2}} \quad (9)$$

Thus, at 100 mmHg P_{O_2} the CO saturation S_{CO} at 1 mmHg P_{CO} is about 60%, while at 500 mmHg P_{O_2} it is about 30% at the same P_{CO}. On the other hand, when the P_{CO} is equally 50 mmHg, S_{CO} is constantly about 100%, regardless of the P_{O_2}. Therefore, it seems possible that the O_2 replacement reaction by CO is dependent on P_{CO} as well as P_{O_2}. Hence, we attempted to measure the reaction rate at a P_{CO} range lower than 10 mmHg in a laked Hb solution.

In order to measure the rate at such a low P_{CO} the Hb

concentration in the solution must also be reduced to a range comparable with P_{CO}, otherwise the change in S_{CO} will become indiscernible. When the Hb concentration is reduced, the light path in a sample cell by which the light is absorbed must be lengthened. Namely, structures of a mixing chamber and an observation tube should be improved to fulfil the requirements arising from the dilution. On the other hand, in order to increase the signal to noise ratio a selection of a wave length at which the change in admittance appears most significantly is also a very important factor. Roughton, Forster and Cander (1957) and Sirs and Roughton (1963) measured the change in admittance due to the reaction at two wave lengths of about 560 and 580 mμ, or at 520 and 460 mμ. The authors (1969) also measured the change in S_{CO} by using a pair of filters having a transmission peak at 470 and 513 mμ. However, as to the wave length showing the most significant change in admittance no knowledge is obtainable at present. Thus, in this study we attempted to show the change in transmission spectrum due to the reaction by means of a rapid scanning spectrometer.

Method

The change in S_{CO} was determined photometrically by using a stopped flow method. The change in transmission due to the reaction is generally very small, therefore, it was technically necessary to take the difference between the two spectra of the reacting solution and a reference solution of 100% S_{O_2}. That is to say, two observation cells were prepared for these solutions and the difference between the two transmission spectra was measured. The structure of the optical device is shown schematically in Fig. 10.

The light from a tungsten lamp is dispersed by a grating mirror and after passing through a compensating mirror which was specially made to render the output signal from a photomultiplier flat, over a wave length range of about 400 to 600 mμ, it was reflected to Mirror-I by a rotating scanning mirror driven by a synchronous motor. The revolution of the rotating mirror is 1500 R.P.M., thus, the scanning is repeated 25 times per sec. Since the mirror surface is not rectangular to the rotating axis, the light arriving at Slit-II after reflection at Mirror-I changes its color as the mirror rotates. The duration for which the dispersed light is injected into Slit-II, namely, the scanning duration is 3 msec. In other words, a scanning of 3 msec is repeated every 40 msec. After passing through Slit-II, the light is divided into two parts by a half-mirror, and each part is injected into a mixing chamber of an O_2Hb reference solution or a reacting sample. Then, the transmitted lights are led into

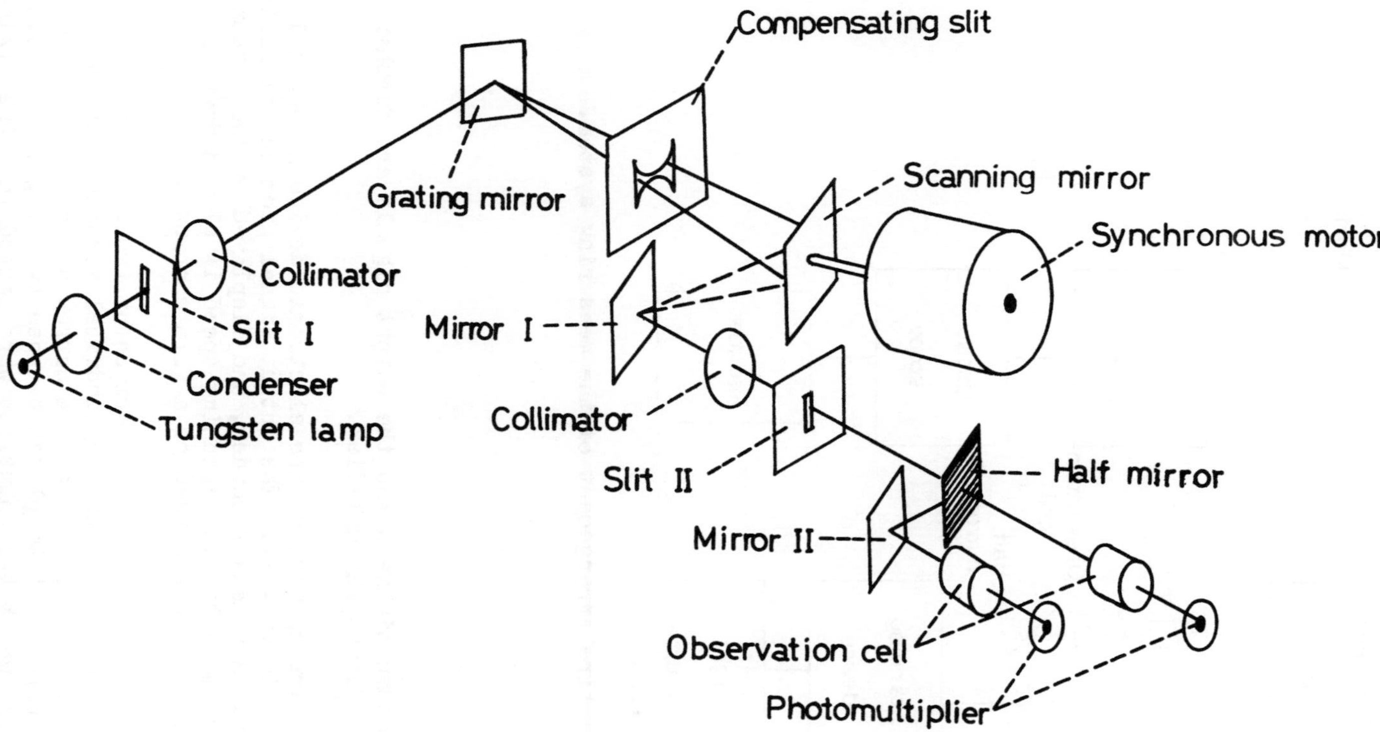

Fig. 10 Schematic diagram of the optical device of a rapid scanning spectrometer.

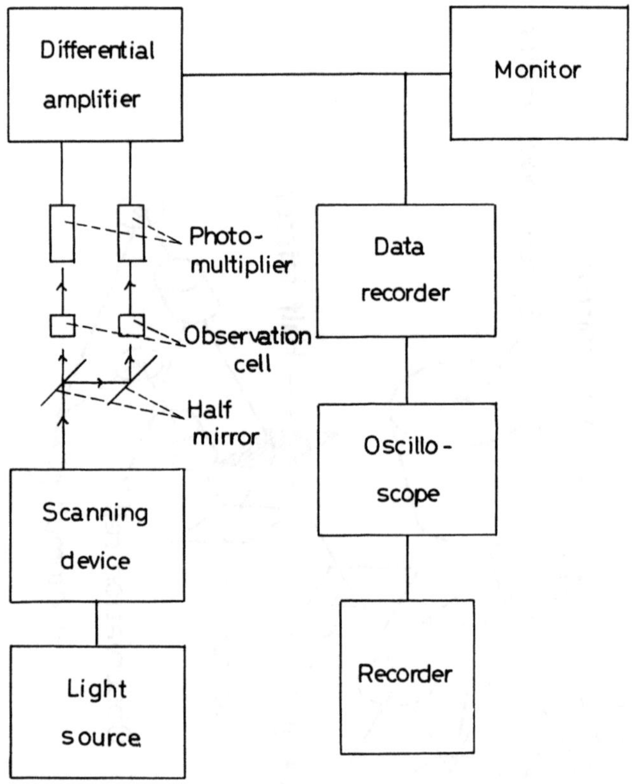

Fig. 11 The entire arrangement of the measuring system.

respective photomultipliers and the output signals are further led to a DC differential amplifier.

The whole arrangement of the electronic device is shown in a block diagram in Fig. 11. The output signal from the amplifier is first stored in a data recorder, and displayed on the screen of an oscilloscope after the scanning speed is slowed down to 1/5 of the original and is recorded on a film continuously.

Hb and buffer solutions were driven by two syringes of 10 ml into an observation cell with a cylindrical structure as shown in Fig. 12. At the inlet of the cell a narrow space in ring form was attached; this was interplaced between two ring plates with 24 holes through which two solutions were jetted into the mixing

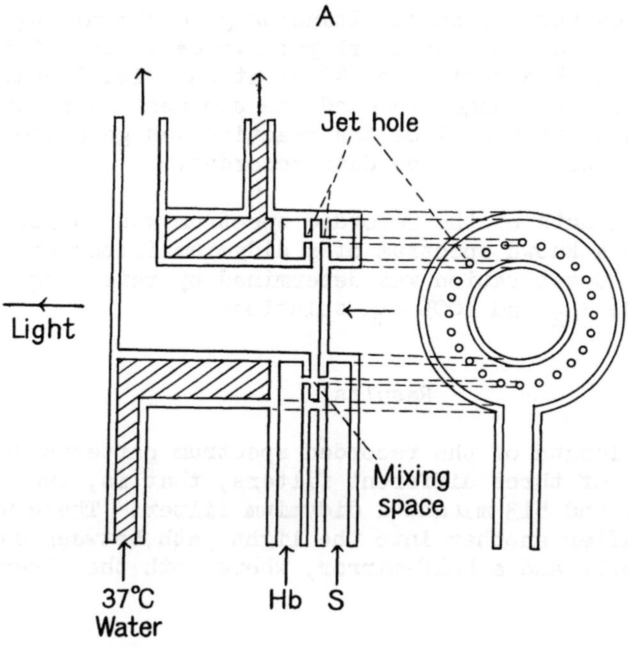

Fig. 12 A. Structure of the observation cell.

space. The cylinder of the cell has a 7 mm inner diameter, and the light penetrates through the cell in the axial direction of the cylinder. The light path has a length of 15 mm.

The Hb solution was prepared using freshly drawn blood, which was processed by freezing and thawing and thereafter was diluted by a phosphate buffer solution to a 1,500 fold dilution. The pH was adjusted by $NaHCO_3$ to become 7.26 at 40 mmHg P_{CO_2}. The CO containing gas was prepared by adding a certain amount of CO to the gas mixture of a known volume, which was used for tonometry of the CO free solution. The CO concentration was measured after the measurement by burning a gas sample using a heated Pt wire.

Tonometry was performed by bubbling solutions in five vessels: Three vessels were used for tonometry of the Hb solution and the other vessels were used for that of the buffer solution. Each vessel was connected with a respective syringe through a three-way cock. After 30 minutes tonometry, the solutions were transferred to four syringes and driven simultaneously to the mixing chambers by an air pump at a speed of 10 ml per 1.5 sec. Outlet tubing was connected with a syringe of 40 ml at the distal end; this was pushed up until the plunger reached the stopper. When the plunger was stopped, a start signal of the reaction was generated at the stopper and transmitted to the data recorder.

The wave length of the recorded spectrum was calibrated by comparing with a known spectrum of a didymium filter or interference filters. The CO saturation was determined by referring to the two spectra of 100% S_{O_2} and 100% S_{CO} solutions.

Results

The wave length of the recorded spectrum patterns was calibrated by use of three different filters, that is, two interference filters of 471 and 513 mμ and a didymium filter. These were inserted one after another into the light path between one of the observation cells and a half-mirror, where both the observation

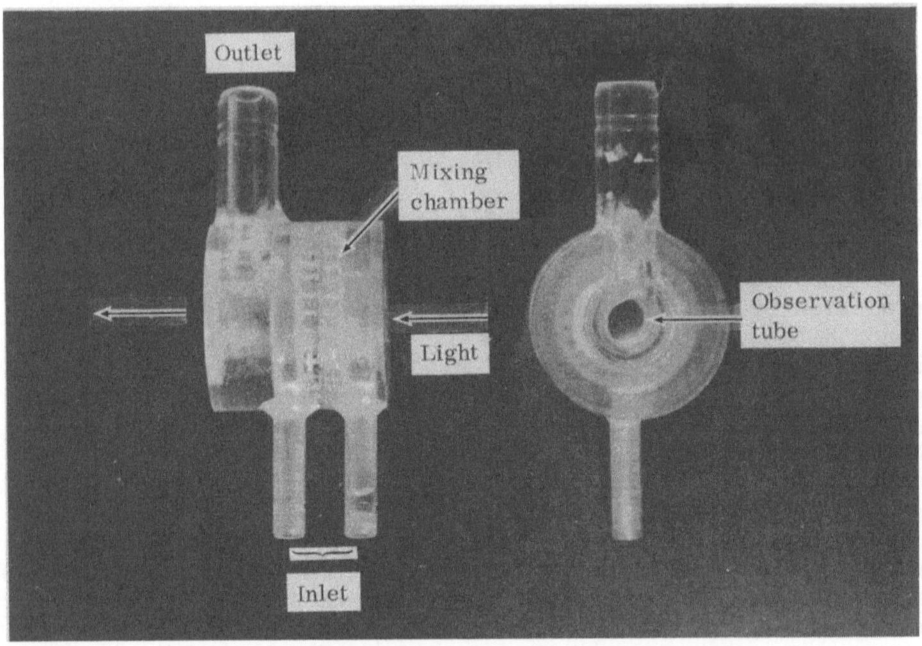

Fig. 12 B Actual photograph of the observation cell.

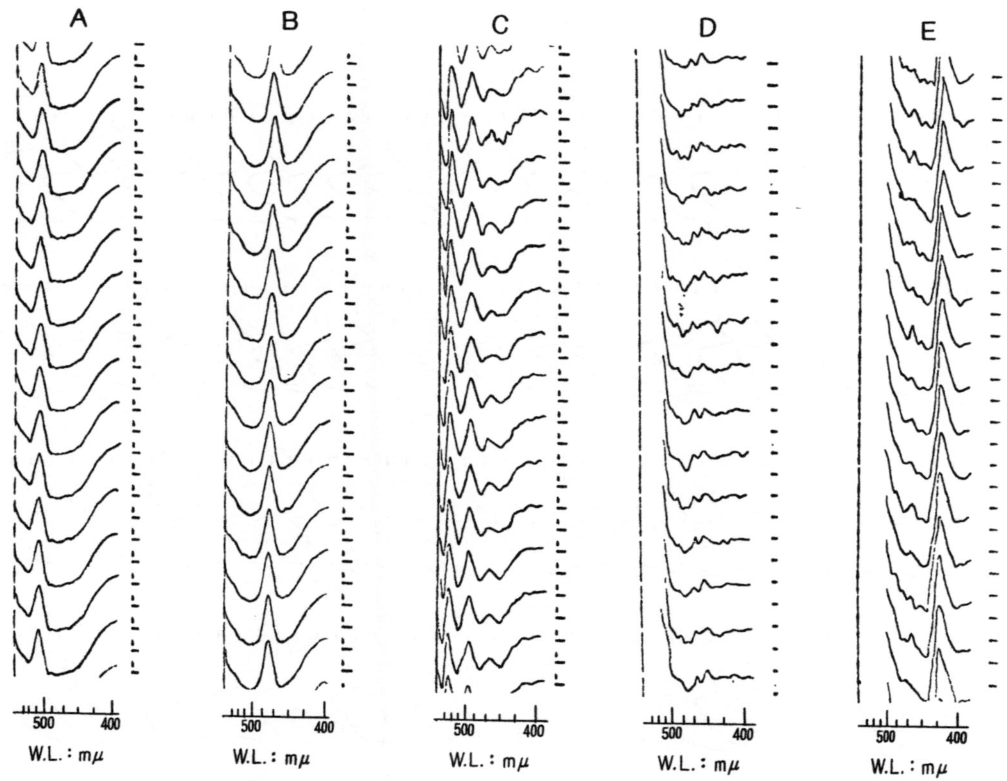

Fig. 13 Spectrum patterns obtained with interference filters of 513 mμ(A) and 471 mμ(B), a didymium filter (C), in O_2Hb (D) and COHb (E) solutions. In cases of A, B and C, the filters were set just in front of one of two observation cells, which were filled with distilled water. The pattern of D was obtained by filling both observation cells with O_2Hb solution and that of D, by filling the cells with O_2Hb and COHb solutions respectively. W.L. is an abbreviation of the wave length.

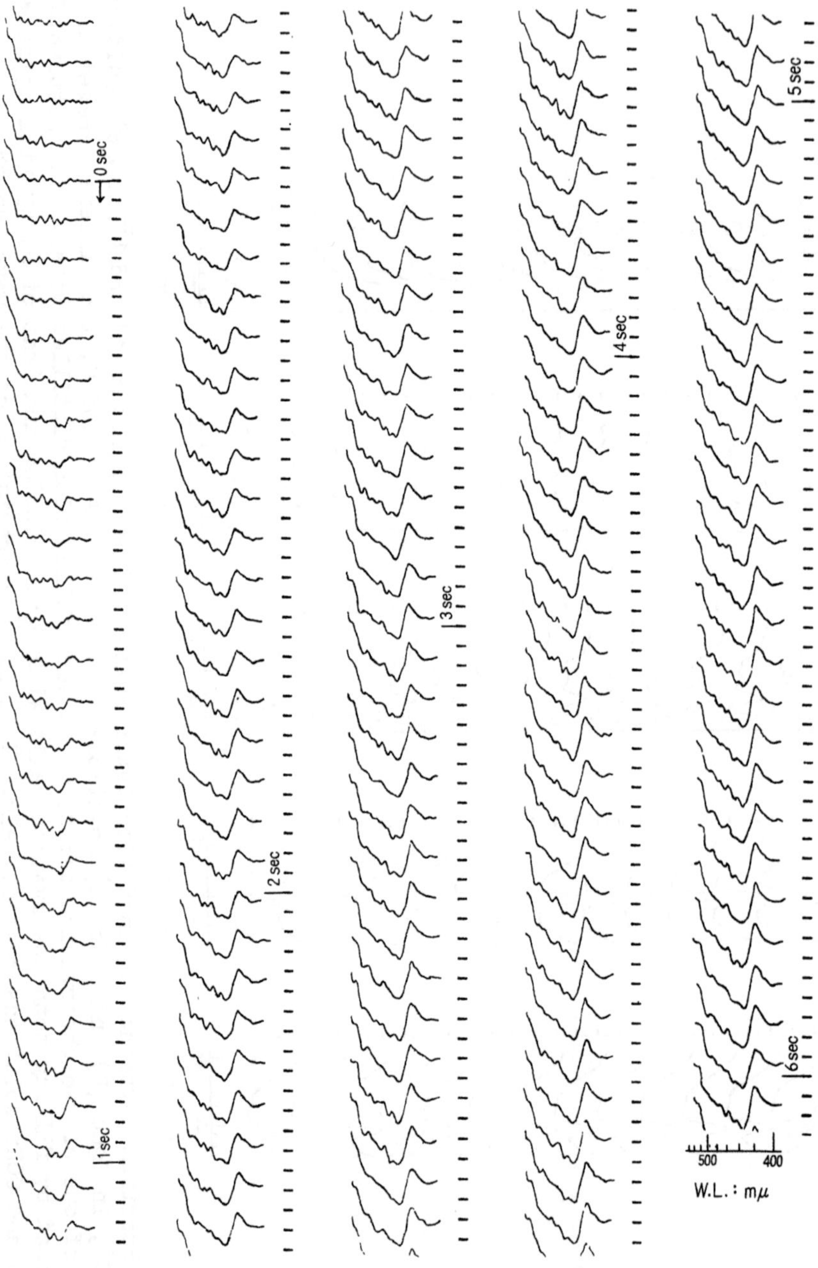

Fig. 14 Change of spectrum pattern during the course of O_2 replacement reaction by CO from O_2Hb. P_{O_2} was 180 mmHg and P_{CO} was 6 mmHg.

cells were filled with distilled water. Spectrum patterns A, B and C in Fig. 13 were obtained by using the filters of 513, 471 mμ and didymium respectively. As shown by a calibrated scale at the bottom of each pattern, the space of the abscissa was not linearly proportional to the wave length. At a shorter range of the wave length the space was greatly expanded. Patterns D and E in Fig. 12 were obtained when both the observation cells were filled with the same O_2Hb solution and when one of them was filled with O_2Hb and the other, with COHb solution, respectively. In the former case a fair balance could be attained over a range of 41 to 520 mμ. In the latter case the light absorption by COHb was observed as a sharp peak at a range around 420 mμ and on the other hand over a range of 430 to 500 mμ transmission was rather increased.

An example of a change in spectrum due to the reaction of CO and O_2Hb is seen in Fig. 14. The recorded pattern was obtained when the Hb solution with 181 mmHg P_{O_2} was mixed with an equal amount of buffer solution with 12 mmHg P_{CO} and the P_{O_2} of 179 mmHg. Thus, the P_{O_2} after mixing was 180 mmHg and the P_{CO} was 6 mmHg. The time mark on the record commenced with the start of reaction. The interval of the time mark was about 20 msec, and the spectrum was recorded for 6 sec continuously. The distance between the two peaks of 420 and 440 mμ extended as the reaction proceeded, at 6 sec the CO saturation became about 50%.

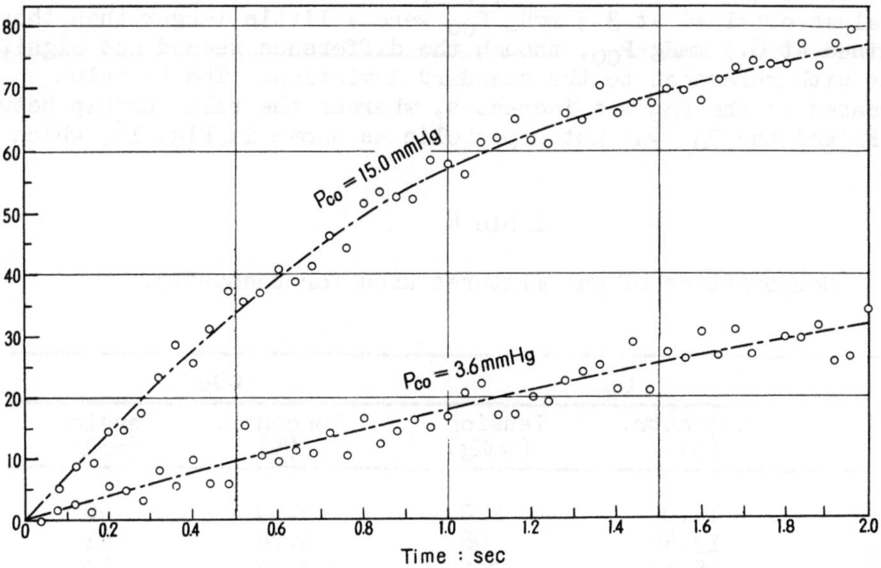

Fig. 15 Change in S_{CO} due to the replacement reaction obtained from transmission difference between two wave lengths of about 420 and 440 mμ. P_{CO} was 15.0 and 3.6 mmHg, and P_{O_2} was about 104 mmHg in both cases.

Two examples of the S_{CO}-time curve obtained at P_{CO}'s of 15.0 and 3.6 mmHg are shown in Fig. 15. The points plotted are the average values of four serial records measured under the same experimental conditions. The points are scattered around the mean smoothed curve with a width of 5% S_{CO} or thereabouts. Although the S_{CO}-time curve at P_{CO} of 15 mmHg was upwards convex, that of P_{CO} of 3.6 mmHg was almost linear during the reaction time recorded. In other words, the time rate of S_{CO} change was constant, regardless of the S_{CO}. This fact was generally confirmed in all the measurements at a low P_{CO}.

Since it was troublesome to obtain the mean smoothed S_{CO}-time curve for each experiment, Fukui and Kakiuchi (1969) improved the display system of Fig. 11 by using an electronic averaging device and measured the reaction rates at 3.3 and 6.5 mmHg P_{CO}'s and at five different P_{O_2}'s of about 100, 200, 343, 500 and 670 mmHg. The compositions of gas mixtures used for tonometry are shown in Table 4. In order to show the reaction velocity numerically, the velocity factor Fs was calculated according to the following formula:

$$Fs = (0.4 \times 10^{-2}\ dS/dt)/P_{CO} \qquad (10)$$

The Fs values obtained in 5 normal subjects are tabulated in Table 5. These were generally smaller than those obtained hitherto by Roughton et al (1957) and by us (1969) in a higher P_{CO} range. The Fs values obtained at 3.3 mmHg P_{CO} were a little larger than those obtained at 6.5 mmHg P_{CO}, though the difference seemed not significant with reference to the standard deviation. The Fs value decreased as the P_{O_2} was increases, whereas the relationship between the Fs and the P_{O_2} was not hyperbolic as shown in Fig. 16, which

Table 4

Compositions of gas mixtures used for tonometry.

No.	O_2		CO_2	
	Concentr. (%)	Tension (mmHg)	Concentr. (%)	Tension (mmHg)
1	13.95	100	5.59	40
2	13.48	96	5.74	41
3	28.12	201	5.63	40
4	48.12	343	5.54	39
5	69.91	499	5.76	39
6	94.43	673	5.54	40

Table 5. Velocity factors of CO obtained in a Hb solution at five different P_{O_2}'s and at 3.3 and 6.5 mmHg P_{CO}'s.

	P_{O_2}: mmHg	100		200		340		500		675	
	P_{CO}: "	3.3	6.5	3.3	6.5	3.3	6.5	3.3	6.5	3.3	6.5
Subject	T.K.	.0185	.0192	.0121	.0108	.0088	.0078	.0068	.0067	.0047	.0047
	Y.K.	.0155	.0155	.0109	.0111	.0075	.0077	.0074	.0066	.0048	.0043
	K.F.	.0150	.0143	.0120	.0125	.0100	.0084	.0061	.0054	.0043	.0038
	N.S.	.0229	.0187	.0135	.0134	.0127	.0103	.0076	.0065	.0049	.0042
	K.N.	.0217	.0204	.0136	.0109	.0122	.0098	.0067	.0064	.0050	.0042
	Mean	.0187	.0176	.0124	.0117	.0102	.0088	.0069	.0063	.0047	.0042
	S.D.	.0032	.0023	.0010	.0010	.0020	.0011	.0005	.0005	.0002	.0003

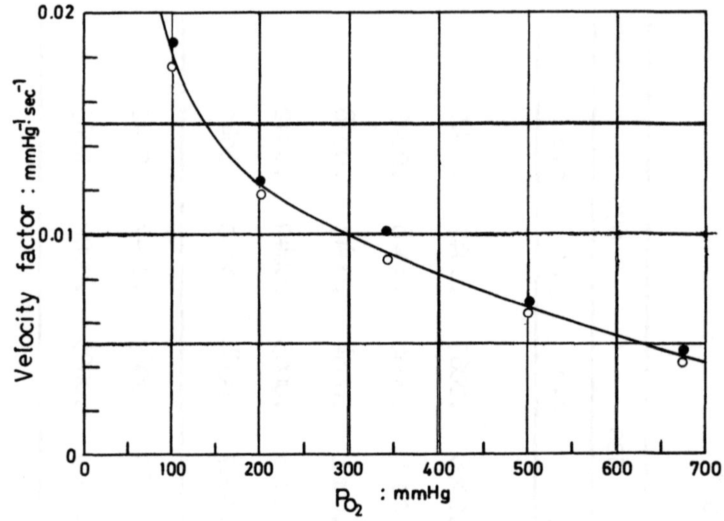

Fig. 16 Fs values plotted against the P_{O_2}. P_{CO} was 3.3 (closed circle) and 6.5 (open circle) mmHg.

was different from the description of Roughton and others (1957).

Discussion

Recently we have measured th O_2 replacement reaction by CO from O_2Hb by the use of a rapid flow method, and found that the reaction rate had a very high value at the initial stage of the reaction and rapidly decreased as the S_{CO} increased. In detail, the Fs value in a range of 0 to 10% of S_{CO} was about 0.12 sec^{-1}. $mmHg^{-1}$, but soon after it decreased with the increase of S_{CO} to comparable values as those shown in Table 5.

Koyama and Mochizuki (1967, 1968, 1969) showed that the oxygenation velocity of the red cell measured in a rapid flow method increased with flow velocity through an observation tube. However, it is doubtful that the reaction rate in a laked Hb solution may also be influenced by the flow. In the previous method the mixing chamber had only two jets as shown in Fig. 5; one jet was for the buffer solution and the other for the Hb solution, while in the present mixing chamber each of the solutions was jetted through 24 holes and mixed in a very narrow space. Therefore, it is additionally doubtful whether the mixing of two sample solutions was still incomplete in the present measurement.

Sirs and Roughton (1963) showed that the difference between two transmission signals at two wave lengths should linearly correlate with S_{CO}, regardless of the range of wave length. However, it may be possible that during the reaction process the changing ratio of transmission may not always be identical over the entire wave length range. For the purpose of analysing $D_{L_{CO}}$, the clarification of the relationship between the velocity factor and P_{CO} is absolutely necessary, especially in a low P_{CO} range under 3 mmHg. Hence, before the clarification of the above relationship an investigation of the reason, as to how the difference in the Fs values arises, is now keenly demanded.

IV. THE TREATMENT OF THE DIFFUSION BARRIER AROUND THE RED CELL IN THE LUNG CAPILLARY

As shown in the foregoing papers, the replacement reaction of O_2 by CO from O_2Hb at a low P_{CO} range is independent on the S_{CO} at least in a low S_{CO} range of 0 to 20%, that is to say, the back pressure of COHb need not be taken into account in the treatment of the reaction, which is different from the oxygenation reaction. Hence, the diffusion rate through the barrier in the lung may be calculated more easily by using the data on the $D_{L_{CO}}$ than that on the $D_{L_{O_2}}$.

The experimental fact that the Fs value was almost independent of the P_{CO} indicates that the CO quantity taken up by Hb molecules is linearly proportional to the P_{CO}. When the red cell is assumed to be a solid plate of a thickness 2d with the origin at the center, the diffusion equation in the red cell is given as follows:

$$\frac{\partial P}{\partial t} = \alpha D \frac{\partial^2 P}{\partial x^2} - Fs \cdot P, \qquad (11)$$

Let the transfer coefficient across the diffusion barrier which includes also the alveolar membrane be η. Then, at the boundary between the red cell and the plasma, x = d, the CO quantity penetrating through the barrier may be denoted as $\eta(P_a - P_{x=d})$, where P_a is the alveolar gas P_{CO}. The diffusion quantity entering into the red cell, on the other hand, is expressed as $\alpha D(\frac{dP}{dx})_{x=d}$. Therefore, at any time,

$$\eta \cdot (P_a - P_{x=d}) = \alpha D(\frac{dP}{dx})_{x=d}. \qquad (12)$$

Moreover, since CO molecules do not diffuse across the plane at the center, the P_{CO} gradient at x = 0 should be 0, namely,

$$\left(\frac{dP}{dx}\right)_{x=0} = 0. \tag{13}$$

At the initial stage, on the other hand, the P_{CO} in the red cell is 0, that is,

$$P_{t=0} = 0. \tag{14}$$

Thus, the differential equation (11) can be solved under the initial and boundary conditions of Eqs (12), (13) and (14). Especially, since the early transition stage of the reaction calculated is very short as compared with the contact time in the lung, the CO quantity entering into the red cell is calculated by taking only the steady state term of the solution as follows:

$$\alpha D\left(\frac{dP}{dx}\right)_{x=d} = Pa\,\frac{\eta\cdot X}{\eta + X}, \tag{15}$$

where

$$X = \sqrt{\alpha\cdot D\cdot Fs}\quad \tanh\left(\sqrt{\frac{Fs}{D}}\cdot d\right) \tag{16}$$

The X value of Eq (16) has already been obtained by Roughton (1932). Thus, from Eq (15) the concentration increase (ml) in one ml red cells during a unit time may be given by,

$$\frac{d[CO]}{dt} = \frac{Pa}{d}\;\frac{\eta\cdot X}{\eta + X}. \tag{17}$$

When Q ml blood with the hematocrit Ht flows through the entire lung capillary during tc sec, the CO uptake $\dot{V}_{CO}$ may be given as follows:

$$\dot{V}_{CO} = Q\cdot Ht\cdot tc\cdot Pa\,\frac{\eta\cdot X}{(\eta + X)\,d}. \tag{18}$$

Since, $D_{L_{CO}} = \dot{V}_{CO}/Pa$, the $D_{L_{CO}}$ may be given by the following;

$$1/D_{L_{CO}} = d/(Q\cdot Ht\cdot tc)(\,1/\eta + 1/X). \tag{19}$$

Thus, by using the two relationships of $D_{L_{CO}}$ against P_{O_2} and X against P_{O_2}, $d/(Q \cdot Ht \cdot tc)$ and $1/\eta$ of Eq (19) may easily be determined according to a similar method described by Forster and others (1957). Fig. 17 shows the Fs value obtained by Fukui and Kakiuchi and the $Fc = X/d$ calculated from it by Eq (16), where the following values were used for α, D and d : $\alpha = 0.24 \times 10^{-4}$ $mmHg^{-1}$, $D = 0.5 \times 10^{-5}$ $cm^2 \cdot sec^{-1}$ and $d = 0.8 \times 10^{-4}$ cm.

The reciprocal of the Fc is shown in Fig.18 with that of θ value derived by Forster and others (1957). Different from the $1/\theta$ value, $1/Fc$ was not linearly correlated with P_{O_2}. When the reciprocals of the D_{LCO}'s obtained by Kruhoffer (1954) and Forster et al (1957) at various P_{O_2} were plotted against that of the Fc's at the same P_{O_2}, however, the relationships between them became fairly linear as shown in Fig. 19. By referring to the $1/D_{LCO}$ values at $1/Fc = 0$, the value for $(Q \cdot Ht \cdot tc \cdot \eta/d)$ can be evaluated as 46 and 71 $ml \cdot sec^{-1} \cdot mmHg^{-1}$ for both the cases of Kruhoffer and Forster et al respectively. Moreover, by taking the ratio of η to X, the transfer coefficient η in both cases can be obtained as 0.08×10^{-5} and 0.13×10^{-5} $cm \cdot sec^{-1} \cdot mmHg^{-1}$. These values for CO may correspond to 0.10×10^{-5} and 0.15×10^{-5} $cm \cdot sec^{-1} \cdot mmHg^{-1}$ of the η values for O_2 respectively, if the difference between the solubilities of O_2 and CO is taken into account. Such η values for O_2 as above seem compatible with the value which may be considered as reasonable in the appreciation of the Fc value measured by a stopped flow method as shown in Table 1 and in Fig. 19.

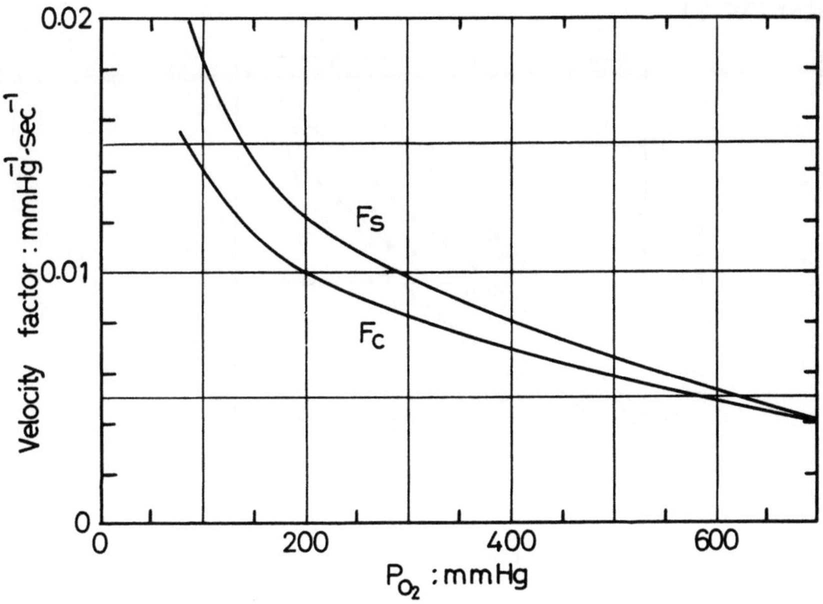

Fig. 17 Fs values measured by Fukui and Kakiuchi and Fc values calculated by using the above Fs values.

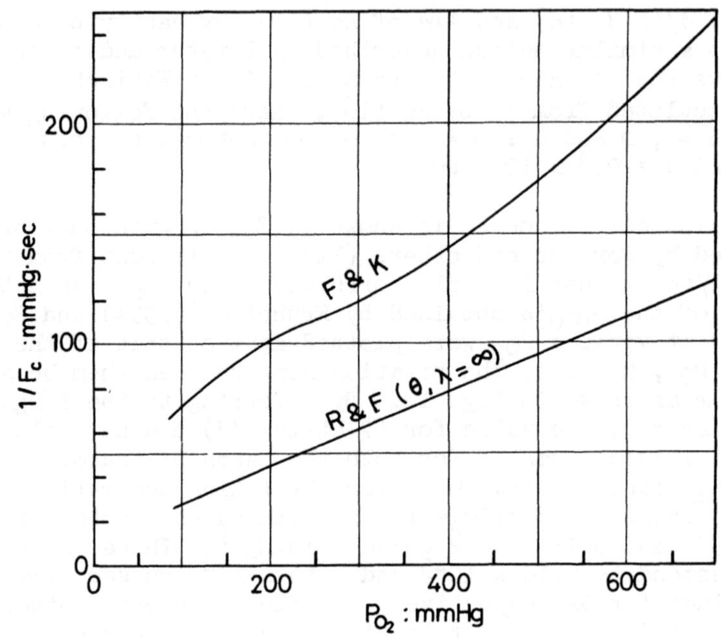

Fig. 18 Reciprocals of the Fc values plotted against the P_{O_2}. The curve F & K was calculated from the Fc value in Fig. 17 and the curve R & F was calculated from the θ values obtained by Roughton and Forster (1957)

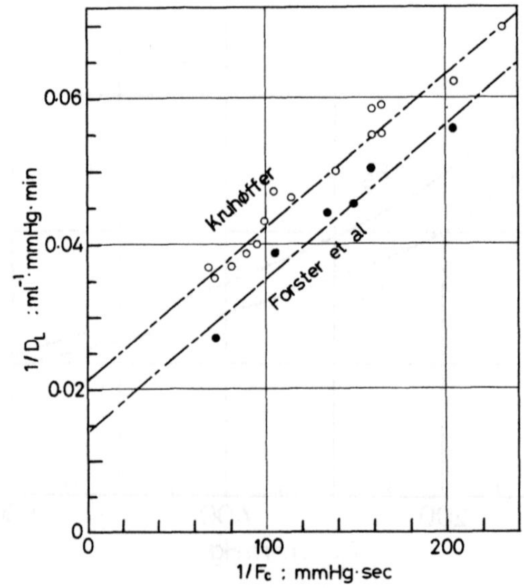

Fig. 19 Reciprocals of D_{LCO} obtained by Kruhoffer and Forster et al plotted against those of the Fc.

References

1. Bartels, H., Beer, R., Fleischer, E., Koepchen, H.P., Krall, J. Rodewald, G., Wenner, J. and Witt, I.: Bestimmung von Kurzschlußdurchblutung and Diffusionskapazitat der Lunge bei Gesunden und Lungenkranken. Pflugers Arch. 261, 99-132, 1955.

2. Bohr, C.: Uber die spezifische Tatigkeit der Lungen bei der respiratorischen Gasaufnahme und ihr Verhalten zu der durch die Alveolarwand stattfindenden Gasdiffusion. Skand. Arch. Physiol. 23, 221-280, 1909.

3. Douglas, C.C., Haldane, J.S. and Haldane, J.B.S.: The law of combination of hemoglobin with carbon monoxide and oxygen. J. Physiol. 44, 275-304, 1912.

4. Forster, R.E., Roughton, F.J.W., Kreuzer, F. and Briscoe, W.A.: Photometric determination of rate of CO and O_2 by reduced human red cell suspensions at 37°C. J. Appl. Physiol. 11, 260-268, 1957.

5. Forster, R.E.,Roughton, F.J.W., Cander, L., Briscoe, W.A.: Apparent Pulmonary diffusing capacity, for CO at varying alveolar O_2 tensions. J. Appl. Physiol. 11, 277-289, 1957.

6. Fukui, K. and Kakiuchi, Y.: The kinetics of the reaction of $CO \quad O_2Hb = O_2 \quad COHb$ in human hemoglobin solution at a low P_{CO} range. Jap. J. Physiol. in print.

7. Gibson, Q.H., Kreuzer, F., Meda, E. and Roughton, F.J.W.: The kinetics of human hemoglobin in solution and in the red cell at 37°C. J. Physiol. 129, 65-89, 1955.

8. Hartridge, H. and Roughton, F.J.W.: The rate of distribution of dissolved gas between the red blood corpuscle and its fluid environment. J. Physiol. 62, 232-242, 1927.

9. Holland, R.A.B. and Forster, R.E.: The effect of size of red cells on the kinetics of their O_2 uptake. J. Gen. Physiol. 49, 727-742, 1966.

10. Koyama. T. and Mochizuki, M.: Application of the Pt-polarography to the stopped flow method for measurement of the oxygen uptake by the red blood cell. Rev. Polarography. 14, 405-411, 1967.

11. Koyama, T., Furuse, T., Arai, T. and Mochizuki, M.: A study on the oxygenation velocity factor of the red blood cell by use of the rapid flow method combined with a Pt-electrode as the oxygenation detector. Bul. Res. Inst. Appl. Electr. 20, 144-152, 1968.

12. Koyama, T. and Mochizuki, M.: A study on the relationship between the oxygenation velocity of the red blood cell and the flow velocity in a rapid flow method. Jap. J. Physiol. 19, 534-546, 1969.

13. Kruhoffer, P.: Studies on the lung diffusion coefficient for carbon monoxide in normal human subjects by means of $C^{14}O$. Acta Physiol. Scand. 32, 106-123, 1954.

14. Miyamoto, Y. und Moll, W.: Gestalt, Anordnung und Dichte der Erythrozyten in Blutcapillaren. Pflugers Arch. 300, 21, 1968.

15. Mochizuki, M. and Fukuoka, J.: The diffusion inside the red cell. Jap. J. Physiol. 8, 206-223,1958.

16. Mochizuki, M., Nakamura, K. and Oyama, Y.: Study on the oxygenation velocity of human hemoglobin. Jap. J. Physiol. 16, 519-527,1966.

17. Mochizuki, M.: Study on the oxygenation velocity of the human red cell. Jap. J. Physiol. 16, 635-648, 1966.

18. Mochizuki, M.: A theoretical study on the velocity factor of oxygenation of the red cell. Jap. J. Physiol. 16, 658-666, 1966.

19. Mochizuki, M.: Transport of the respiratory gases. The relationship between O_2 diffusing capacity and the oxygenation velocity of the red cell. Jap. Cir. J. 32, 1817-1820, 1968.

20. Mochizuki, M., Furuse, T. and Fujisawa, Y.: A kinetic study on the oxygenareplacement reaction by carbon monoxide from oxygenated hemoglobin in human hemoglobin solution and red cell suspension. Bul. Res. Inst. Appl. Electr. 21, in print.

21. Mochizuki, M., Furuse, T., Arai, T. and Kakiuchi, Y.: An application of rapid scanning spectrophotometry for the measurement of the reaction rate of CO and O_2Hb. Bul. Res. Inst. Appl. Electr. 21, in print.

22. Nicolson, P. and Roughton, F.J.W.: A theoretical study of the influence of diffusion and chemical reaction velocity on the rate of exchange of carbon monoxide and oxygen between the red blood corpuscle and the surrounding fluid. Proc. Roy. Soc. B. 138, 241-264, 1951.

23. Roughton, F.J.W.: Diffusion and chemical reaction velocity as joint factors in determining the rate of uptake of oxygen and carbon monoxide by the red blood corpuscle.

Proc. Roy. Soc. B. 111, 1-36, 1932.

24. Roughton, F.J.W.: The kinetics of the reaction $CO + O_2Hb \rightleftharpoons O_2 + COHb$ in human blood at body temperature. Am. J. Physiol. 143, 609-620, 1945.

25. Roughton, F.J.W.: The average time spend by the blood in the human lung capillary and its relation to the rates of CO uptake and elimination in man. Am. J. Physiol. 143, 621-633, 1945.

26. Roughton, F.J.W., Forster, R.E. and Cander, L.: Rate at which carbon monoxide replaces oxygen from combination with human hemoglobin solution and in the red cell. J. Appl. Physiol. 11, 269-276, 1957.

27. Roughton, F.J.W. and Forster, R.E.: Relative importance of diffusion and chemical reaction rates in determining rate of exchange of gases in the human lung, with special reference to true diffusing capacity of pulmonary membrane and volume of blood in the lung capillary. J. Appl. Physiol. 11, 290-302, 1957.

28. Saito, Y.: A sputtered Pt film electrode for polarographic O_2 measurement. J. Appl. Physiol. 23, 979-983, 1967.

29. Sirs, J.A. and Roughton, F.J.W.: Stopped flow measurement of CO and O_2 uptake by hemoglobin in sheep erythrocytes. J. Appl. Physiol. 18, 158-165, 1963.

30. Sirs, J.A.: Influence of metabolism on uptake of CO and O_2 by hemoglobin in erythrocytes. J. Appl. Physiol. 18, 175-178, 1963.

31. Staub, N.C., Bishop, J.M. and Forster, R. E.: Velocity of O_2 uptake by human red blood cells. J. Appl. Physiol. 16, 511-516, 1961.

32. Staub, N.C. and Schultz, E.L. : Pulmonary capillary length in dog, cat and rabbit. Res. Physiol. 5, 371-378, 1968.

MEMBRANE RESISTANCE AND DIFFUSION COEFFICIENTS: THE INTACT RED CELL

Robert Franco and Daniel Hershey

Department of Chemical and Nuclear Engineering

University of Cincinnati, Cincinnati, Ohio 45221

The transport of oxygen through the interior of the red cell is characterized by a diffusion coefficient and oxygen transport through the membrane is described in terms of a mass transfer coefficient. In order to determine these coefficients, a flow apparatus is used in which blood is mixed with isotonic saline solution and flows down a glass tube. The basic data obtained is extracellular oxygen partial pressure versus time. The results are independent of hemoglobin reaction or diffusion since at all times the hemoglobin is kept fully saturated with oxygen. Therefore, there is no need for hemoglobin-oxygen rate constants and hemoglobin diffusivity. Also, the controlling diffusion equation is linear because of the absence of reaction terms and a straight forward analytical solution is possible.

EXPERIMENTAL METHODS

A schematic diagram of the equipment is shown in Figure 1. A stream of oxygenated blood (P_{O_2} = 715) is mixed with deoxygenated isotonic saline (P_{O_2} = 0) in a saline:blood ratio of 2:1. This insures that the partial pressure of oxygen in the red cells is always above the value necessary for complete conversion of hemoglobin to oxyhemoglobin. The blood-saline system flows down the tube with a velocity which assures mixing due to turbulent flow ($N_{RE} > 2500$). In our first attempts, a microelectrode (diameter= 0.05 cm., length=18.7 cm.) was inserted in the tube and the P_{O_2} at any point in the tube was measured by sliding the electrode in and out. The elapsed time from mixing is proportional to the distance from the mixing chamber to the tip of the electrode and therefore a time-dependent curve for extracellular oxygen partial pressure is generated. However, the readings from the microelectrode were not stable enough. In addition, the electrode limited how small the diameter and length of the tube could be made. With the microelectrode system, the following observations were made:

1. At $N_{RE} < 2000$, a large P_{O_2} gradient was observed across the diameter of the tube. Apparently, streamlines were being set up in the mixing chamber and continued down the tube.

2. At $N_{RE} > 2500$ this gradient disappeared and the readings were constant across the tube diameter. It was therefore apparent that mixing due to turbulent flow was taking place.

To overcome the shortcomings of the microelectrode, we tried a system employing a macroelectrode, which is shown in Figure 2. In this case the fluid from the tube impinges directly on the membrane of an oxygen macroelectrode. The time of measurement is varied by changing the flow rate while maintaining a constant blood-saline flow ratio. The experiment yields a curve of P_{O_2} in the saline-plasma mix (extracellular P_{O_2}) versus time.

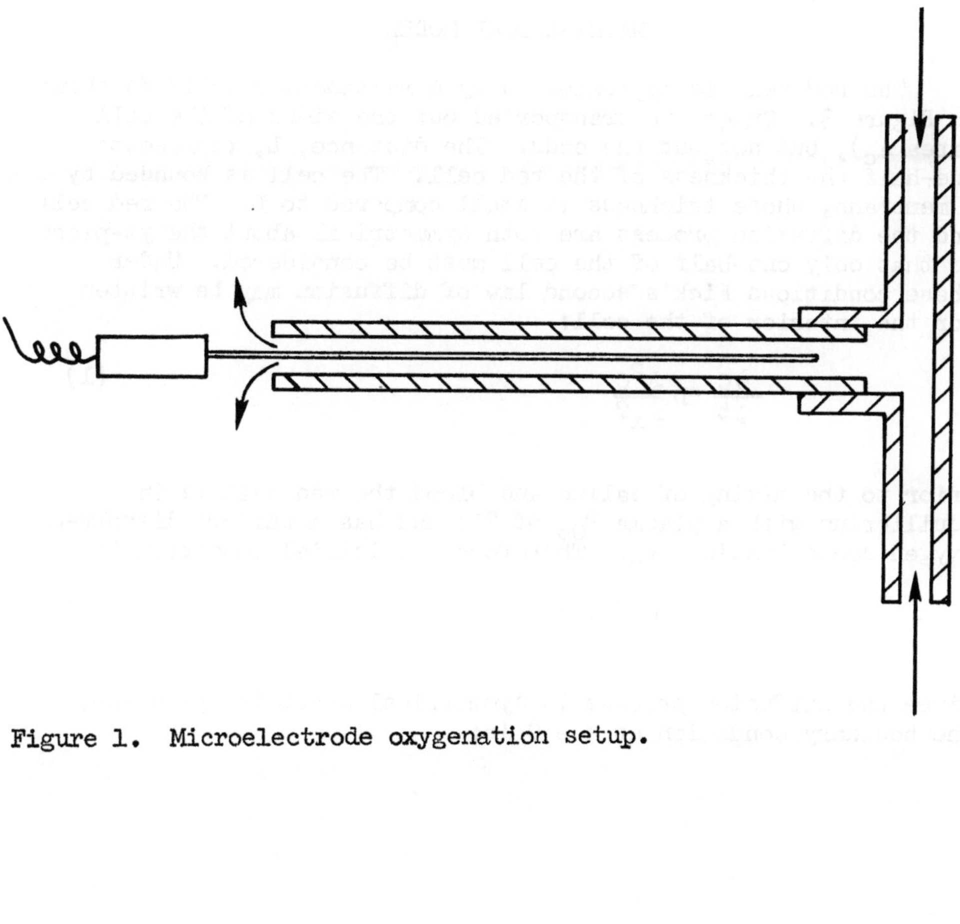

Figure 1. Microelectrode oxygenation setup.

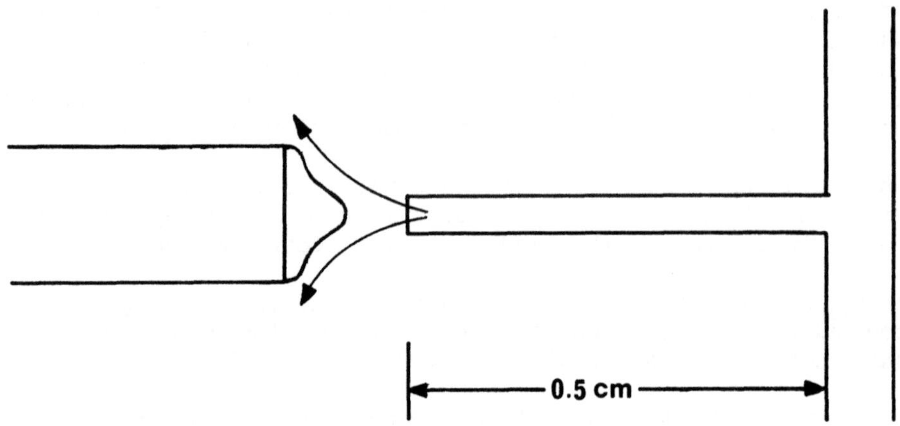

Figure 2. Macroelectrode oxygenation setup.

MATHEMATICAL MODEL

The red cell is approximated by a rectangular solid as shown in Figure 3. Oxygen is transported out the sides of the cell (area A_c), but not out the ends. The distance, L, represents one-half the thickness of the red cell. The cell is bounded by a membrane, whose thickness is small compared to L. The red cell and the diffusion process are both symmetrical about the yz-plane so that only one-half of the cell must be considered. Under these conditions Fick's second law of diffusion may be written for the interior of the cell:

$$\frac{\partial c}{\partial t} = D \frac{\partial^2 c}{\partial x^2} \tag{1}$$

Prior to the mixing of saline and blood the red cell is in equilibrium with a plasma P_{O_2} of 715 and has a uniform dissolved oxygen concentration, v_o. Therefore the initial condition is:

$$c\ (x,o) = v_o \tag{2}$$

Since the diffusion process is symmetrical about the yz-plane, the boundary condition at $x = 0$ is:

$$\left[\frac{\partial C(x,t)}{\partial x}\right]_{x=0} = 0 \tag{3}$$

There remains to be found a boundary condition at x = L. This boundary condition introduces the effect of the red cell membrane on the deoxygenation process. Assuming that the flux of oxygen through the membrane is passive, there are at least two ways of representing the process.

1. The diffusion equation may be written for the membrane and coupled with the diffusion equation for the cell interior. This is relatively difficult mathematically and it is necessary to specify the thickness of the membrane.

2. A mass transfer coefficient, h, across the membrane may be defined by

$$N(t) = hA_c\left[C(L,t)-C_p(t)\right] \tag{4}$$

Essentially, this assumes that the flux through the membrane is proportional to the area of the membrane and also to the concentration driving force across the membrane. In this case it is not necessary to assume a thickness for the cell membrane.

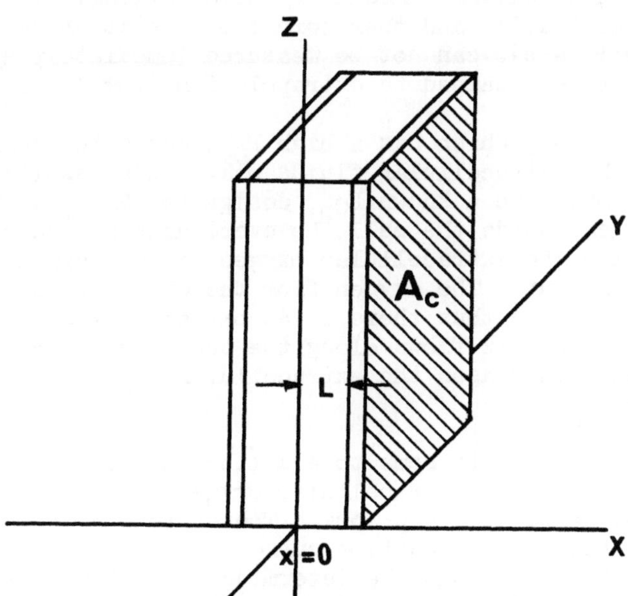

Figure 3. Schematic representation of the red cell.

The second of these approaches was chosen for our work. Equation (4) may be rearranged to give the boundary condition at x = L:

$$C(L,t) = C_p(t) + \left[\frac{N(t)}{hA_c}\right] \quad (5)$$

The system of equations (1, 2, 3, 5) may be solved analytically for the concentration of oxygen at any point in the cell at any time. It is of the form:

$$C = C(x,t,D,h) \quad (6)$$

There are two unknown constants in equation (6), D and h. Comparison of the experimental results for molar flux out of the cells with the theoretically predicted results allows the direct determination of the diffusion coefficient once the membrane coefficient, h, has been determined.

DETERMINATION OF h

Since the concentration driving force across the membrane is known immediately after mixing, h may be calculated directly from equation (4) at t = 0. The assumption is that the membrane resistance is controlling immediately after mixing. The change in plasma concentration and therefore the rate at which oxygen leaves the cell, N(t), can not be measured immediately after mixing and must be obtained be extrapolation from later times.

When the blood, which has a high P_{O_2}, meets the saline stream with low P_{O_2} the extracellular fluids mix. This results in a plasma-saline mix with a known P_{O_2}, designated $P_{O_{2t=0}}$. The dissolved oxygen inside the cell, however, does not come to equilibrium with the extracellular oxygen immediately after mixing but must diffuse out. The oxygen from the outer portion of the cell leaves first and thus there is an oxygen flux out of the cell. As the blood-saline mix flows along the tube the oxygen diffusing out of the cells increases the extracellular P_{O_2}. This value is designated as $P_{O_{2t}}$.

Eventually the cells come to equilibrium with the surrounding fluid. The equilibrium extracellular oxygen partial pressure, $P_{O_{2eq.}}$, may be calculated a priori. Therefore, the starting and end points of the P_{O_2} versus time curve are known beforehand, but the intermediate values must be determined experimentally. The P_{O_2} data is presented in terms of the fraction of expected change in extracellular P_{O_2}:

$$\frac{(P_{O_2})_{eq} - (P_{O_2})_t}{(P_{O_2})_{eq} - (P_{O_2})_{t=0}}$$

which has a range of zero $\left((P_{O_2})_t = (P_{O_2})_{eq}\right)$ to one $\left((P_{O_2})_t = (P_{O_2})_{t=0}\right)$. Figure 4 shows the data available from the experiments. With the present equipment we can not make P_{O_2} measurements for times less than 0.5 milliseconds. Therefore the data must be interpolated between known points at 0 and 0.5 milliseconds. The first interpolation attempted is a straight line on the semilong plot. With this line the molar flux out of the cells may now be calculated

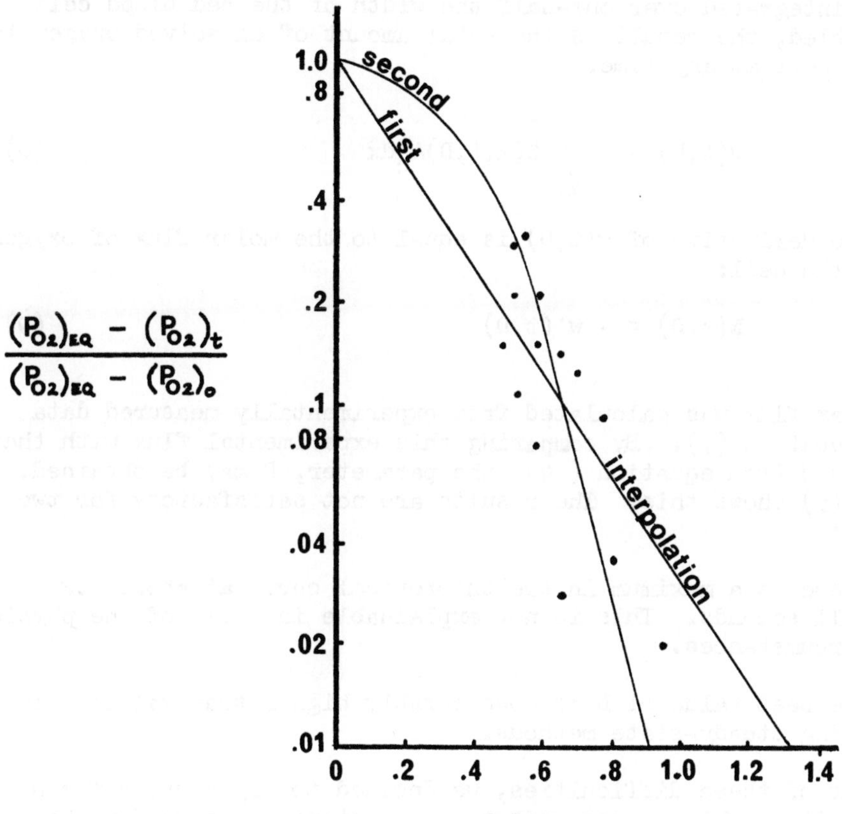

Figure 4. Actual oxygenation data and extrapolation.

directly from equation (7);

$$N(t) = -V_p \alpha \left[(P_{O_2})_{EQ} - (P_{O_2})_0 \right] \frac{d}{dt} \left[\frac{(P_{O_2})_{EQ} - (P_{O_2})_t}{(P_{O_2})_{EQ} - (P_{O_2})_{t=0}} \right] \quad (7)$$

From equations (4) and (7) the mass transfer coefficient across the membrane, h, is calculated from the molar flux at zero time. Note that the initial molar flux and therefore the coefficient, h, are highly dependent upon the slope of the experimental curve at zero time. Thus it would be highly desirable to obtain data for smaller times so that a more accurate interpolation could be made. In any case, with this first fit we obtain a value for h of 0.155 $\frac{cm}{sec}$. This h may now be used in equation (5) and to calculate the diffusion coefficient, D. Another look at the experimental data, (figure 4), shows that a second interpolation may be used. This approach is not reported upon here. If equation (6) is integrated over one-half the width of the red blood cell and doubled, the result is the total amount of dissolved oxygen in the red cell at any time:

$$w(t,D) = 2 \int_0^L C(x,t,D) A_c \, dx \quad (8)$$

The time derivative of w(t,D) is equal to the molar flux of oxygen out of the cell:

$$N(t,D) = -w'(t,D) \quad (9)$$

The molar flux was calculated from experimentally measured data using equation (7). By comparing this experimental flux with that determined from equation (9), the parameter, D may be obtained. Figure (5) shows this. The results are not satisfactory for two reasons:

(1) There is a maximum in the theoretical curve at about 0.2 milliseconds. This is not explainable in terms of the physical circumstances.

(2) The best value of D is considerably higher than values obtained using steady-state methods.

In light of these difficulties, we decided to try a value for h of infinity. This corresponds to a resistance of zero for the membrane. As shown in figure 6 using the same procedures as before the theoretical curve corresponds to the experimental curve

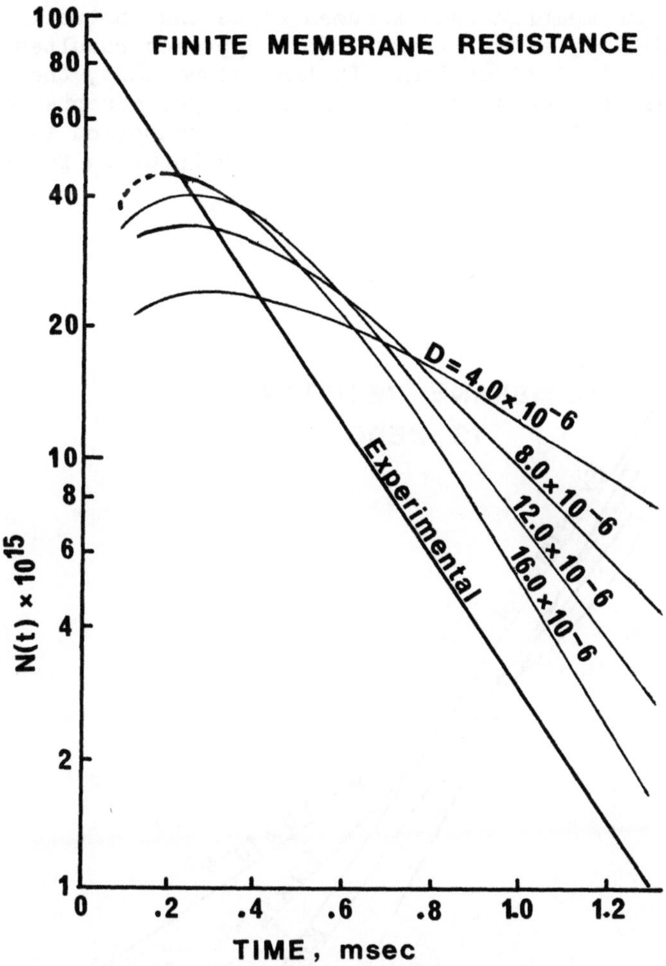

Figure 5. Comparison of experimental results and predictions from the membrane model.

for a diffusion coefficient of 0.8×10^{-5}. At zero time, however, the theoretical curves do not converge. This is understandable since with h equal to infinity the oxygen at the surface of the red cell sees no resistance but is subjected to a driving force. Therefore, the flux must be infinite.

In summary, therefore, two extreme cases were investigated. In the first case, the membrane was assumed to be the controlling resistance immediately after mixing. This approach resulted in a questionable physical situation. In the second case, the membrane resistance was assumed equal to zero which yielded acceptable values for the diffusion coefficient. While neither assumption is strictly true the results give insights which will be helpful in further work.

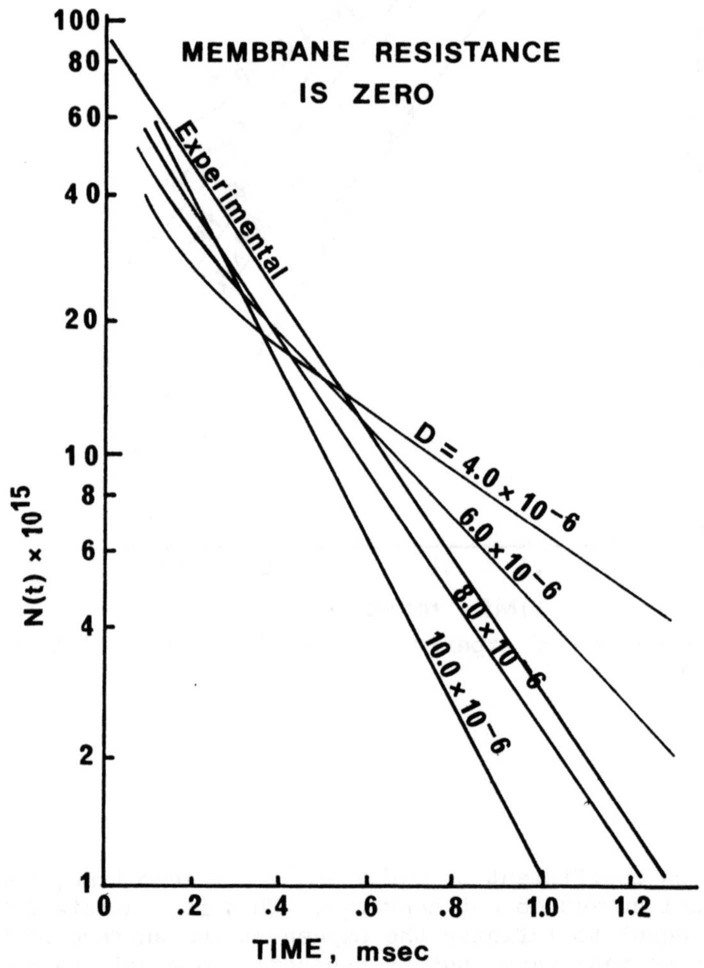

Figure 6. Comparison of experimental results and predictions from the membraneless model.

NOMENCLATURE

C - concentration of dissolved oxygen inside the red cell, moles/cc

t - time, secs.

D - diffusion coefficient of dissolved oxygen inside the red cell. cm^2/sec.

x - interior dimension of the red cell, cm.

L - one-half the effective thickness of a red cell, cm

A_c - one-half the effective area of a red cell, cm^2

V_o - initial concentration of free oxygen, moles/cc

$N(t)$ - flux of oxygen leaving a red cell, moles/sec.

C_p - concentration of extracellular oxygen, moles/cc

h - mass transfer coefficient across the cell membrane, cm/sec.

$(P_{O_2})_t$ - extracellular oxygen partial pressure, mm Hg

$(P_{O_2})_{t=0}$ - extracellular oxygen partial pressure at time of mixing

$(P_{O_2})_{eq}$ - extracellular oxygen partial pressure at time = ∞

V_p - volume of extracellular fluid associated with one red cell, cc

α - conversion factor, moles/(cc)(mm Hg)

$w(t,D)$ - total amount of oxygen dissolved in cell, moles

MODELS FOR SIMULTANEOUS DIFFUSION AND CHEMICAL REACTION OF OXYGEN WITHIN THE INTACT RED CELL OF WHOLE BLOOD

Ronald C. Menke and Daniel Hershey

Department of Chemical and Nuclear Engineering

University of Cincinnati, Cincinnati, Ohio 45221

A great amount of work has been done in the area of the oxygenation of blood. Studies have been devoted to the diffusion of ligand gases through plasma, hemoglobin solutions, and whole blood. Chemical reaction studies have been made testing various kinetic models. Much work has been done with hemoglobin solutions and millipore membranes to simulate the environment within the red cell. Fleischman (7) has presented a survey of some of the pertinent work done in this field. Of more interest to this discussion are the studies that have involved the simultaneous considerations of oxygen diffusion and chemical reaction with hemoglobin in the intact red cell of whole blood.

One of the initial approaches (16, 19, 33) to the problem was to use an effective diffusivity which would account for oxygen diffusion in both plasma and the red cell. Thus, blood was treated as a homogenous solution, and very little physical insight into the actual cell oxygenation mechanism was gained from the studies. Roughton (25) described the process for the transfer of oxygen from the plasma across the cell membrane and into the cell. He then (26) set up the equations for the simultaneous diffusion and reaction of oxygen within the red cell. He assumed that the kinetics for hemoglobin solutions would be the same as for the intact red cell, as shown in equation (1).

$$Hb + O_2 \underset{k}{\overset{k'}{\rightleftharpoons}} HbO_2 \qquad (1)$$

Roughton used several models for the cell consisting of layers and spheres of hemoglobin solution with and without membranes.

He also assumed that there would be no hemoglobin diffusion. His area of interest was the limiting case near zero saturation where the concentration of oxyhemoglobin is approximately zero and the concentration of hemoglobin may be assumed constant. He used the kinetic model of equation (1) with the reverse reaction eliminated and the forward reaction approximately first order to obtain analytic solutions.

Mochizuki and Fukuoka (20) also considered the simultaneous diffusion and reaction of oxygen within the intact red cell. Their model employed a disc-shaped cell, one dimensional diffusion, and the kinetics of equation (1). By expressing the equilibrium relation between oxygen and hemoglobin, $K = k'/k$, where K is the equilibrium constant at a given saturation, and by finding K values at different saturations from the equilibrium-dissociation curve, they were able to eliminate the reverse rate constant, k. Their equations were written in terms of the partial pressure of oxygen instead of oxygen concentration, and they used literature data based on studies of homogeneous hemoglobin solutions. What they considered, in effect, was a uniform concentration of oxyhemoglobin within the cell at any time, t. This suggests a model which, within the cell, allows for either the diffusion of hemoglobin or rapid, facilitated oxygen diffusion, or both. The concentration of oxyhemoglobin would be a function of cell distance, x, as well as time, t, for the case where oxygen, by means of ordinary diffusion, enters the cell to react with hemoglobin held stationary within the cell. By numerical methods they solved the conservation equation for oxygen. Their solution produced the forward rate constant, $k' = 1.0$ 1/sec-mm Hg, which was then used to find the product of the oxygen solubility in the cell, α_c, and the diffusion coefficient of oxygen in the cell, D. They found this value of $\alpha_c D$ to be equal to 0.755×10^{-10} cc/cm-sec-mm Hg for a fixed saturation of 10% oxyhemoglobin.

Fleischman (7) employed a well-stirred batch reactor to eliminate diffusion in the plasma in his study of the simultaneous diffusion and reaction of oxygen in the red cell of whole blood. He solved the resulting equations numerically for a spherical model of the cell. He assumed no hemoglobin diffusion in the cell and, examining five simplified models for the kinetics, concluded that the model given by equation (1) was in best agreement with experimental data. The reported results of his work include the developing of an indirect method for measuring free oxygen concentration in the cell, an effective diffusion coefficient, D_{eff}, for oxygen in the red cell equal to 1×10^{-8} cm^2/sec at 25°C, and forward and reverse rate constants equal to 100 (millimoles/liter-sec)$^{-1}$ and 2 sec^{-1}, respectively.

Moll (21) considered the additional possibility of facilitated diffusion of oxygen by means of hemoglobin and oxyhemoglobin within the cell. He reported a diffusion coefficient for hemoglobin of 5 to 6 x 10^{-8} cm^2/sec. Moll (22) then studied the simultaneous reaction and diffusion of both hemoglobin and oxygen within the intact red cell. His approach was to solve the simultaneous differential equations by the calculus of finite differences. He used a slab model for the cell with one-dimensional diffusion and insignificant membrane resistance, and he considered the kinetics to follow the model of equation (1). His results indicated that hemoglobin diffusion does not accelerate the initial one-third of the oxygen uptake process. However, it does speed up the last two-thirds of the saturation process by as much as 20-30%.

The objective of this study was to set up several models for the simultaneous diffusion and reaction of oxygen within the intact red cell and to approach the solution of the differential equations analytically. The model which best described experimentally observed behavior could then be determined.

THEORETICAL CONSIDERATIONS

In order to develop meaningful overall models for the oxygenation process occurring within the intact red cell of whole

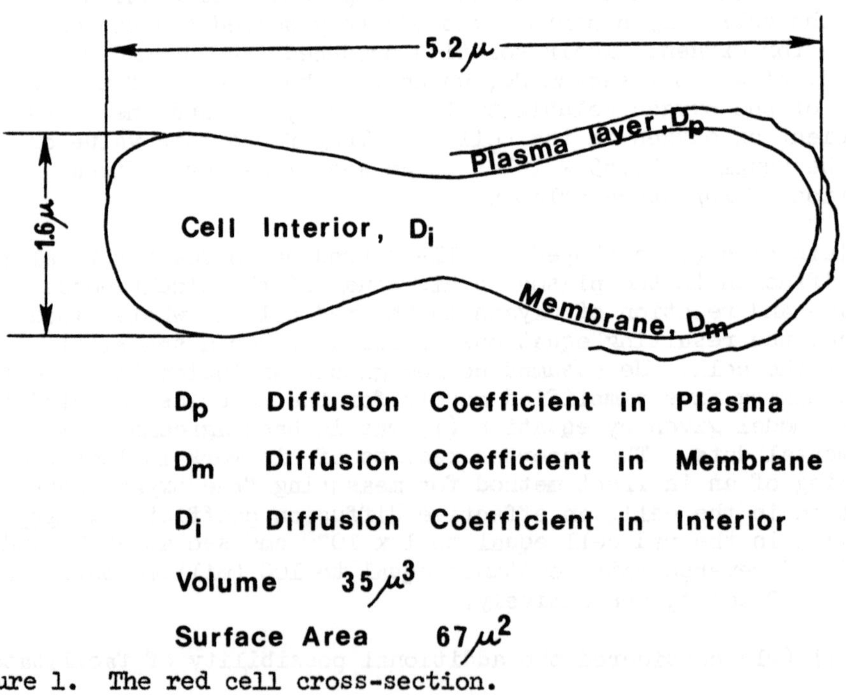

Figure 1. The red cell cross-section.

blood, it is necessary, first to describe the intact cell and the kinetics of the oxygen-hemoglobin reaction, and then, to consider the relative importance of the roles played by diffusion and chemical reaction. Considering a cross-section of a cell shown in Figure 1, oxygen must be transported from the plasma across the stagnant film adhering to the surface of the cell, through the cell membrane, and into the cell interior where it reacts with hemoglobin to form oxyhemoglobin. Diffusion of oxygen in the plasma has been eliminated by the use of a well-stirred reactor.

The cell seldom exists in this undistorted shape under conditions in which oxygen transfer to the cell occurs (2, 22). The turbulence of the flow in a stirred reactor would cause some deformation. The cylindrical disc model for the cell depicted in Figure 2 is considered to be a reasonable approximation to the cell geometry. The cell width has been calculated for this model, and the half-cell thickness of 0.888μ agrees well with Forster's (8) cell thickness calculation of 1.6μ.

For the purpose of this study, the cell membrane is not viewed as offering significant resistance to the transfer of oxygen (15). An effective diffusivity is employed to describe collectively the diffusion of oxygen in the plasma boundary layer, the cell membrane, and the cell interior. In view of the high concentration of hemoglobin within the cell and the paracrystalline nature of the cell interior (23, 34), it is assumed that a homogenous distribution of hemoglobin exists initially within the cell, and that hemoglobin diffusion is negligible as the oxygenation of the hemoglobin proceeds.

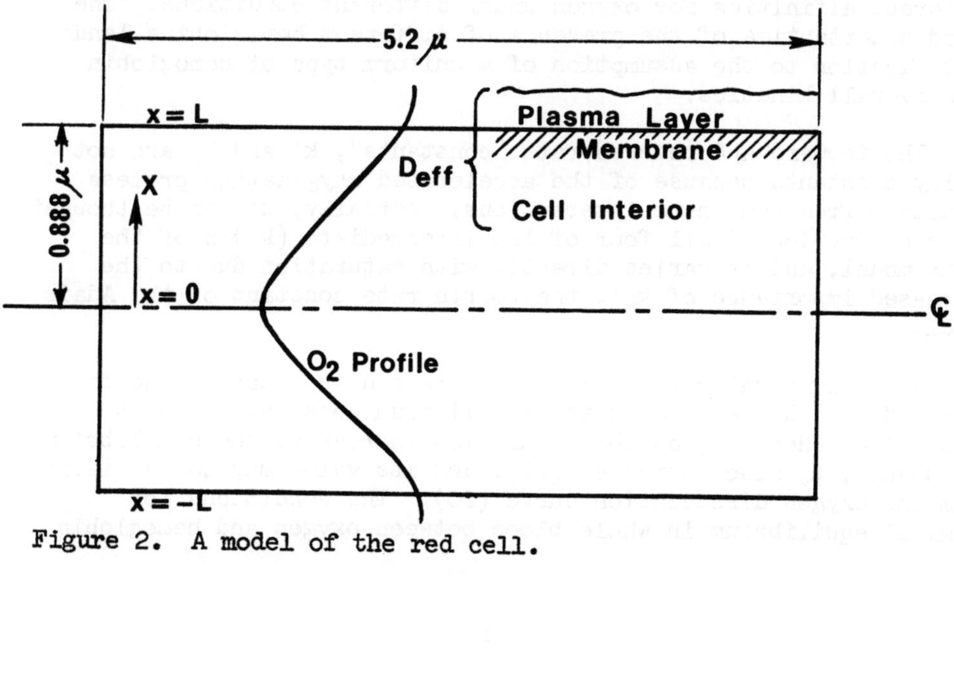

Figure 2. A model of the red cell.

The kinetics model for the reaction of oxygen with hemoglobin is assumed to be expressed by equation (1). There are several reasons for selecting this simplified overall approach. It has been shown (7, 10, 22) to hold empirically for the overall reaction and to be a good approximation to the Adair Intermediate Compound Hypothesis (10) which is the generally accepted model for the oxygenation of the hemoglobin molecule. The Adair model is based on a four-step reaction:

$$Hb_4 + O_2 \underset{k_1}{\overset{k_1'}{\rightleftharpoons}} Hb_4O_2 \qquad (2)$$

$$Hb_4O_2 + O_2 \underset{k_2}{\overset{k_2'}{\rightleftharpoons}} Hb_4O_4 \qquad (3)$$

$$Hb_4O_4 + O_2 \underset{k_3}{\overset{k_3'}{\rightleftharpoons}} Hb_4O_6 \qquad (4)$$

$$Hb_4O_6 + O_2 \underset{k_4}{\overset{k_4'}{\rightleftharpoons}} Hb_4O_8 \qquad (5)$$

However, the intermediate compounds of the Adair model have not been detected, and it is believed that hemoglobin exists predominantly in either the completely unsaturated (Hb_4) or the completely saturated (Hb_4O_8) state. Also, Kernohan (14) has reported two electrophoretically distinct hemoglobins in sheep cells which have different affinities for oxygen under different conditions. The added uncertanties of the presence of different hemoglobins lends justification to the assumption of a uniform type of hemoglobin with overall kinetics.

The forward and reverse rate "constants", k' and k, are not really constants because of the accelerated oxygenation process resulting from heme-heme interactions. Actually, k' can be thought of as a function of all four of the intermediate (k')'s of the Adair model, and it varies directly with saturation due to the increased importance of k_4', the fourth rate constant of the Adair model.

As a given saturation, k' and k are true constants, and their ratio, $K = k'/k$, represents the equilibrium constant for that saturation. However, as the saturation increases, the equilibrium "constant", K, also increases (22), and its value may be estimated from the oxygen dissociation curve (20). The conditions for chemical equilibrium in whole blood between oxygen and hemoglobin are usually expressed in terms of the dissociation curve, as shown in Figure 3. This curve gives the equilibrium relationship between the partial pressure of oxygen in the plasma and the percent saturation of the hemoglobin in the red cells. Saturation refers to the fraction of the hemoglobin that has been oxygenated to form oxyhemoglobin. Since the solubility of oxygen in the red cell is known (27), for a given partial pressure of oxygen in the plasma it is possible to find the concentration of dissolved oxygen in the cell which is in chemical equilibrium with hemoglobin of a given saturation. The dissociation curve is a function of temperature and pH and has been reported by Karhan (13) and Hesselberth (12) at 25°C for pH values of 8.6 and 8.1, respectively, as shown in Figure 3.

Three models have been considered in an effort to determine the relative importance of the phemomena of diffusion and chemical reaction to the overall oxygenation process.

MODEL 1. The first model considers simultaneous diffusion and reversible chemical reaction of oxygen within the cell depicted in Figure 2. The rates of change of oxygen and oxyhemoglobin

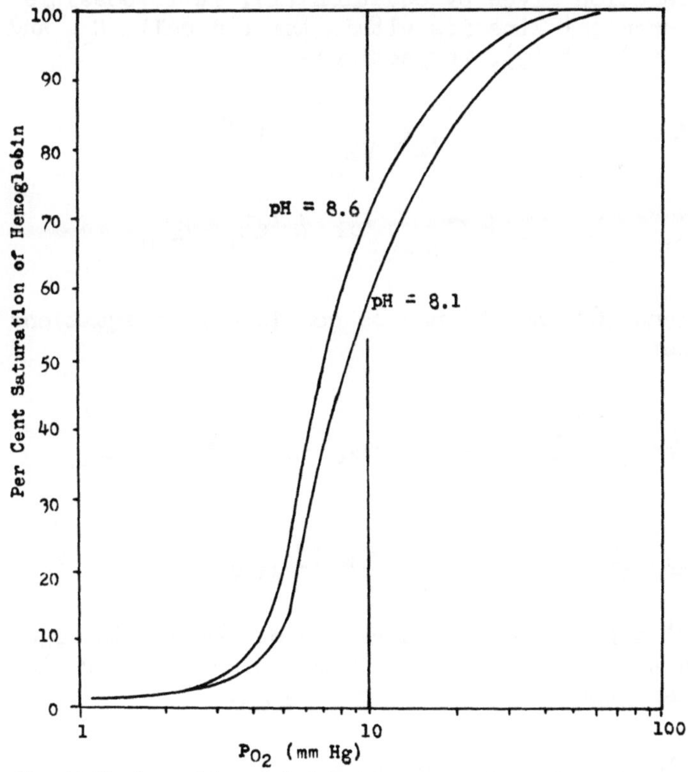

Figure 3. Equilibrium dissociation curves for sheep's blood at 25°C. (12, 13).

concentrations, C_{O_2} and C_{HbO_2} within the intact red cell are expressed by the unsteady state equations (6) and (7), respectively:

$$\frac{\partial C_{O_2}}{\partial t} = \nabla \cdot D_{O_2} \nabla C_{O_2} + R_{O_2} \tag{6}$$

$$\frac{\partial C_{HbO_2}}{\partial t} = \nabla \cdot D_{HbO_2} \nabla C_{HbO_2} + R_{HbO_2} \tag{7}$$

where the first term in each equation is a molecular diffusion term, D_{HbO_2} and D_{O_2} are the effective diffusion coefficients of oxyhemoglobin and oxygen within the cell, and the second term in each equation is a chemical reaction term. As shown in Figure 2, the model of the cell is a cylindrical disc in which oxygen diffuses into the cell across the face of the cell. This unidirectional diffusion is considered for $0 \leq x \leq L$, where L is the half-cell thickness, since the cell is symmetrical. From the oxygen-hemoglobin reaction given by equation (1), it is possible to express the chemical kinetics within the red cell, R_{O_2} and R_{HbO_2}, by equations (8) and (9), respectively:

$$R_{O_2} = -k' C_{O_2} C_{Hb} + k C_{HbO_2} \tag{8}$$

$$R_{HbO_2} = -k C_{HbO_2} + k' C_{O_2} C_{Hb} \tag{9}$$

Thus, equations (6) and (7) may be combined with equations (8) and (9) to get:

$$\frac{\partial C_{O_2}}{\partial t} = D_{EFF} \frac{\partial^2 C_{O_2}}{\partial x^2} - k' C_{O_2} C_{Hb} + k C_{HbO_2} \tag{10}$$

$$\frac{\partial C_{HbO_2}}{\partial t} = k' C_{Hb} C_{O_2} - k C_{HbO_2} \tag{11}$$

Equation (12) allows the concentration of hemoglobin, C_{Hb}, to be expressed in terms of the initial hemoglobin concentration, C_{Hb_o}, and the oxyhemoglobin concentration, C_{HbO_2}.

$$C_{Hb_o} = C_{Hb} + C_{HbO_2} \tag{12}$$

Equation (13) expresses the reverse rate "constant", k, in terms of the forward "constant", k', and the ratio of the rate "constants", K:

$$K = k'/k \quad ; \quad k = k'/K \tag{13}$$

The ratio k'/k = K may be evaluated by means of the oxygen dissociation curve and has been shown (7, 20) to be equal to:

$$K = \frac{S_{O_2}}{(100 - S_{O_2})(\alpha_c P_{O_2}/17.024)} \tag{14}$$

Here, S_{O_2} is the percent saturation of the cell, α_c is the Bunsen cell solubility, and P_{O_2} is the partial pressure of oxygen in the plasma.

The following initial and boundary conditions are used:

1) $$t = 0 \; ; \; C_{O_2} = 0 \; ; \; C_{HbO_2} = 0 \; ; \; C_{Hb} = C_{Hb_o} \tag{15}$$

2) $$x = 0 \; ; \; C_{O_2} = \text{FINITE} \; ; \; \frac{\partial C_{O_2}}{\partial x} = 0 \tag{16}$$

3) $$x = \pm L \; ; \; C_{HbO_2}(L,t) = K' P_{O_2} + C_1 \tag{17}$$

The system is to be well stirred so that there is no diffusion in the plasma. Initially the blood is completely deoxygenated. As oxygenation proceeds equilibrium is assumed to exist at all times between the oxyhemoglobin concentration at the surface of the cell and the surrounding plasma. The equilibrium relationship of equation (17) can be obtained by breaking up the sigmoidal equilibrium curve into sections which can each be linearized in terms of a slope, K', and an intercept, C_1. This approximation is more managable, mathematically, but a particular set of values, K' and C_1, is applicable only over a given range of the saturation curve. The values K' and C_1 must be changed as the saturation proceeds from one linearized section of the dissociation curve to the next.

The experimental P_{O_2} data is also a function of time and may be expressed as in equation (18):

$$P_{O_2}(t) = k^* t + C_2 \tag{18}$$

Here again the P_{O_2} vs. time curve has been broken into linear sections with each section having a slope, k*, and an intercept, C_2.

It now becomes a problem of solving the equations (10) through (18) simultaneously. Since an analytic approach is desired, a method has been developed to handle the second-order, non-linear differential equations. This method is discussed elsewhere (18). The result is an expression for the concentration of oxyhemoglobin as a function of cell distance, x, and time, t, $C_{HbO_2}(x, t)$. Using the mean value theorem (29), an average cell concentration of oxyhemoglobin dependent only on time can be obtained:

$$\langle C_{HbO_2} \rangle = K'k^*t - K'k^*L^2\left(\frac{Q + k'C_{Hbo}}{3DQ}\right) + (K'C_2 + C_1) + \frac{K'k^*}{L}\sum_{n=1}^{\infty}\left[\frac{2e^{P_n t} D(Q+P_n)^2}{P_n^2 L(P_n^2 + Q^2 + 2QP_n + k'QC_{Hbo})}\right] + \frac{(K'C_2+C_1)}{L}\sum_{n=1}^{\infty}\left[\frac{2e^{P_n t} D(Q+P_n)^2}{P_n L(P_n^2+Q^2+2QP_n + k'C_{Hbo}Q)}\right] \tag{19}$$

where:

$$Q = k''/K + k'C_{O_2}(\bar{x}, T) \tag{20}$$

$$P_n = -\frac{1}{2}\left[Q + k'C_{Hbo} + \frac{\pi^2 D}{L^2}(n-\tfrac{1}{2})^2\right] \pm \frac{1}{2}\left[Q^2 + 2Qk'C_{Hbo} - \frac{2\pi^2 DQ}{L^2}(n-\tfrac{1}{2})^2 + (k'C_{Hbo})^2 + \frac{2k'C_{Hbo}}{L^2}\pi^2 D(n-\tfrac{1}{2})^2 + \frac{\pi^4 D^2}{L^4}(n-\tfrac{1}{2})^4\right]^{1/2} \quad n = 1, 2, 3 \ldots \tag{21}$$

$$B_n = \left(\frac{P_n^2 + P_n Q + P_n k'C_{Hbo}}{D(Q+P_n)}\right)^{1/2} \quad n = 1, 2, 3 \ldots \tag{22}$$

and $C_{O_2}(\bar{x}, T)$ is an empirical constant arising from the linearization technique. Its significance is discussed later.

Equation (19) will predict analytically the oxyhemoglobin curve which can be measured and recorded experimentally. Examining the equation, it can be seen that the average cell-concentration of oxyhemoglobin is a function only of known values, measurable constants, time, and $C_{O_2}(\bar{x}, T)$. The model can be verified by its ability to generate, analytically, curves which will agree with experimentally recorded curves of oxyhemoglobin as a function of time. The adjustable parameter is $C_{O_2}(\bar{x}, T)$. Equation (19) is particularly cumbersome to evaluate, and a digital computer solution was developed. Details are given elsewhere (18).

MODEL 2. The second model considers diffusion and simultaneous irreversible chemical reaction. The reverse reaction in equation (1) is de-emphasized in favor of the diffusion and the forward chemical reaction of oxygen with hemoglobin within the red cell. This model is considered in light of experimental observations (7, 18) which indicate such a rapid formation of oxyhemoglobin that the reverse reaction appears to be completely negligible.

The kinetics for the present model become:

$$Hb + O_2 \xrightarrow{k'} HbO_2 \qquad (23)$$

Using an approach identical to that for the first model, the equations for the conservation of oxygen and oxyhemoglobin are equations (10) and (11):

$$\frac{\partial C_{O_2}}{\partial t} = D_{EFF}\frac{\partial^2 C_{O_2}}{\partial x^2} - k' C_{Hb} C_{O_2} \qquad (24)$$

$$\frac{\partial C_{HbO_2}}{\partial t} = k' C_{Hb} C_{O_2} \qquad (25)$$

Solving equations (24) and (25), along with equations (12) and (15) through (18) for the same initial and boundary conditions yields a solution of the same form as equation (19):

$$\begin{aligned} \langle C_{HbO_2} \rangle = {} & K'k^* t - K'k^* L^2 \left(\frac{Q + k' C_{Hb_0}}{3DQ}\right) + (K'C_2 + C_1) \\ & + \frac{K'k^*}{L} \sum_{m=1}^{\infty} \left[\frac{2e^{P_m t} D (Q + P_m)^2}{P_m^2 L (P_m^2 + Q^2 + 2QP_m + k'QC_{Hb_0})}\right] \\ & + \frac{(K'C_2 + C_1)}{L} \sum_{m=1}^{\infty} \left[\frac{2e^{P_m t} D (Q + P_m)^2}{P_m^2 L (P_m^2 + Q^2 + 2QP_m + k'QC_{Hb_0})}\right] \end{aligned} \qquad (26)$$

where:

$$Q = k' C_{O_2}(\bar{x}, T) \qquad (27)$$

and the other terms remain unchanged. The method of solution is shown elsewhere (18).

MODEL 3. The third model represents a significant departure from the first two, in that, now, the diffusion is de-emphasized in favor of the reversible chemical reaction. For this case, the governing equation would be expressed by equation (28), where the diffusion term has been neglected:

$$\frac{\partial C_{HbO_2}}{\partial t} = -\frac{\partial C_{O_2}}{\partial t} = k' C_{Hb} C_{O_2} - k C_{HbO_2} \tag{28}$$

An analytic solution of equation (28) with equations (12) and (13) and the initial condition (15) results in equation (29):

$$C_{HbO_2} = \left(k' C_{O_2}(T)\, C_{Hb_0} \right) \left(\frac{1 - e^{-[k'/K + k' C_{O_2}(T)]t}}{k'/K + k' C_{O_2}(T)} \right) \tag{29}$$

where $C_{O_2}(T)$ is a parameter arising from the linearization technique, and it is comparable to the $C_{O_2}(\bar{x}, T)$ value of the first two models. The development is shown elsewhere (18).

If the diffusion of oxygen within the cell is facilitated by the diffusion of oxyhemoglobin and/or by some additional mechanism such as the transfer of oxygen from one hemoglobin molecule to another as suggested in the bucket-brigade description (4, 5, 6, 32) for hemoglobin solutions, there would be little or no gradient of oxygen or oxyhemoglobin actually existing within the cell and model 3 would be implied. Oxygen coming into the cell would find hemoglobin available for reaction as soon as it entered the cell membrane.

EXPERIMENTAL PROCEDURE AND EQUIPMENT

Whole sheep's blood was used, since it has the attraction of being inexpensive and readily available in a fresh state. The fibrinogen was removed and heparin was added to prevent clotting during the experimental measurements. After bring the blood to 25°C, the following two measurements were made on each sample:

a) Measurement of hematocrit, the volume fraction of cells in the blood, usually expressed in ml cells/100 ml blood, by means of centrifugation (17).

b) Measurement of the total hemoglobin content. The total hemoglobin is the sum of the hemoglobin and the oxyhemoglobin at any time, t:

$$C_{Hb_o} = C_{Hb} + C_{HbO_2} \tag{30}$$

A haemoglobinometer which operates on an optical technique (the cyanomethemoglobin method) was used to measure the total hemoglobin content (24, 35).

PROCEDURE

The blood was deoxygenated. This was accomplished by placing 1400 CC samples of the blood in a plastic reactor (which would not cause clotting) and by bubbling nitrogen gas through the stirred blood. After about two hours, the blood was completely deoxygenated. At the same time that the oxygen was being driven off, the carbon dioxide of the blood was being removed, raising the pH from 6.8-7.2 to 8.3-8.7 where it remained constant thereafter.

Foaming was a problem, and a silicone anti-foaming agent was used to coat the surface of the blood. There was concern over the possibility of causing hemolysis from the stirring and sparging. Checks were made to determine the degree of hemolysis, and it was not found to be significant.

After deoxygenation in the reactor, the blood was ready to be oxygenated. The equipment is shown in Figure 4. The oxygenation was performed by passing a metered stream of gas of a particular oxygen-nitrogen composition through gas spargers immersed in the blood. The saturation process was followed by monitoring the change in the concentration of oxyhemoglobin in the red cells and the change of the partial pressure of oxygen in the plasma. This data was to be measured and recorded as a function of time. Nine different experimental runs were made.

The oxygen partial pressure in the plasma was measured by means of a polarographic P_{O_2} macro-electrode inserted directly into the plasma. The electrode was connected to an amplifier for direct P_{O_2} readings, and the amplifier output was connected to a recorder which gave a continuous plot of the P_{O_2} vs. time curve. Saline solutions of the same salt content (0.9% by weight NaCl) as plasma were used to calibrate the electrode before each run. This was found to be necessary since membrane aging and protein buildup caused the calibration to vary over a period of time.

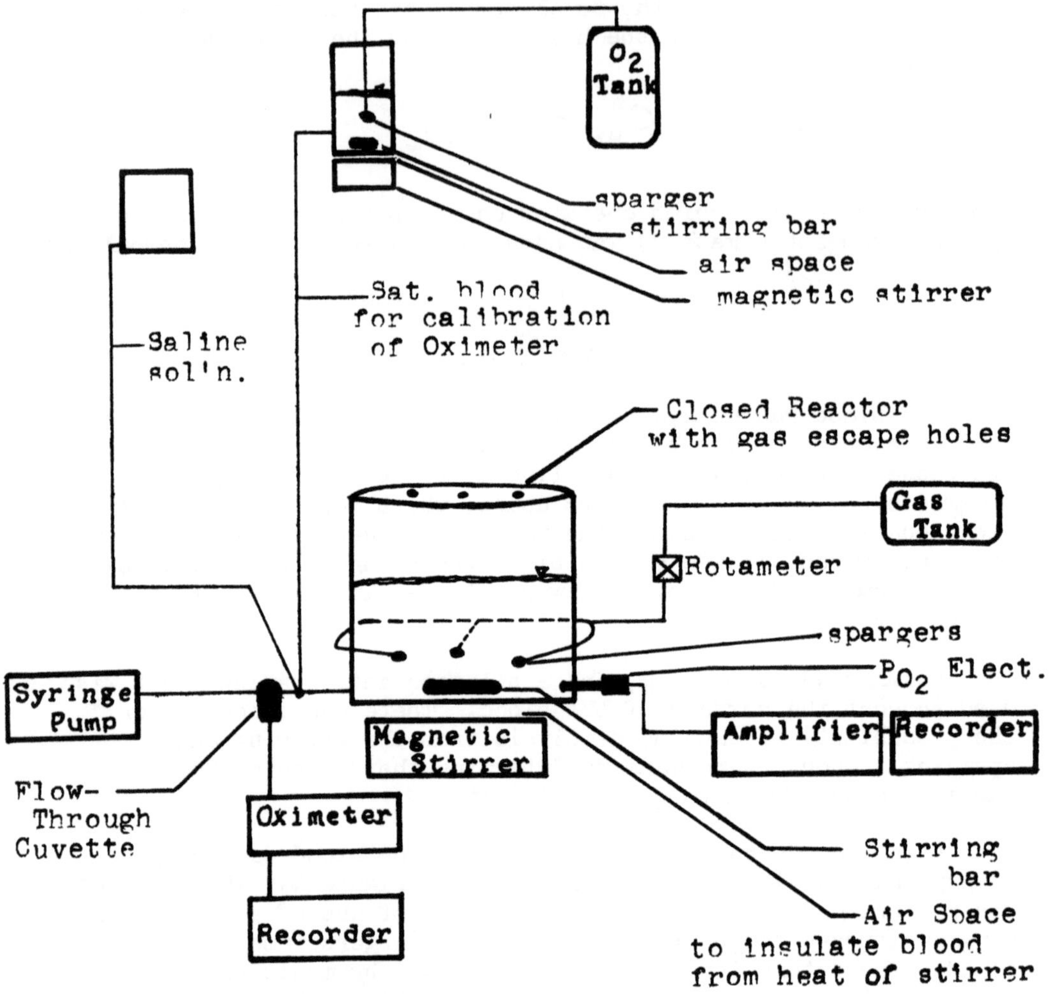

Figure 4. Experimental apparatus.

An oximeter was used to measure the change in the concentration of oxyhemoglobin with time. Based on an optical technique, this instrument gives a rapid, continuous reading of the percent saturation, S_{O_2}, of the hemoglobin of the red cells of whole blood. The saturation is expressed by equation (31) as the ratio of the concentration of the oxyhemoglobin, C_{HbO_2}, to the concentration of total hemoglobin, reacted and unreacted.

$$S_{O_2} = \frac{C_{HbO_2}}{C_{Hb} + C_{HbO_2}} = \frac{C_{Hb}}{C_{Hb_0}} \tag{31}$$

The flow rate of the blood through the oximeter was fast enough to make the flow time lag negligible in comparison to the response time of the electrode assembly. The output of the oximeter was connected to a recorder which gave a continuous plot of the percent saturation vs. time curve. The oximeter was calibrated before each run as given in (18). The samples used for calibration of the oximeter were taken from the blood to be oxygenated in a given run to insure complete similarity of hematocrit, hemoglobin content, etc.

EXPERIMENTAL RESULTS

Nine sets of experimental data were collected and used for verification of the proposed models. Some of the experimental curves of percent saturation versus time and P_{O_2} vs. time are presented graphically in Figures 5-8. The data of run #4, shown in Figure 8, has been replotted in Figure 9 in order to show the complete development of the P_{O_2} vs. time curve. This, then, represents a typical example of the type of data collected, and a general discussion of Figure 9 would be applicable to all of

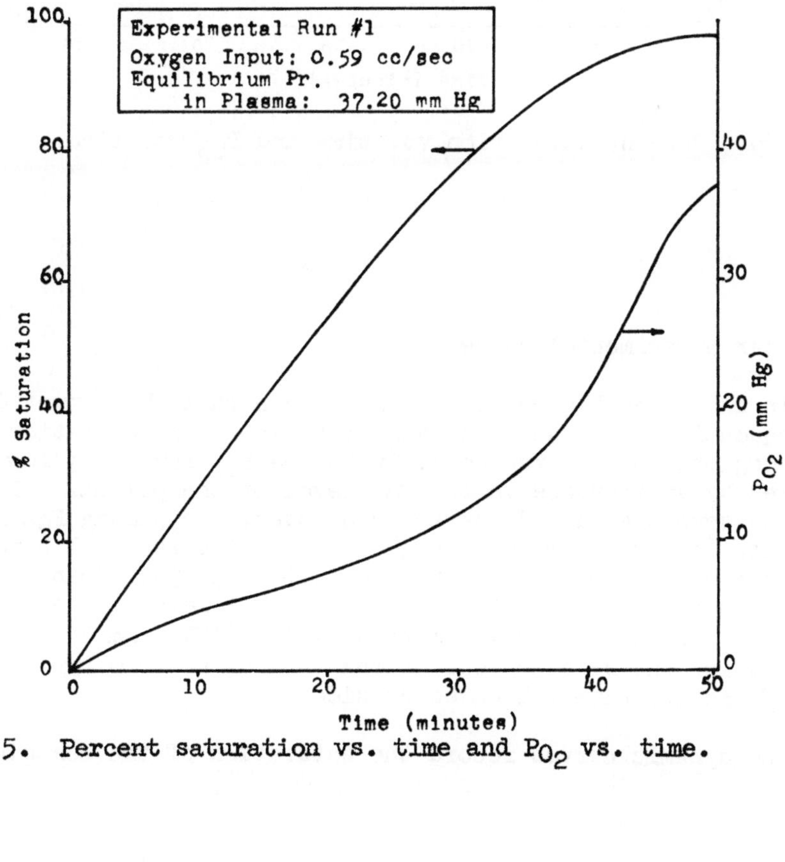

Figure 5. Percent saturation vs. time and P_{O_2} vs. time.

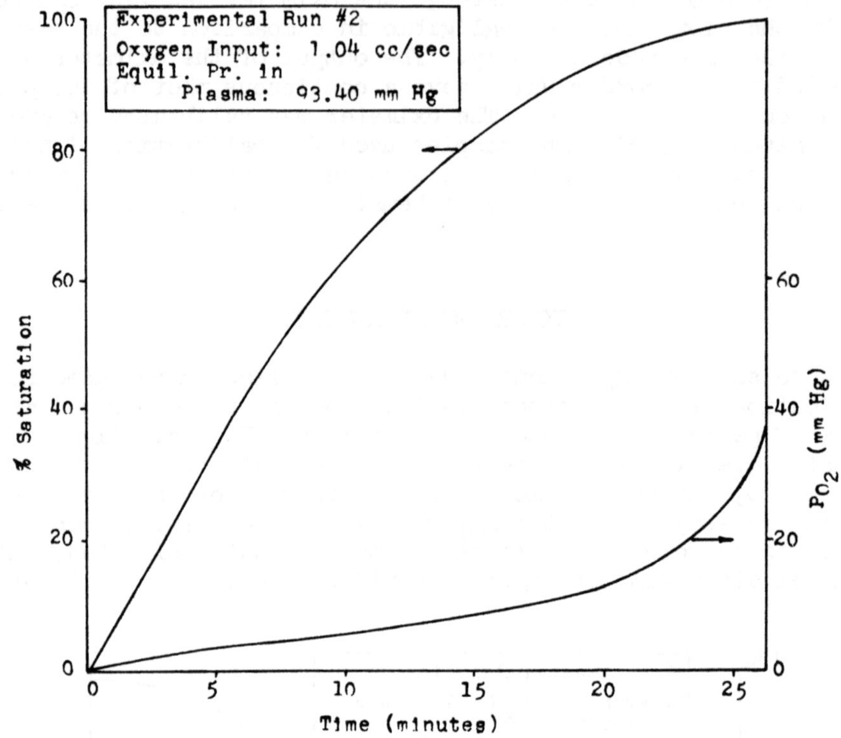

Figure 6. Percent saturation vs. time and P_{O_2} vs. time.

the other experimental curves.

In examining Figure 9, it can be seen that the curves depict two sequential processes: a rapid rise in the % saturation of the hemoglobin of the red cells (with the P_{O_2} remaining low), followed by an increase in the P_{O_2} level of the plasma. Although there is some overlap of the two processes, the plasma P_{O_2} remains quite low (below 10 mm Hg in Fig. 9) until the red cell hemoglobin reaches a high degree of saturation (90%, Fig. 9). Then, as the saturation curve tapers off to 100%, the P_{O_2} level of the plasma increases rapidly to the equilibrium value (177.7 mm Hg, Fig. 9). The behavior and relationship of the two curves were typical of each of the experimental cases considered.

The dynamic curves record the saturation of the hemoglobin

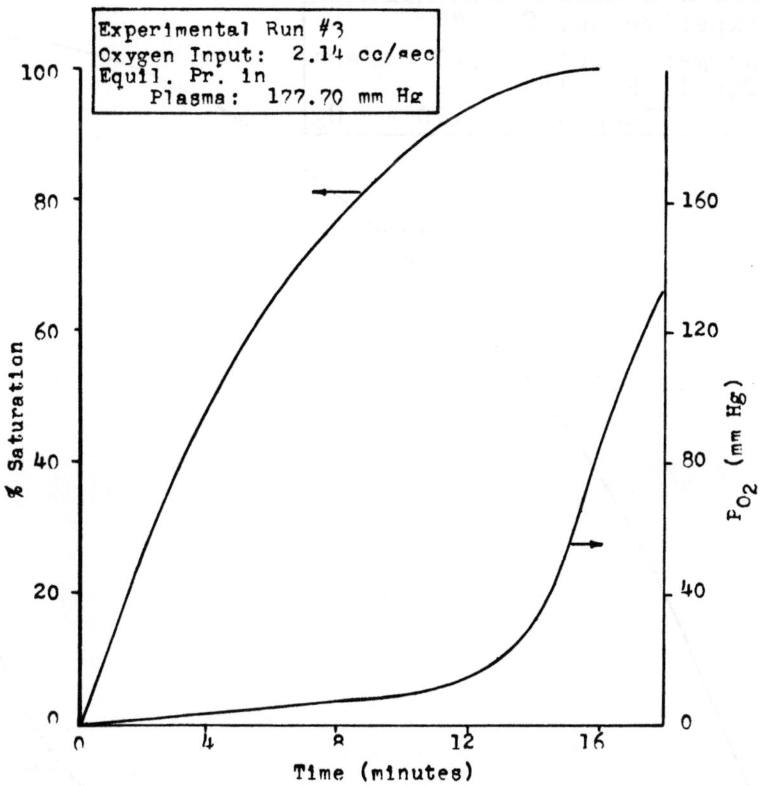

Figure 7. Percent saturation vs. time and P_{O_2} vs. time.

and the plasma P_{O_2} as they exist during the oxygenation process without sufficient time for equilibrium conditions to be established. If the blood to be oxygenated were contacted with a very low volumetric flow rate of oxygen, the saturation would progress at a pace slow enough such that the system would not be far removed from equilibrium and might actually approach equilibrium conditions.

In Figure 3, the equilibrium curves of Karhan (13) and Hesselberth (12) are presented for sheep's blood at 25°C for pH's of 8.6 and 8.1, respectively. The equilibrium curve, also referred to as the dissociation curve, expresses the relationship that exists between the % sat. of the hemoglobin and the plasma P_{O_2}

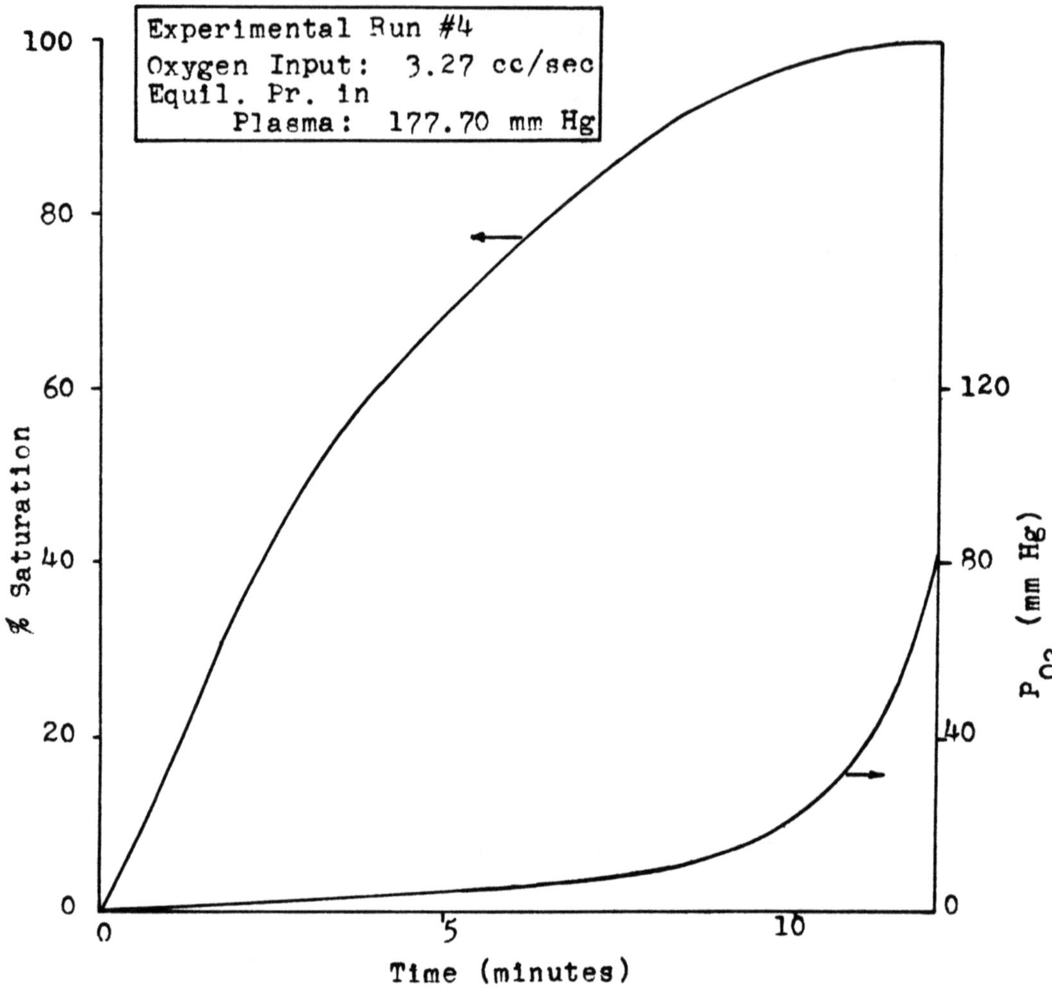

Figure 8. Percent saturation vs. time and P_{O_2} vs. time.

under equilibrium conditions. In order to gauge the approach of an experimental run to equilibrium, it can be compared to the equilibrium situation by plotting the dynamic results and the dissociation curve on the same graph. This has been done in Figure 10, and this is typical of the other results.

On examining Figure 10, it can be seen that the equilibrium curve serves as an upper boundary for the dynamic curves. The oxygen input rate for run #1 was very low, 0.59 cc O_2/sec, and this dynamic curve made the closest approach to equilibrium

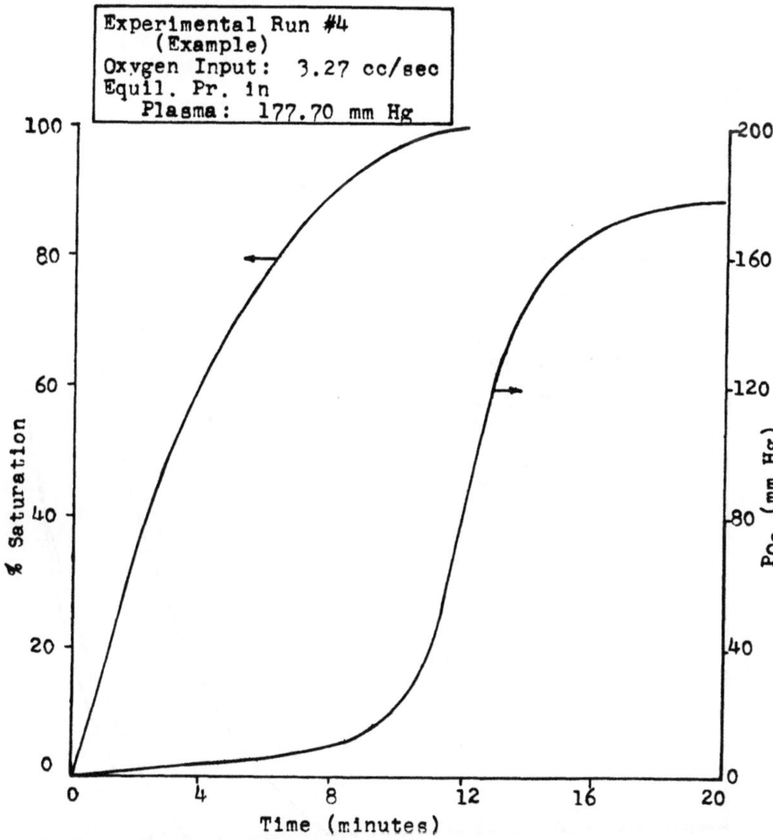

Figure 9. Percent saturation vs. time and P_{O_2} vs. time.

conditions of the nine cases studied. Although the oxygen input rate was high enough, at first, to allow the dynamic oxygen uptake to be larger than that prescribed by the equilibrium situation, at about 40% saturation the dynamic curve converges with the equilibrium curve. The oxygen flow rate for run #2 was about twice as much as for run #1, 1.04 cc O_2/sec, and it can be seen in Figure 10 that the dynamic curve is farther removed from equilibrium. The dynamic curve converges with the equilibrium curve as the saturations nears 100% and at a much higher P_{O_2}

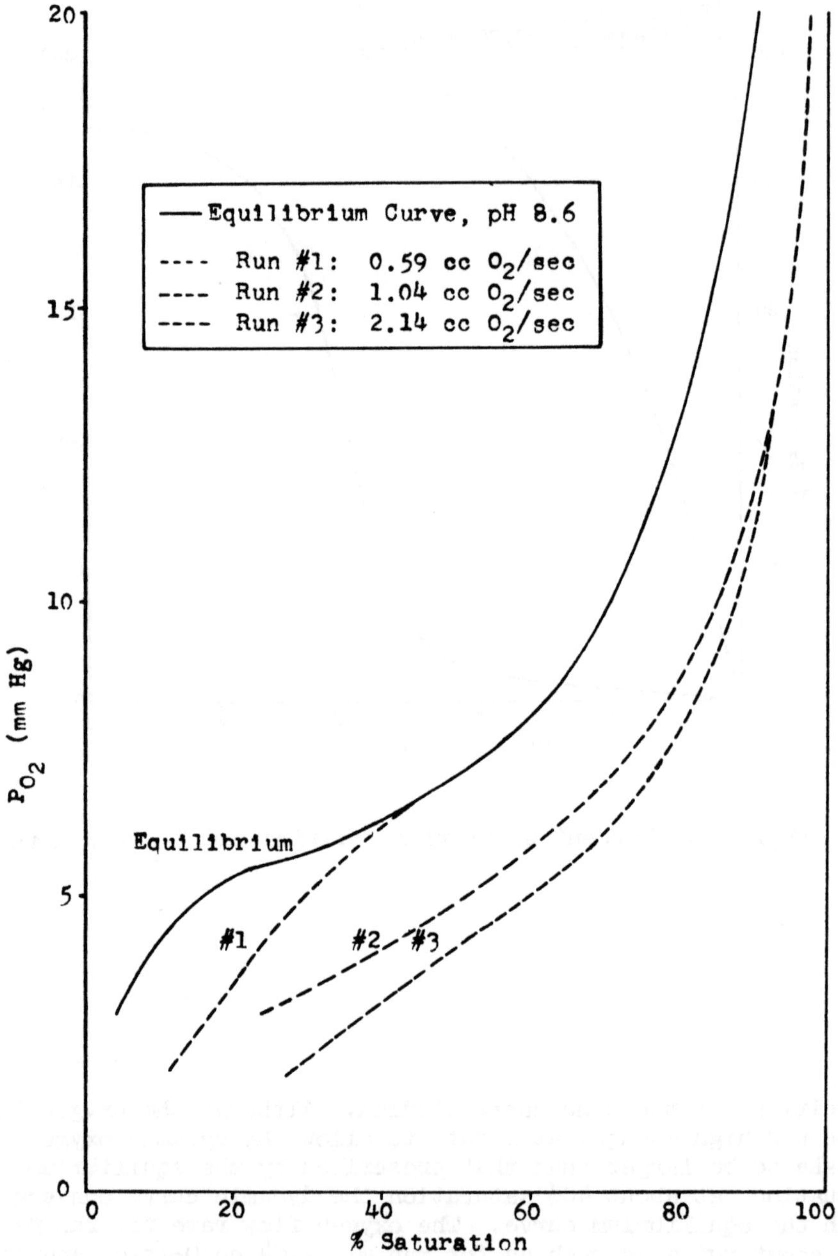

Figure 10. Comparison of dynamic curves to the equilibrium dissociation curve.

than shown in Figure 10. The oxygen flow rate for run #3 was about twice that of run #2, 2.14 cc O_2/sec, and the dynamic curve is shifted farther from equilibrium. As the oxygen input rates become even larger, the dynamic curves being to overlap, indicating that they are all behaving similarly. It can be seen that the rate of oxygen admission to the blood sample is an important consideration for describing how far a system is removed from equilibrium.

RESULTS OF MODEL TESTING

The experimental data presented in the last section was used to test the three models proposed to describe the oxygenation process for whole blood.

MODEL 1. Diffusion and Reversible Chemical Reaction

Analytic curves of % sat. vs. time were generated by using equation (19) with $C_{O_2}(\bar{x}, T)$ as a parameter. These analytic curves were then compared to the experimental curves, some of which are shown in Figures 5-8. The results for the first two runs are shown in Figures 11 and 12.

The $C_{O_2}(\bar{x}, T)$ is the average value of the free oxygen concentration within the intact red cell. It arises from the method of solution of the unsteady-state governing equations, as discussed by Menke (18). As oxygen enters the blood sample, the free oxygen concentration within the cells increases from zero initially for a completely deoxygenated blood sample to a maximum value, C_{O_2} (max.), achieved when saturation is complete and the cells are in equilibrium with the plasma P_{O_2}. Thus, for $C_{O_2}(\bar{x}, T)$ to represent an average value of the free oxygen concentration within the cell during the oxygenation process, it would have to lie between these limits for each run; i.e. $0 \leq C_{O_2}(\bar{x}, T) \leq C_{O_2}$ (max.).

It can be seen in Figure 11 that the model is unsuccessful in describing the initial phase of the oxygenation process. The analytic curve, even with the maximum value of $C_{O_2}(\bar{x}, T)$, predicts a much lower rate of oxygen uptake at first, and only at about 40% saturation does the analytic curve converge with the experimental curve. The analytic model is only successful when an equilibrium situation has been achieved.

The second run (Figure 12) had an oxygen input rate high enough to cause the experimental curve to be substantially removed from the equilibrium curve until complete saturation was

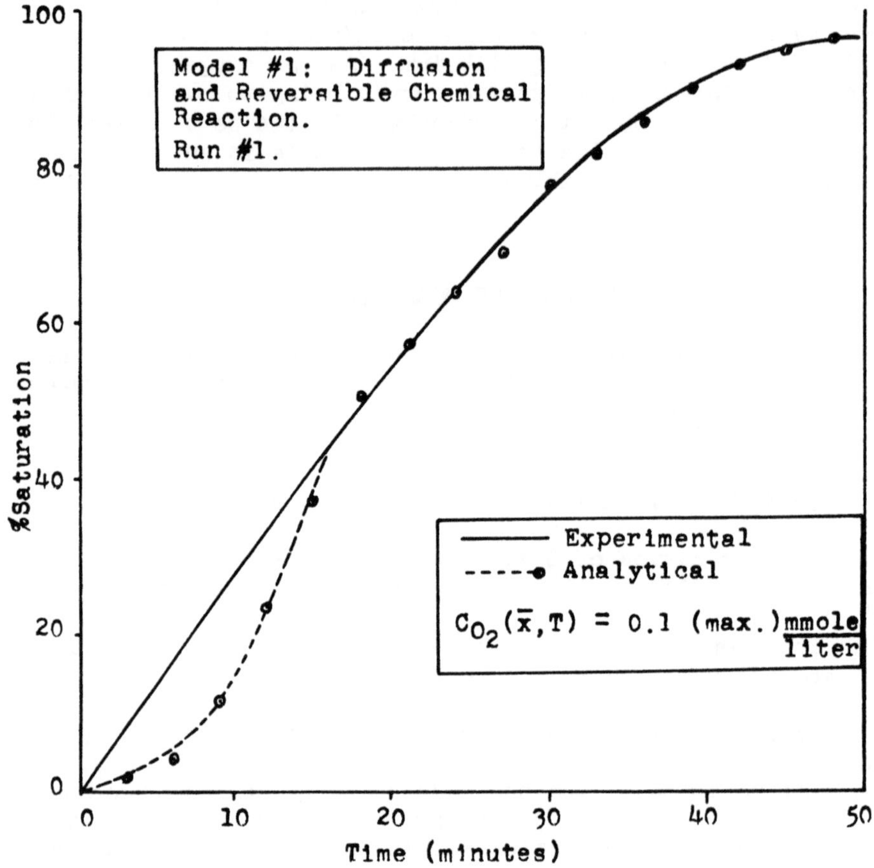

Figure 11. Percent saturation vs. time.

approached. The prediction of model 1 (analytical curve) does not converge with the experimental curve until saturation is near 100%. The other runs are all similar in that the analytic model of diffusion and reversible chemical reaction does not describe the oxygenation process for any of the runs. Model 1 seems only to apply when the oxygenation is proceeding slowly enough to approach equilibrium conditions.

The values of the effective diffusion coefficient, D_{eff}, and the forward rate constant, k', reported by Fleischman (7) were used in the analytic equation (20). He found an average effective diffusion coefficient under equilibrium conditions to be 1.0×10^{-8} cm^2/sec. His forward rate constant value was 100 l/mmole-sec. Other values of D_{eff} and k' have been reported

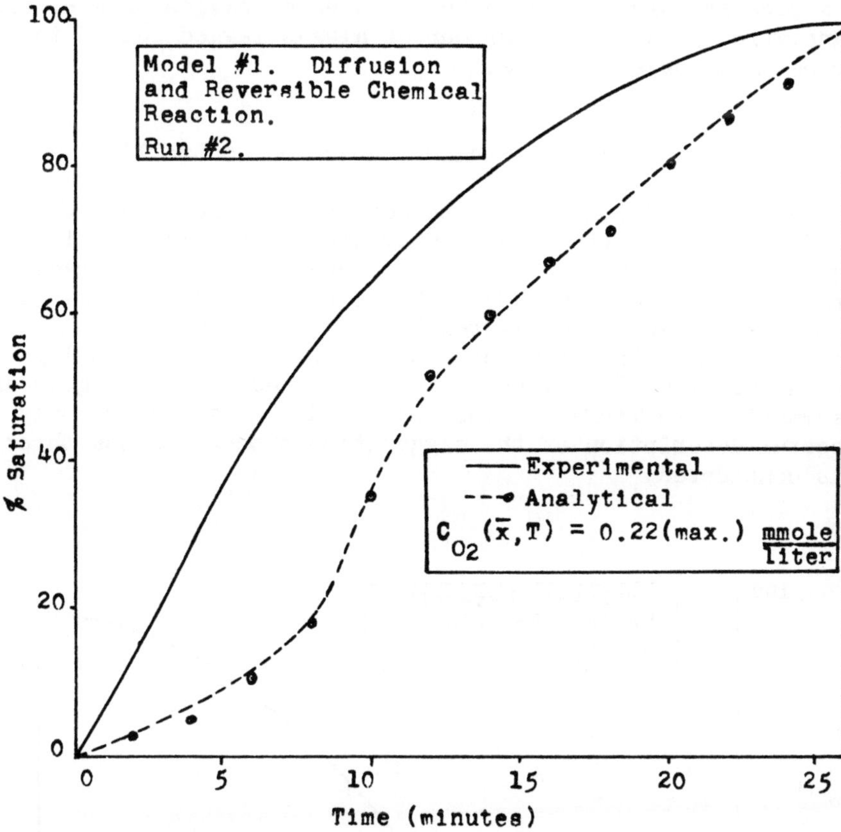

Figure 12. Percent saturation vs. time.

which do not agree with Fleischman's values. Grote and Thews (11) reported an effective diffusion coefficient equal to 0.8×10^{-5} cm^2/sec, and a forward rate constant, k', was reported by Gibson, et. al. (9) to be equal to 3.5×10^3 l/mmole-sec. Because of the large differences between these values for D_{eff} and k', the analytic curves were recalculated using Gibson's value for k' and Grote and Thews' value for D_{eff} in equation (20). The results were no more satisfactory. The diffusion model still failed to describe the oxygenation process.

MODEL 2. Diffusion and Irreversible Chemical Reaction

This model was no more successful than Model 1 in predicting the experimental curves. Whenever the diffusion terms were included as controlling steps in the oxygenation process, the models consistently predicted lower rates of oxygen uptake, and the analytic % sat. vs. time curves always lagged behind the experimentally observed ones.

MODEL 3. Reversible Chemical Reaction Controlling

The third model was tried to see if diffusion might be omitted as a controlling step. Only the reversible reaction of hemoglobin to oxyhemoglobin was considered. Analytic curves were generated using equation (29) with C_{O_2} (T) as a parameter. The results are presented in Figures 13-21. The values of C_{O_2} (T) which resulted in the best description of the experimentally recorded curves of % sat. vs. time are listed on each figure and are summarized in Table 1. The third model achieves the only successful description of the oxygenation process of the three models considered.

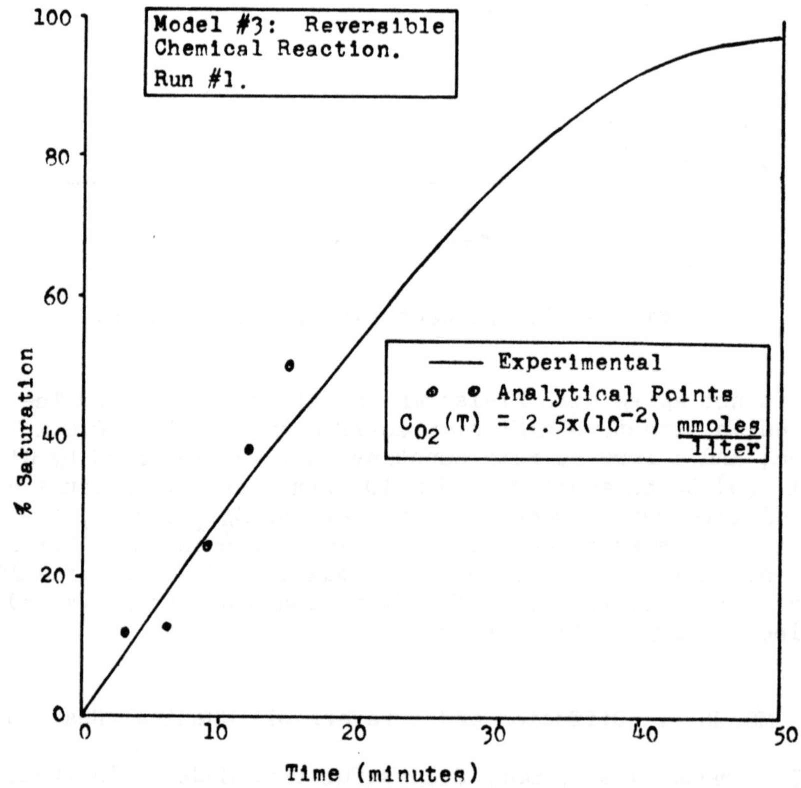

Figure 13. Percent saturation vs. time.

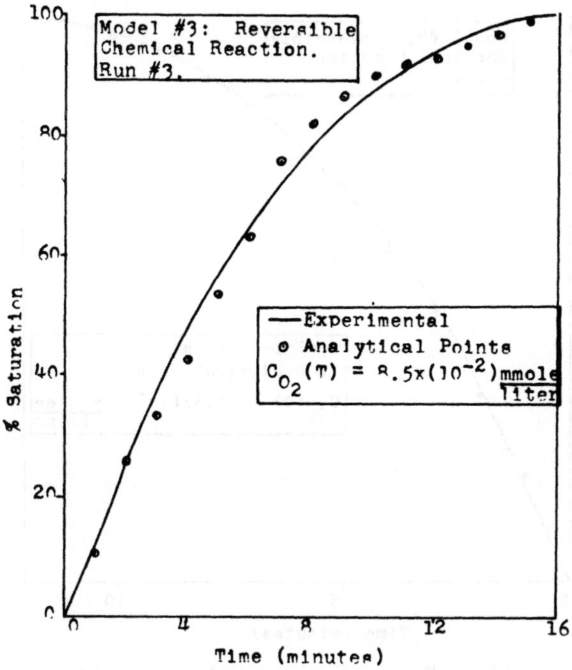

Figure 14. Percent saturation vs. time.

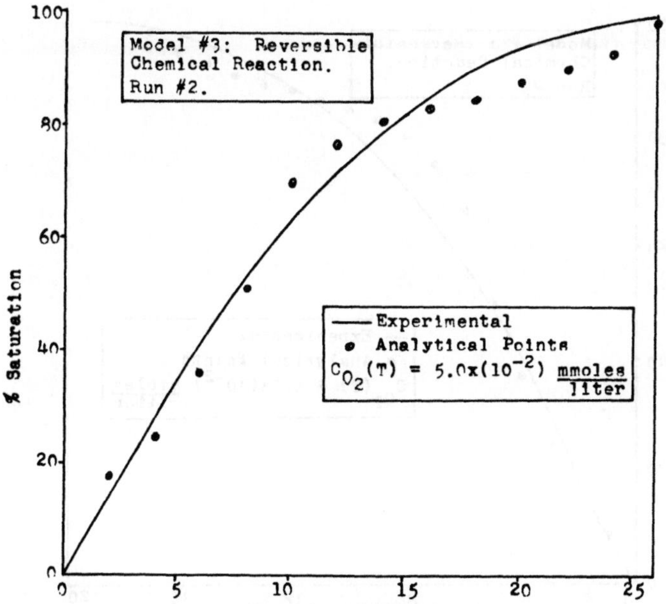

Figure 15. Percent saturation vs. time.

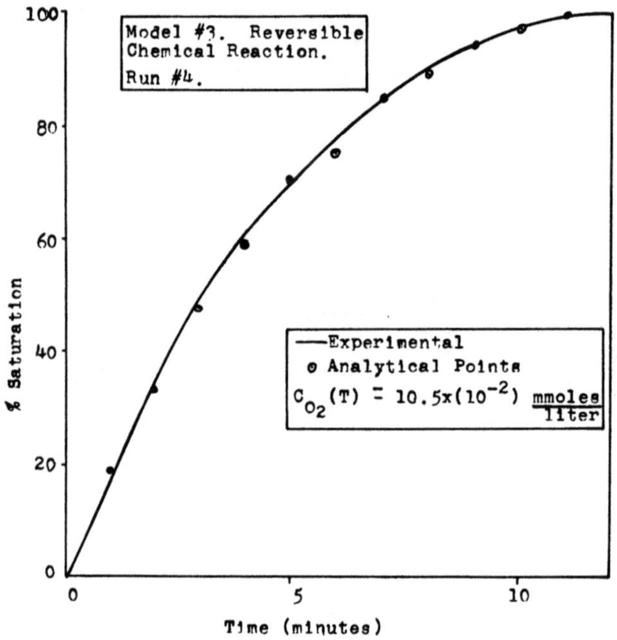

Figure 16. Percent saturation vs. time.

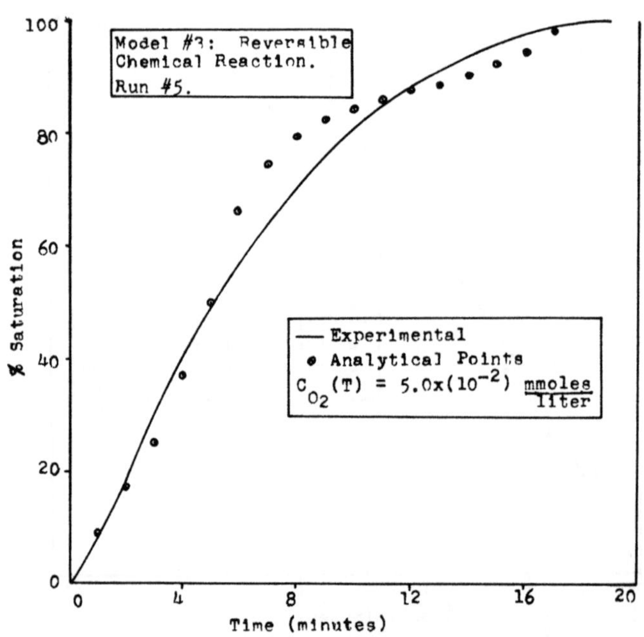

Figure 17. Percent saturation vs. time.

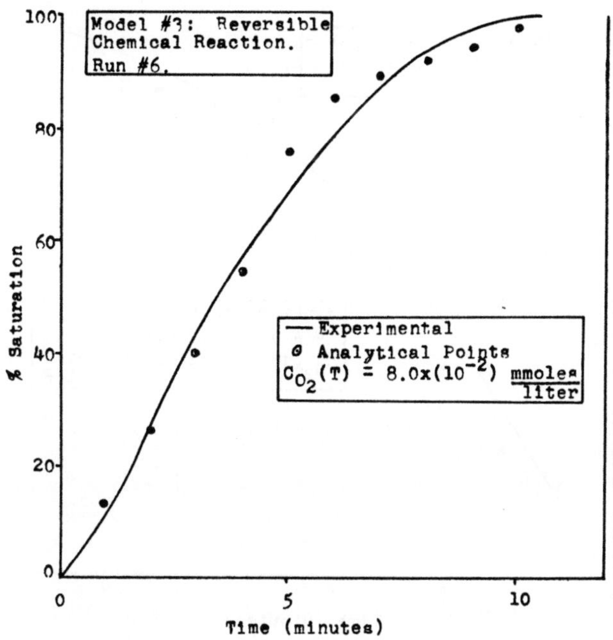

Figure 18. Percent saturation vs. time.

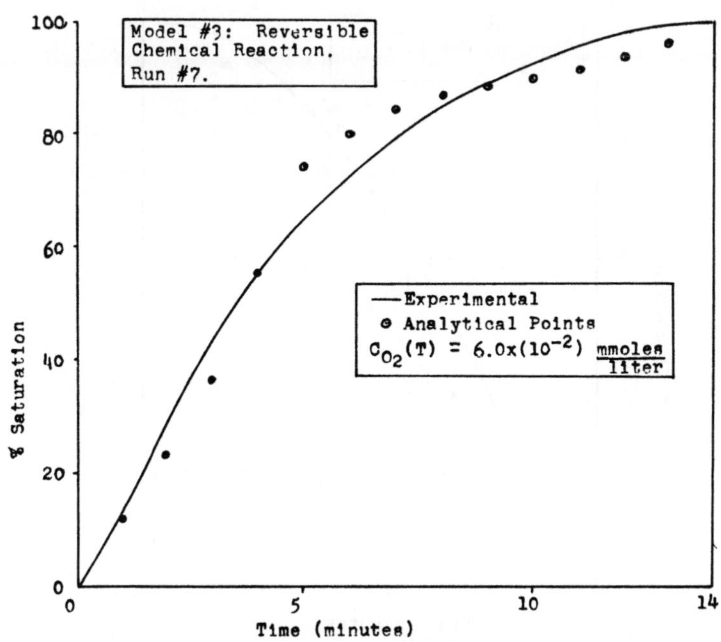

Figure 19. Percent saturation vs. time.

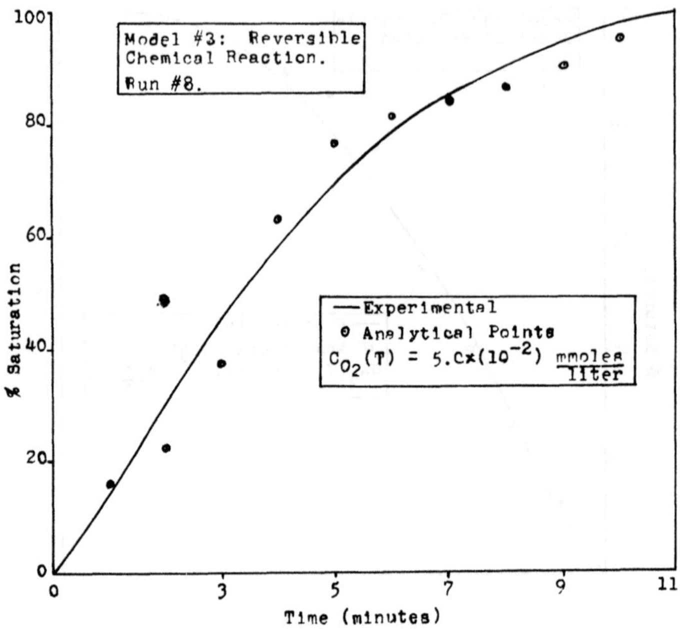

Figure 20. Percent saturation vs. time.

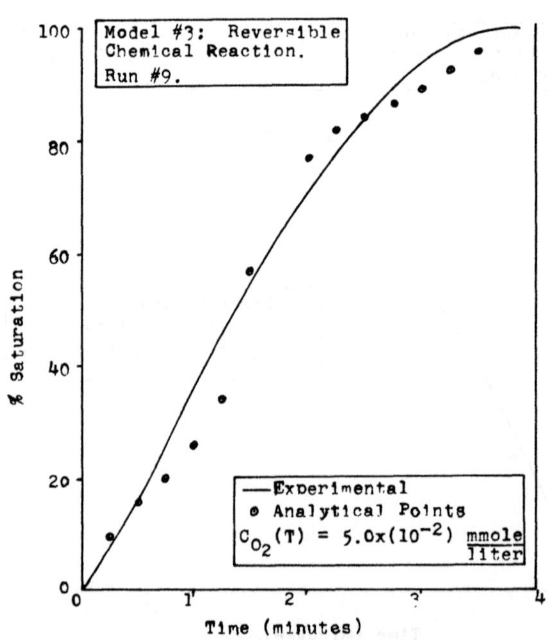

Figure 21. Percent saturation vs. time.

TABLE 1.

Average Free Oxygen Values, $C_{O_2}(\bar{T})$, Predicted by Model #3.

Run #	Oxygen input rate	Avg. Free O_2 Conc. $C_{O_2}(\bar{T})$	Max. Free O_2 Conc. $C_{O_2}(\max.)$	Ratio of Avg./Max.
#	cc O_2/sec	mmoles/liter $x(10^{-2})$	mmoles/liter $x(10^{-2})$	%
1	0.59	2.5	8.74	28.6
2	1.04	5.0	21.95	22.8
3	2.14	8.5	41.70	20.4
4	3.27	10.5	41.70	25.2
5	1.61	5.0	41.45	12.1
6	2.33	8.0	41.85	19.1
7	3.27	6.0	41.95	14.3
8	3.14	5.0	84.40	5.9
9	5.80	5.0	169.00	3.0

DISCUSSION OF RESULTS

Since the oxygen entering the system must first pass through the plasma and then into the red cell, the expected behavior for the process was a gradual buildup of oxygen in the plasma creating an oxygen gradient for transport into the cell. The concentration of oxyhemoglobin in the cell at the outer boundary was expected to be in equilibrium with the plasma P_{O_2}, and an oxyhemoglobin profile was expected to exist within the cell decreasing from the maximum oxyhemoglobin concentration at the cell wall to the minimum concentration at the center of the cell. An oxygen gradient was also expected to exist from the cell membrane to the cell interior. As the plasma oxygen content built up, the cell oxygen content would have built up simultaneously.

The experimental results do not support these ideas. Instead of having an oxygen build up in the plasma, the oxygen is continually removed by the cells, keeping the P_{O_2} of the plasma very low until the cell oxygen demand is largely satisfied. Only then does the P_{O_2} of the plasma start to increase until it reaches its equilibrium value. The oxygen is also rapidly used up as it enters the cell.

These observations tend to de-emphasize the importance of diffusion in the oxygenation process. Thus, it is not surprising that the first two models did not describe the observed behavior. The third model de-emphasized the importance of diffusion by considering only the reversible reaction as the controlling mechanism. It was successful in describing the oxygenation process.

The relatively constant values of $C_{O_2}(T)$ (the average free oxygen concentrations within the cells) predicted by the third model should be noted. These values were of about the same size for each of the nine runs studied, inspite of the differences in blood samples, experimental conditions, and overall times for the oxygenation processes. Table 1 contains the list of the values of the average free oxygen concentrations within the cells for each run. This value is the one found to give the best agreement between the experimental and the analytical curves. Also in Table 1 are the values of the maximum free oxygen concentrations which will be achieved when the saturation is completed and the cells are in equilibrium with the equilibrium plasma P_{O_2}. The percent of the maximum is calculated, and it can be seen that this value ranges from 28.6% for run #1 to 3.0% for run #9.

CONCLUSIONS

From the results of the modeling of the oxygenation process, it can be concluded that the oxygenation process in whole blood is not diffusion controlled but that the controlling mechanism is the reaction kinetics. The reaction-controlling model describes the process well for the nine experimental cases examined.

A relatively constant value of the free oxygen concentration within the cells, C_{O_2} (T), can be reported for the nine cases considered. This value results from the method of solution of the unsteady-state governing equation. It is empirically determined from data to be the value which gives the best agreement between the analytically derived equation and the experimentally recorded curves. For the nine cases examined, the experimental range of this value of C_{O_2} (T) was 2.5 to 10.5 x 10^{-2} mmoles/l, and the average for the nine runs was 6.4 x 10^{-2} mmoles/l. Although it might be argued that this term is of questionable physical significance, it does seem to represent an average value of free oxygen within the intact red cell.

Based on oxygenation Model 3 and the experimental results, a mechanism for oxygenation of whole blood with intact red cells is suggested as follows: Oxygen enters the system by being physically dissolved in the plasma. However, an oxygen gradient from the plasma to the cell exists such that most of the oxygen is drawn out of the plasma into the red cell. Inside the red cell the oxygen concentration stays at a relatively constant level for much of the oxygenation process. The oxygen entering the cells reacts with hemoglobin to form oxyhemoglobin. As the hemoglobin becomes saturated, the forward reaction rate is increased and the reverse reaction rate decreased due to heme-heme interactions. There is also the possibility of facilitated oxygen transport due to hemoglobin diffusion as the oxyhemoglobin concentration gradient increases (21). Thus, the oxygen gradient into the cell would be maintained, and the level of oxygen within the cell would remain relatively constant until a sufficiently large amount of the hemoglobin had been oxygenated to cause the diffusion terms to become significant. Once this point is reached, the oxygen uptake by the cells begins to slow down, and the level of oxygen is allowed to build up in the plasma.

EDITORS NOTE

The original paper contained an appendix section in which were shown the detailed mathematical derivations. It is suggested that the authors be contacted if this information is desired.

NOMENCLATURE

B - a collection of terms; $B = \dfrac{p^2 + pQ - pk'C_{Hb_o}}{D(Q + p)}$

B_n - B as described above with $p = p_n$ (Models #1 and #2)

B' - A collection of terms; $B' = k'C_{O_2}(T) + k'/K$ (Model #3)

C_1 - intercept value for straight-line segments of the equilibrium dissociation curve, % saturation

C_2 - intercept value for straight-line segments of the P_{O_2} vs. time curve, mm Hg

C_{O_2} - concentration of oxygen, mmoles/liter

C_{Hb} - concentration of hemoglobin, mmoles/liter or gm Hb/ 100 ml

C_{Hb_o} - concentration of hemoglobin initially, a constant for each case, mmoles/liter or gm Hb/100 ml

C_{HbO_2} - concentration of oxyhemoglobin, gm Hb/100 ml

$C_{O_2}(\bar{x}, T)$ - an empirical free oxygen concentration averaged over cell thickness and time, mmoles/liter (Models #1 and #2)

$C_{O_2}(T)$ - an empirical cell free oxygen concentration averaged over the time for saturation process, mmoles/liter (Model #3)

$C_{O_2}(\text{max.})$ - maximum free oxygen concentration in cell at equilibrium, mmoles/liter

$C_{O_2}(x, T)$ - an empirical cell free oxygen concentration averaged over the time for the saturation process, mmoles/liter

$\langle C_{HbO_2}(x,t) \rangle$ - concentration of oxyhemoglobin averaged over cell thickness by the Mean Value Theorem, mmoles/liter

D - diffusion coefficient, sometimes used in place of D_{eff}, cm^2/sec

D_m - diffusion coefficient in the membrane, cm^2/sec

D_I - diffusion coefficient in the cell interior, cm^2/sec

D_p - diffusion coefficient in the plasma film, cm^2/sec

D_{eff} - effective diffusion coefficient for plasma film, membrane, and cell interior, cm^2/sec

D_{O_2} - diffusion coefficient of oxygen, cm^2/sec

D_{HbO_2} - diffusion coefficient of oxyhemoglobin, cm^2/sec

f(t) - a function of t (general)

f(T) - the function of t evaluated at T, some intermediate value of t

g(t) - a function of t, (general)

Hb - hemoglobin heme unit- $\frac{1}{4}$ of molecule

Hb_4 - the hemoglobin molecule - four heme units

HbO_2 - oxygenated hemoglobin heme unit

Hb_4O_8 - oxygenated hemoglobin molecule

k - reverse rate constant, (constant only at a given saturation), sec^{-1}

k' - forward rate constant, (constant only at a given saturation), liters/mmoles-sec

K - equilibrium constant, (k'/k), (constant only at a given saturation), liters/mmole

K' - slope of the straight-line segment of the equilibrium dissociation curve, % saturation/mm Hg

k* - slope of the straight-line segment of the P_{O_2} vs. time curve, mm Hg/minute

L - half-cell thickness for cylindrical disc cell model, cm

n (k) - integers, n = 1, 2, 3, 4 ...

O_2 - oxygen

P_n - poles p_1, p_2, p_3 ... p_n, where n = 1, 2, 3 ...

P_{O_2} - the oxygen tension of plasma; the partial pressure of the gas which would be equilibrated with the plasma

Q - a collection of terms; $Q = k'/K + k'C_{O_2}(\bar{x}, T)$, (Model #1)
$Q = k'C_{O_2}(\bar{x}, T)$, (Model #2)

R_{O_2} - rate of change of oxygen by means of reaction

R_{HbO_2} - rate of change of oxyhemoglobin by means of reaction

S_{O_2} - the percent saturation of the oxyhemoglobin of the red cells

t - time, sec or min

x - cell distance; x = 0 at cell center; x = L at cell wall, cm

α_c - (Bunsen) cell solubility, cc O_2/cc cell-atm

ABBREVIATIONS

Equil. - equilibrium

Hemat. - hematocrit

% - per cent

Pr. - pressure, atm. or mm Hg

Temp. - temperature, ^{o}C

BIBLIOGRAPHY

1. Bird, R.B.; W.E. Stewart; E.N. Lightfoot: Transport Phenomena, John Wiley (1960).

2. Bloch, E.H.: "A Quantitative Study of the Hemodynamics in the Living Microvascular System," Am. J. Anat. 110, pp 125-45 (1962).

3. Churchill, R.V.: "Complex Variables and Applications," McGraw-Hill, 2nd Edition, New York, N. Y., (1960).

4. Collins, R.E.: "Transport of Gases Through Hemoglobin Solutions," Science 133, pp 1593, (1961).

5. Enns, T.: "Molecular Collision-Exchange Transport of O_2 by Hemoglobin," Proc. Nat. Acad. Sci. U.S.A. 51, pp 247, (1964).

6. Fatt, I.: R.C. LaForce; "Theory of Oxygen Transport Through Hemoglobin Solutions," Science 133, pp 1919, (1961).

7. Fleischman, M. and Hershey, D.: "Simultaneous Diffusion and Chemical Reaction of Oxygen in Whole Blood," Chemical Engineering Symposium Series, 1970 (in Press)

8. Forster, R.E.: "Rate of Gas Uptake by Red Cells," in: Handbook of Physiology, Section 3, Respiration, Vol. 1, Edited by W.O. Fenn and H. Rahn. American Physiological Society, Washington D.C., pp 827-37, (1964).

9. Gibson, Q.H.; F. Kreuzer; E. Meda; F. Roughton: "The Kinetics of Human Hemoglobin in Solution and in the Red Cell at 37°C," J. Physiol. (London) 129, pp 65-89, (1955).

10. Gibson, Q.H.: "The Kinetics of Reactions Between Hemoglobin and Gases," Ch. 1 in Progress in Biophysics and Biophysical Chemistry, J.A.V. Butler and B. Katz, Eds., Pergammon Press, (1959).

11. Grote, J.; g. Thews: "Die Bedingunger fur die Sauerstoff-

versorgung des Herzmuskelgewebes," Pflugers Arch. ges. Physiol. 276, pp 142-165, (1962).

12. Hesselberth, J.F.: "Simultaneous Diffusion and Chemical Reaction of Oxygen in Whole Blood in a Wetted-Wall Column," PhD Thesis, Univ. of Cincinnati, (1968)

13. Karhan, T.L. and Hershey, D. : "Equilibrium and Diffusion of the O_2-Blood System," A.I. CH.E. Journal, 14, 969-973 (1968).

14. Kernohan, J.C.: "Kinetics of the Reactions of Two Sheep Hemoglobins with O_2 and CO," J. Physiol. 155, pp 580-88, (1961).

15. Kreuzer, F.; W.Z. Yahr: "Influence of the Red Cell Membrane on the Diffusion of Oxygen," J. Appl. Physiol. 15, pp 1117, (1960).

16. Marx, T.I.; et. al.: "Diffusion of Oxygen into a Film of Whole Blood," J. Appl. Physiol. 15, pp 1123-9, (1960).

17. Meale, J.B.: Laboratory Medicine-Hemotology, C.V. Mosby, St. Louis, (1962).

18. Menke, R.C.: "Models for Simultaneous Diffusion and Chemical Reaction of Oxygen Within the Intact Red Cell of Whole Blood," PhD Thesis, Univ. of Cincinnati, (1970)

19. Miller, C.; et. al.: "Mass Transfer of Oxygen into Blood Using a Wetted-Wall Column," Chemical Engineering in Medicine and Biology, D. Hershey, Editor, Plenum Press, N. Y. 1967.

20. Mochizuki, M.; J.I. Fukuoka: "The Diffusion of Oxygen Inside the Red Cell," Japan. J. Physiol. 8, pp 206-24, (1958).

21. Moll, W.: "Measurements of Facilitated Diffusion of Oxygen in Red Blood Cells at 37°C," Pflugers Arch. 305, pp 269-278, (1969).

22. Moll, W.: "The Influence of Hemoglobin Diffusion on Oxygen Uptake and Release by Red Cells," Respiration Physiology 6, pp 1-15, (1968/69)

23. Perutz, M.F.: "Submicroscopic Structure of the Red Cell," Nature 161, pp 204, (1948).

24. Peters, J.P.; D.D. van Slyke: Quantitative Clinical Chemistry, Williams and Wilkins, Baltimore, Vol. II, (1932).

25. Roughton, F.J.W.: "Diffusion and Chemical Reaction velocity as Joint Factors in Determining the Rate of Uptake of Oxygen and Carbon Monoxide by the Red Blood Corpuscle," Proc. Roy. Soc. B 111, pp 1-36, (1932).

26. Roughton, F.J.W.: "Diffusion and Simultaneous Chemical Reaction Velocity in Hemoglobin Solutions and Red Cell Suspensions," Ch. 2 in reference 10.

27. Sendroy, J.; R.T. Dillon; D.D. van Slyke: "Studies of Gas and Electrolyte Equilibrium in Blood," J. Biol. Chem. 105, pp 597-632, (1934).

28. Spiegel, M.R.: Theory and Problems of Laplace Transforms, Schaum's Outline Series, Schaum Pub. Co. N. Y., (1965).

29. Taylor, A.E.: Advanced Calculus, pp 51-52, Ginn and Co., New York, N. Y., (1955).

30. Taylor, A.E.: Ibid. pp 118-120.

31. Taylor, A.E.: Ibid. pp 619.

32. Wang, J.H.: "Transport of Oxygen Through Hemoglobin Solutions," Science 133, pp 1770, (1961).

33. Weissman, M.H.; L.F. Mockros: "Oxygen Transfer to Blood Flowing in Round Tubes," 19th Annual Conference on Engineering in Medicine and Biology, San Francisco, (1966).

34. Whittam, R.: Transport and Diffusion in Red Blood Cells, Williams and Wilkins, Baltimore, (1964).

35. Wintrobe, M.M.: Clinical Hematology, Lea and Febger, Philadelphia, (1961).

IN VITRO MEASUREMENT OF UNCOMBINED OXYGEN CONCENTRATION IN INTACT RED CELLS

Marvin Fleischman, University of Louisville, Louisville Kentucky and Daniel Hershey, University of Cincinnati Cincinnati, Ohio 45221

Oxygen transfer in blood is receiving increasing attention because of current activities in the areas of heart transplants and oxygenator development. In the course of our work on simultaneous diffusion and chemical reaction (1) of oxygen within red cells, it became necessary to measure the concentration of uncombined oxygen in intact cells. Since no method was available, a technique using polarographic PO_2 electrodes was developed and is described in this paper. The data thus obtained was used in the governing equations for intracellular oxygen transport to evaluate the overall diffusion coefficient of oxygen in the red cell at 25°C.

METHODS FOR MEASURING INTRACELLULAR FREE OXYGEN

The free oxygen concentration in the intact red cell is very low, on the order of millimoles per liter. Because any penetration of the red cell by measuring instruments, e.g., an electrode, would damage the cell, intracellular oxygen must be measured indirectly. Though mathematical models for intracellular oxygen transport have used free oxygen as the dependent variable (7), nobody has reported the actual measurement of free oxygen in intact red cells. Sendroy and Van Slyke (8) did report a Bunsen solubility coefficient for intracellular oxygen, though free oxygen was not measured. They found values of the oxygen solubility in suspensions of varying hematocrit, and extrapolated to 100% cells.

Two Electrode Method

A successful method for measuring intracellular free oxygen concentration using polarographic oxygen (PO_2) electrodes was developed. A measured volume of whole blood equilibrated with a gas of sufficient PO_2 to completely saturate the hemoglobin, was sealed in a vessel of known volume. The gas phase in the closed bottle was at a higher PO_2 than that of the equilibrated blood. Therefore, when stirring was started, oxygen was transferred from the gas to the blood. Polarographic electrodes in the gas phase and in the blood continuously monitored the PO_2 in both phases. The experimental system is shown in Figure 1, and is discussed in detail in the next section.

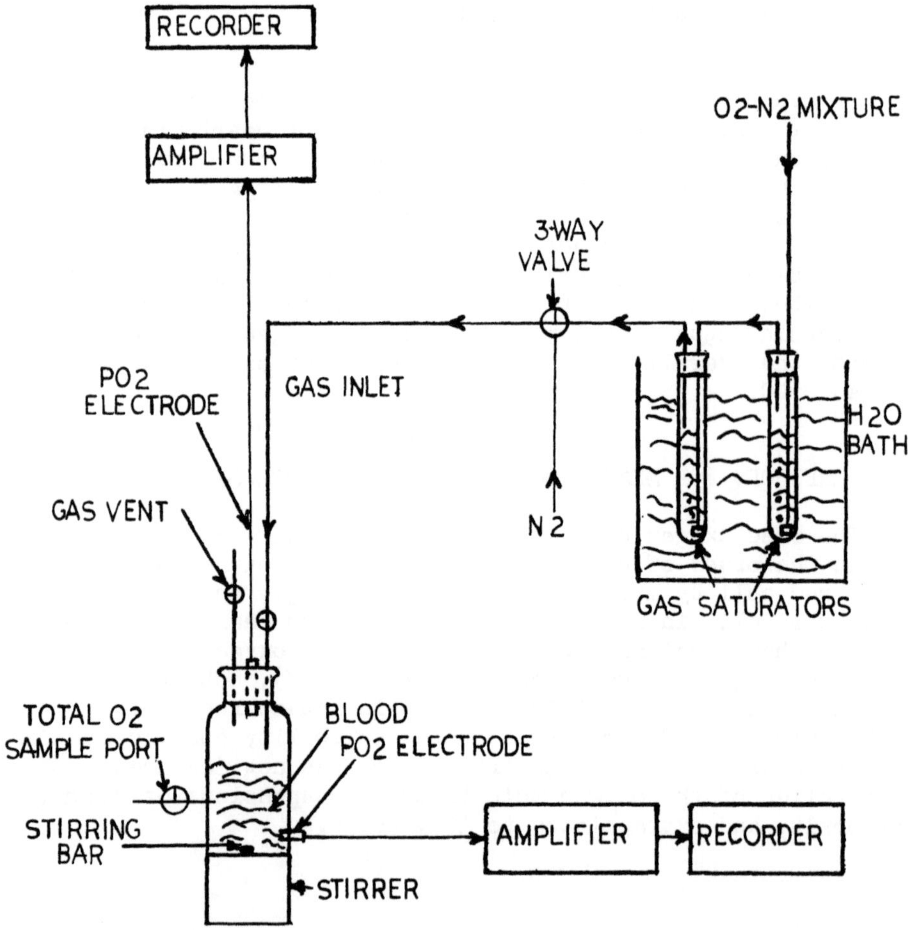

Figure 1. Experimental system for determining intracellular oxygen concentration.

In the "Two Electrode" method the change in average intracellular free oxygen concentration from zero time to any given time t, i.e., $\Delta\bar{c}(t)$, is found from the measured gas and blood PO_2 values by a total oxygen balance. Since the vessel is closed, the total amount of oxygen in the vessel is constant with time. This is shown by equation 1 in which Vg is the volume of the gas phase in ml, T the temperature in °K, and R is the gas constant, 62.4 cm^3-mm Hg/millimole-°K. The f in equation 1 is an empirical electrode membrane correction factor to convert the measured PO_2 of liquid samples to the equivalent gas PO_2. As supplied by the electrode manufacturer, the value of f is 1.02

$$\left.\begin{aligned}&\underbrace{\alpha_p f V_L(1-Ht) P_{O_2}BL(1)}_{\text{initial millimoles } O_2 \text{ in plasma}} + \underbrace{c_1 V_L Ht}_{\text{initial millimoles } O_2 \text{ in cells}} + \underbrace{V_g P_{O_2}GAS(1)/RT}_{\text{initial } O_2 \text{ in gas phase}} \\ &= \alpha_p f V_L(1-Ht) P_{O_2}BL(t) + \bar{c}(t) V_L Ht + V_g P_{O_2}GAS(t)/RT\end{aligned}\right\} \quad (1)$$

$PO_2BL(1)$ is the initial blood PO_2, equal to the PO_2 of the gas with which the blood was initially equilibrated. C_1 is the initially uniform intracellular oxygen concentration as determined by equation 2.

$$c_1 = \alpha_c P_{O_2}BL(1)/17.024 \qquad \text{for all r at t=0} \qquad (2)$$

The experimental change in intracellular oxygen concentration $\Delta\bar{c}(t)_{exp.}$ is found by rearranging equation 1.

$$\Delta\bar{c}_{EXP} = \bar{c}(t) - c_1 = \frac{\dfrac{V_g\left[P_{O_2}GAS(1) - P_{O_2}GAS(t)\right]}{RT} - \alpha_p f V_L(1-Ht)\left[P_{O_2}BL(t) - P_{O_2}BL(1)\right]}{17.024} \qquad (3)$$

The change in intracellular oxygen is determined instead of the absolute concentration at time t, to eliminate the necessity of calculating c_1. This in turn makes the experimental results independent of α_c, the oxygen solubility in the cells, and reduces experimental error.

The oxygen balance assumes that the gas in the sealed bottle obeys the ideal gas law and that the plasma obeys Henry's law with respect to oxygen uptake. Because the compressibility factors and fugacity coefficients of the gas mixtures at the pressures involved were unity, the ideal gas law can be used. Karhan (10) showed that oxygen absorption into sheep plasma follows Henry's law.

Metabolism Correction Used in Two Electrode Method

Equation 3 doesn't account for oxygen uptake by blood metabolism. Because of the low intracellular oxygen concentrations involved in the "Two Electrode" experiments, metabolic oxygen consumption is significant, and was accounted for empirically using Kelman's (11) data.

Metabolic oxygen consumption of shed blood results from uptake by white cells, platelets, and plasma and cell components, mainly lipids. Based on Kelman's work (12), the average metabolic consumption rate at 25°C is 3.7×10^{-5} cc O_2(STP)/min/ml. of blood. To correct for metabolism in the "Two Electrode" method, an equivalent change in gas phase PO_2, $\Delta PO_2GAS_{met.}$, is calculated as shown in equation 4.

$$\underbrace{\Delta P_{O_2}GAS_{MET}}_{MM\ Hg} = \frac{mRT}{V_g} = \frac{3.7\ ML\ (STP)}{MIN - ML\ BLOOD} \times \frac{1\ MMOLE}{22.4\,cc\,(STP)} \times \frac{V_L RT}{V_g} \times t = .031\,t\,\frac{V_L}{V_g} \qquad (4)$$

In equation 4, t is the experimental time over which the change in average intracellular oxygen is measured. The metabolism correction, $\Delta PO_2GAS_{met.}$, is subtracted from the measured change in gas PO_2 over time t, and the corrected ΔPO_2GAS is used in equation 3.

EXPERIMENTAL ASPECTS OF THE TWO ELECTRODE METHOD

Defibrinated sheep blood was used, and wherever possible blood was contained in plasticware. Plasticware is preferable to glass because red cells adhere to glass, and hemolysis can occur when blood is in contact with glass for long periods.

To attain a desired blood PO_2 and/or hemoglobin saturation, gas of the appropriate PO_2 was bubbled through the blood with plastic sparger tubes. Gas mixtures of oxygen and nitrogen of known composition were used for this purpose. Prior to entering the blood, all gases were saturated with water vapor at 25°C to prevent the gas from drying the blood. In addition, saturating the gases eliminates the need for humidity measurement in calibrating the PO_2 electrodes. To reduce foaming of the blood during gas bubbling, teflon shreds coated with Dow Corning Antifoam A Spray were added to the blood.

Hematocrit was determined with a Drummond Microhematocrit Centrifuge, and total hemoglobin by the cyanomethemoglobin method using an EEL Hemoglobinometer. Both hematocrit and hemoglobin were measured before and after each experiment, and both sets of measurements always agreed.

Oxygen partial pressure was measured in both liquids and gases by polarographic electrodes. For measurements in blood, water, and plasma, an Instrumentation Laboratories PO_2 Reservoir Electrode was used with an IL 125B amplifier. A Beckman PO_2 Microelectrode was tried for measuring blood PO_2, but the membranes were quite difficult to change, and the electrode was fragile. To measure PO_2 in gases, a Beckman PO_2 Macroelectrode with a Beckman 160 Gas Analyzer was used.

Procedure For "Two Electrode" Experiments

The experimental system is shown in Figure 1. The bottle used to hold the blood was a rigid one liter plastic bottle. A hole near the bottom held a #0 one hole rubber stopper in which the blood electrode was placed. The bottle was sealed at the top by a #10 stopper which held the gas electrode and the gas inlet and vent tubes. Becton and Dickinson MS20 two way stopcocks were used for the gas inlet and vent valves. Silicone grease was placed on all external points where gas leakage could occur.

The blood was equilibrated prior to the experiment by bubbling water saturated gas of known PO_2 through it for one to two hours. The stopper holding the calibrated blood PO_2 electrode was placed in the bottle, and the equilibrated blood transferred to the vessel under an atmosphere of the equilibrating gas. Without stirring of the blood, the gas electrode was calibrated while in place in the bottle by alternately passing through pure N_2 and a water saturated N_2-O_2 mixture of known PO_2. The inlet and vent valves were then closed, sealing the bottle under an atmosphere of the high PO_2 gas used for calibration.

Stirring of the blood and the blood and gas PO_2 recorders were started simultaneously. The volume of blood used, V_L, was determined at the end of an experiment by measuring the volume of the bottle corresponding to the blood level in the bottle.

Validity of Two Electrode Method Checked

The validity of the two electrode method was checked by using it to determine the solubility of oxygen in distilled water at 25°C. The calculated values shown below in table 1 agree with the published value (12) to within a few percent.

Table 1

Oxygen Solubilities at 25°C Determined By The Two Electrode Method

Number of Experiments	αO_2-H_2O cc(STP)/cc/atm Measured	αO_2-H_2O cc(STP)/cc/atm Literature
4	0.028 ± 0.0009	0.028
	αO_2-Plasma Measured	αO_2-Plasma Literature
5	0.030 0.003	0.026

Although a value of α_p can be estimated from the literature (8), the "Two Electrode" method was used to determine the value for use in equation 3. These results are also given in table 1. Both the literature and our experimental values of α_p are averages only for the blood used in both determinations.

NOMENCLATURE

C- concentration of free oxygen in the intact red cell, millimoles/liter

C_1- initial intracellular oxygen concentration, millimoles/l

$C(R_c)$- oxygen concentration at the cell wall

$\overline{C}$- volumetric average oxygen concentration

$\Delta\overline{C}$- change in $\overline{C}$, millimoles/liter

$\Delta\overline{C}_{exptl}$- experimental value of $\overline{C}$

C_T- total hemoglobin content, grams/100 ml blood

D_c- overall diffusion coefficient of oxygen in the red cell, $cm^2/sec.$

Ht- fractional hematocrit, cc cells/cc blood

MCHC- mean corpuscular hemoglobin concentration, grams/cc cell

O_2- oxygen

PO_2- partial pressure of oxygen, mm Hg.

PO_2BL-oxygen partial pressure in blood, mm Hg.

PO_2GAS-oxygen partial pressure in gas phases, mm Hg.

r- radial distance in cell, cm.

R_c- effective radius of cell

S- fractional saturation of hemoglobin with oxygen

t- time, sec.

T- temperature, ^oK

V_L- volume of blood, liters

v_g- volume of gas phase, cc

y- dimensionless cell radius

α_c- oxygen solubility in red cell, cc O_2(STP)/cc cell/atm.

α_{H_2O}- oxygen solubility in water, cc/cc/atm.

α_p- oxygen solubility in plasma, cc/cc/atm.

θ - dimensionless oxygen concentration

Δ - change in a quantity, e. g., $\overline{C}$

REFERENCES

1. Fleischman, M., "Simultaneous Diffusion and Chemical Reaction

of Oxygen in Whole Blood", Ph.D. Thesis, Univ. of Cincinnati, 1968

2. Keller, K.H., "The Steady State Transport of Oxygen thru Hemoglobin Solutions", Ph.D. thesis, Johns Hopkins Univ., 1965

3. Longmuir, I.S., Roughton, F.J.W., "The Diffusion Coefficients of CO and N_2 in Hemoglobin Solutions", J. Physiol. 118, 264 (1952)

4. Bird, R.B., Stewart, W.E., & Lightfoot, E.N., "Transport Phenomena", p. 545, John Wiley (1960)

5. Kreuzer, F., Yahr, W.Z., "Influence of Red Cell Membrane in Diffusion of Oxygen", J. App. Physiol. 15, 1123-29 (1960)

6. Sirs, J.A., "Uptake of O_2 and CO by Hemoglobin in Sheep Erythrocytes at Various Temperatures", J. App. Physiol. 18, 166-170 (1963)

7. Mochizuki, M., Fukuoka, J.I., "The Diffusion of Oxygen Inside the Red Cell", Japan J. Physiol. 8, 206-224 (1958)

8. Sendroy, J., Van Slyke, D.D., "Solubility and Physical State of Uncombined Oxygen in Blood", J. Biol. Chem. 105, 597-632(1954)

9. Peters, J.P., Van Slyke, D.D., "Quantitative Clinical Chemistry", Vol. II, Williams & Wilkins, Balt. (1932)

10. Hershey, D. and Karhan, T.L., "Equilibrium and Diffusion of the O_2 Blood System," A.I.Ch.E. Journal, <u>14</u>, 969-973 (1968)

11. Kelman, G.R., "Correction Factors to be applied to Measured Blood Gas Tensions and pH", Bio. Med. Eng. 8/67, pp 362-64

12. Handbook of Chemistry and Physics, 39th ed., Chem. Rubber Co., p. 1606, 1957-1958

13. Crank, J., "The Mathematics of Diffusion", p. 87, Oxford Press, 1956

14. Lapidus, L., "Digital Computation for Chemical Engineers", McGraw-Hill, 1962

15. Fasciolo, J.C., Chiodi, H., "Arterial PO_2 During Pure O_2 Breathing", Am. J. Physiol. 147 54-65 (1946)

16. Sirs, J.A., "Influence of Metabolism on Uptake of CO and O_2

by Hemoglobin in Erythrocytes", J. Appl. Physiol. 18, 175-178 (1963)

17. Whyte, J.H., Laver, M.B., "Oxygen Solubility in Blood and Temperature Corrections for PO_2", J. Appl. Physiol. 19, 901-906 (1963)

18. Hesselberth, J.F., "Simultaneous Diffusion and Chemical Reaction of Oxygen in Whole Blood in a Wetted Wall Column", Ph.D. thesis, Univ. of Cincinnati, 1968

19. Perutz, M.F., "Submicroscopic Structure of the Red Cell", Nature 161, 204 (1948)

20. Fleischman, M., Hershey, D., "Estimation of the Rate Constants for the Oxygen-Hemoglobin Reaction Within Intact Red Cells", In Press, CEP Symposium Series, AIChE

21. Anderson, N.M., Sekelj, P., "The Light Transmission of Whole Blood-Determination of Oxygen Saturation", J. Lab. Clin. Med. 65, 153-166 (1965)

THE INTERACTION OF CARBON DIOXIDE WITH THE RATE OF EXCHANGE OF OXYGEN BY RED BLOOD CELLS

John A. Sirs

Department of Physics, St. Mary's Hospital Medical School, University of London, London, W. 2, England

INTRODUCTION

An objective of 'in vitro' studies of the rate of exchange of oxygen by haemoglobin in the red blood cell is to be able to precisely state the rates of these processes occurring 'in vivo' and at the same time provide a quantitative understanding of the biophysical and physicochemical mechanisms involved. It has not yet proven possible 'in vitro' to simulate the 'in vivo' conditions, involving as they do rapid changes of haemoglobin saturation with oxygen, pH, carbon dioxide interactions, Cl^-/HCO_3^- movement, osmotic pressure and many other factors. In order to start to understand this process, the rate of oxygen exchange has mainly been studied in dilute suspensions, maintaining other factors constant, and observing the complete change from zero to 100% saturation of oxyhaemoglobin. A theoretical analysis of this situation, based on a process of diffusion through the cell membrane, followed by diffusion through and chemical reaction with the internal haemoglobin, was first given by Roughton (1). Even then, certain simplifications, such as the representation of a biconcave erythrocyte as equivalent to a thin layer of haemoglobin covered by a membrane, were necessary in order to obtain an analytical solution. More recently, in a full review, Roughton (2) has provided convincing evidence that these simplifications are acceptable and that for the majority of 'in vitro' observations on the rate of uptake of gaseous ligands by erythrocytes, equilibrated with an oxygen-free gas phase, the theoretical analysis is valid.

Since that review, evidence has accumulated to suggest that the 'in vivo' rate of exchange of oxygen can be determined by the chemical reaction velocity constants. The reason for this discrepancy would appear to be due to the presence or absence of factors which influence the flexibility of the cell. Thus the complete removal of oxygen in itself makes the erythrocyte relatively inflexible (Sirs (4)). It is significant that the rates of uptake by cells which have been previously sphered by heating (Carlsen and Comroe (5)) or to which formaldehyde has been added (Sirs (6)), are identical with the rate of uptake of oxygen by untreated cells equilibrated with an oxygen-free media. Both of these treatments are known to make the cells inflexible. Sirs (7) has also obtained similar results with a number of other methods of making the cells rigid. In normal circumstances erythrocytes equilibrated with oxygen remain flexible and the rate of egress is then simply determined by the rate of chemical dissociation of HbO_2 (Sirs (8), (9)). When however these oxygenated cells are treated to make them inflexible, by adding formaldehyde or heating (Sirs (7)), a much slower rate of egress, consistent with the rate being limited by diffusion through the membrane and internal haemoglobin, is observed.

The logical sequence of these studies, in order to approach more closely conditions pertaining 'in vivo', appeared to be to study the rates of oxygen exchange when only partial reduction or oxygenation of haemoglobin occurred with simultaneous variation of the carbon dioxide content of the surrounding media. The experiments which I would like to discuss in this paper provide evidence to support the possibility, previously mentioned by Roughton (10), that complete equilibration of blood does not occur during the time of transit through the lungs and tissues.

EXPERIMENTAL METHODS AND RESULTS

Measurements have been made of the rates of uptake of oxygen by reduced haemoglobin and egress of oxygen from saturated oxyhaemoglobin, within erythrocytes, under conditions which produce only partial oxygen saturation at the final equilibrium. The partial pressures of carbon dioxide, with which the solutions were equilibrated prior to mixing, have also been varied.

Human or sheep's blood was obtained by venepuncture, heparin being used to prevent coagulation. The blood was immediately stored at 4°C. and measurements made within 24 hrs. of collecting the sample. A suspension of red blood cells was prepared just prior to the experiment, by diluting the blood (1:10 to 1:50) in an isotonic solution consisting of 7.6g. NaCl, 2g. $NaHCO_3$, 0.42g. KCl and 0.24g. $CaCl_2$ per litre. The solution was equilibrated with appropriate partial pressures of oxygen and carbon dioxide as required. A second solution, consisting of the isotonic medium without red blood cells, was similarly equilibrated with either N_2, CO or O_2 pressures and a given pCO_2. The cell suspension was then rapidly mixed with the isotonic medium using the stopped-flow technique of Sirs and Roughton (3). The resulting change of HbO_2 with time was observed with a split-beam spectrophotometric detector coupled to both an oscilloscope and a pen recorder. In some of the experiments the cell suspension was simultaneously diluted while being rapidly mixed using the method of Sirs (8). Calibration solutions were also prepared, at a cell concentration corresponding to the mixed suspension, and equilibrated with known partial pressures of oxygen. These were flushed through the stopped-flow apparatus, before and after each experiment, to obtain a calibration curve. This also enabled the final degree of saturation of the reacting mixture to be determined from comparative deflections on the pen recorder. The temperature of the reacting mixture was measured with a thermocouple inserted into the observation tube of the rapid reaction apparatus. The experiments have been restricted to the room temperature, 20-25°C. The gas concentrations in the solutions have been measured using the Van Slyke gasometric technique and by carefully controlling the partial pressure of the gases when equilibrating the solutions. The pH was measured with a standard glass electrode and pH meter. Small samples were taken directly from the cell suspensions and centrifuged to ensure the absence of haemolysis. Microscopic examinations were also routinely made to check no abnormality or crenation of the erythrocytes were present.

In the first series of experiments, the influence of CO_2 on the rate of egress of oxygen from the erythrocytes was investigated. A 1:10 cell suspension, equilibrated with 8cm pO_2 and 3cm pCO_2, to maintain pH 7.4, was rapidly mixed and diluted 1:20 in the isotonic medium equilibrated with 110cm pCO and 3cm pCO_2. In this case the haemoglobin is maintained in a fully saturated state, as HbO_2 is converted to HbCO, and the rate constant, 'r', of this process has been shown (Sirs (9)), to be equivalent to $k_4/4$, the rate constant for the change $Hb_4O_8 \rightarrow Hb_4O_6$. At 21.5°C, 'r' was found to be 11.5 sec^{-1}, for human red blood cells. Repeating this experiment but with the diluting medium equilibrated with 110cm pCO and partial pressures of carbon dioxide from 3cm to 10cm Hg, had no effect on the rate of egress. Similar experiments with no carbon monoxide in the diluting medium, so that simple dissociation of HbO_2 to Hb occurred, indicated that no change of the initial rate of dissociation took place when the dilutant pCO_2 was varied from 3 to 6cm Hg, while the pre-mixed cell suspension pCO_2 remained at 3 cm.

In a further experiment, a suspension of sheep erythrocytes, containing haemoglobin B, was prepared and equilibrated with 7 cm pO_2, 3cm pCO_2 and then nitrogen applied to maintain a positive pressure. This was mixed and diluted 1:20 with the isotonic medium equilibrated with 1cm pCO_2 and oxygen-free nitrogen. The initial rate gave 'r' equal to 11.3 sec^{-1} at 22°C, in agreement with previously published data (Sirs (11)). The HbO_2 does not in these circumstances completely dissociate, and as can be seen in Fig. 1, the final equilibrium of 50% saturation, was only

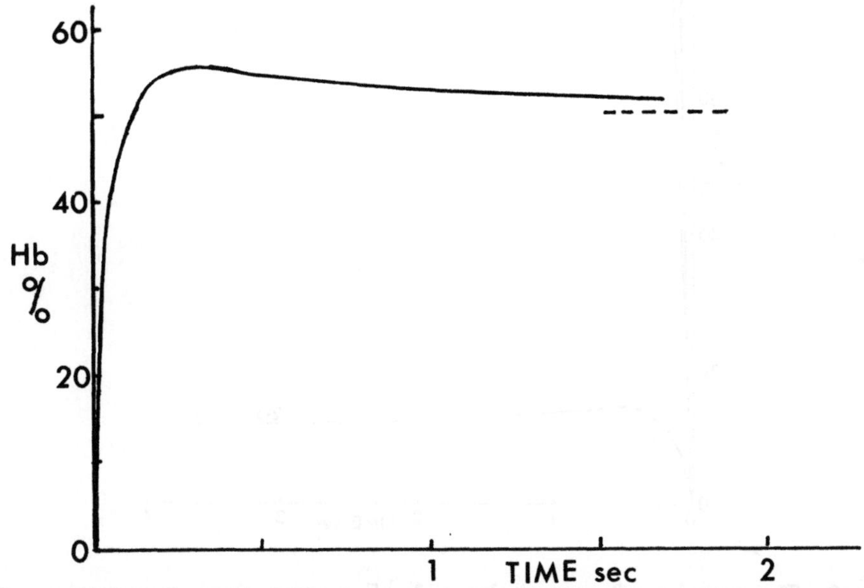

Fig. 1 The egress of oxygen from sheep eyrthrocytes, equilibrated with 7 cm pO_2 and 3 cm pCO_2, after rapid-mixing and 1:20 dilution in an oxygen-free isotonic medium, equilibrated with 1 cm pCO_2. The final mixed suspension corresponds to a 1:150 dilution of whole blood.

achieved after the percentage of haemoglobin had overshot to 55.8% and slowly fallen again.

A similar overshoot phenomena has also been observed with the uptake of oxygen by fresh sheep's blood. In this case the blood was diluted initially 1:45 in the isotonic medium and then reduced by repeated evacuation and flushing with pure nitrogen. The suspension was finally equilibrated with nitrogen. This was rapidly mixed with isotonic solutions equilibrated with 2.5cm, 20cm and 25cm partial pressures of oxygen and 3cm pCO_2. Typical curves of the rate of uptake are shown in Fig. 2. A significant

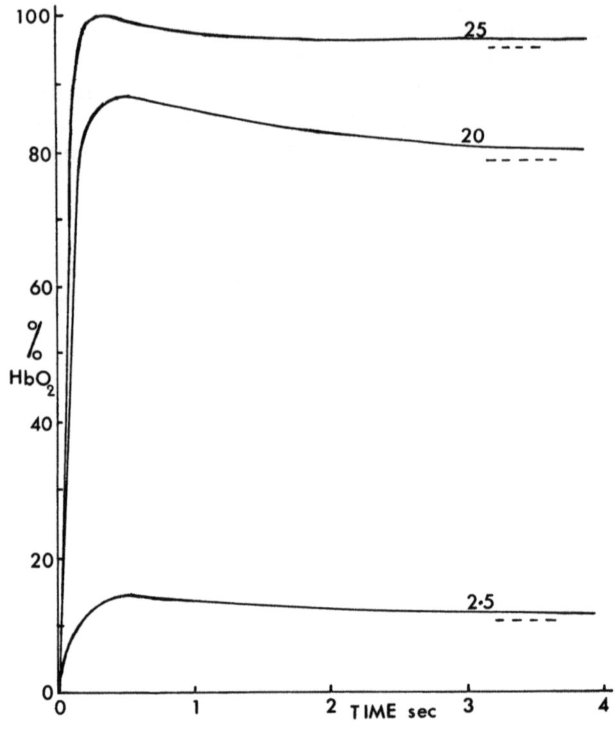

Fig. 2 The uptake of oxygen by a 1:45 oxygen-free suspension of fresh sheep's blood after rapid-mixing with an equal volume of an oxygenated solution. The three curves were obtained at 18°C with solutions equilibrated with 2.5 cm, 20 cm and 25 cm pO_2 and 3 cm pCO_2.

overshoot effect could not be obtained however with an initial cell suspension corresponding to 1:200 dilution of whole blood. There was also no overshoot effect, using some of the same blood, after storage at 4°C for four days.

THEORETICAL ANALYSIS

In order to explain these effects it is necessary to re-examine the nature of the reactions of carbon dioxide in the cell and plasma. Two principal mechanisms are known; the reversible formation of bicarbonate and the reaction with amino-groups to form carbamate. The rate of the overshoot effect, taking several seconds, precludes the carbamate reaction which is relatively fast, and suggests that it must somehow be related to the slower reactions of CO_2 taking place in the plasma. It has been mentioned by Roughton (10), that one of the consequences of carbonic anhydrase being in the red blood cell and not in the plasma might be to produce such an effect. One of the ambiguities of the present understanding of this mechanism however is the statement of Roughton (10, 12) that it would take of the order 250 secs. for 90% dehydration of CO_2 to occur in the plasma, under the conditions prevailing in the lung. This figure is computed on the basis of an analysis given by Brinkman, Margaria and Roughton (13), which involves the assumptions that the hydrogen ion concentration is maintained constant and no back-reaction occurs. In calculating the time of reaction of CO_2 in the cell, as catalysed by carbonic anhydrase, the assumption is also made that if there were no enzyme present the rate in the cell would be the same as that in the plasma. The means by which the carbon dioxide finally comes to equilibrium in the plasma does not appear to have been fully considered.

In the following analysis a more detailed calculation has been made of the rate of formation of bicarbonate from CO_2, which includes the back reaction and allows for variation of the hydrogen ion concentration with varying degrees of buffering. The effect of utilizing an 'open system', in which the carbon dioxide concentration is maintained constant, by the lungs and tissues acting as a 'sink'; and a 'closed system', in which no CO_2 is removed, are also discussed. The effect of introducing carbon dioxide into a system consisting of two compartments, one in which the reaction is fast and the other slow, while the dissolved CO_2 may come to instantaneous equilibrium between them, is then considered. Next the problem of an additional rapid permeability to HCO_3^- and Cl^-, between the compartments, is analysed. Finally an attempt is made to relate the overall process to the oxygenation of the blood as it passes through the lungs.

The Kinetics of the Reaction $CO_2 + H_2O \rightleftharpoons H_2CO_3$ in Buffered Solutions

Faurholt (14) has shown that the dehydration of CO_2 obeys the simple form

$$\underset{(a - x)}{HCO_3^-} + \underset{(h - x/\beta)}{H^+} \rightleftharpoons H_2CO_3 \rightleftharpoons \underset{(c + x)}{CO_2} + H_2O$$

and by applying the Law of Mass Action, the kinetics of the reaction can be expressed by

$$\frac{d\,[CO_2]}{dt} = k\,[H_2CO_3] - k^1\,[CO_2] \quad \ldots\ldots\ldots \quad (1)$$

The combination of bicarbonate with hydrogen ion, being ionic, is relatively instantaneous, so H_2CO_3 is always in equilibrium as given by

$$K = \frac{[H^+]\,[HCO_3^-]}{[H_2CO_3]}$$

Equation (1) can thus be written as

$$\frac{dx}{dt} = \frac{k}{K}\,(a - x)\,(h - \frac{x}{\beta}) - k^1\,(c + x) \quad \ldots\ldots \quad (2)$$

where 'a', 'c' and 'h' are the initial concentrations of HCO_3^-, CO_2 and H^+ respectively. The relative change of H_2CO_3 concentration is negligible. While a change of concentration 'x' of HCO_3^-, at time 't', produces an equivalent change of CO_2, the hydrogen ion concentration only changes by x/β due to the buffering power of the solution. This assumes that the buffer action is effectively instantaneous, due to the ionic nature of the buffer agents in the cells and plasma. The factor $1/\beta$ thus represents the fraction of free hydrogen ions available when x ions are added. Integrating equation (2) gives

$$\exp.\left(-\frac{2Pkt}{\beta}\right) = \left(\frac{Q + P}{Q - P}\right)\left(\frac{x - Q + P}{x - Q - P}\right) \quad \ldots\ldots \quad (3)$$

where

$$2Q = a + \beta h + \frac{\beta k^1 K}{k}$$

$$P = Q - \frac{\beta ah - \beta k^1 Kc/k}{2Q}$$

If the reaction is considered as occurring during the passage of the blood through the lungs, then a state of pseudo-equilibrium is reached, in which the rate of formation of CO_2 is balanced by the rate at which it is being removed via the alveoli, and the carbon dioxide concentration remains constant. Equation (2) then becomes:

$$\frac{dx}{dt} = \frac{k}{K}\left(a - x\right)\left(h - \frac{x}{\beta}\right) - k^1(c). \quad \ldots\ldots\ldots \quad (4)$$

Integrating again gives equation (3), but in this case with new constants P^1 and Q^1 given by

$$2Q^1 = a + \beta h$$

$$P^1 = Q^1 - \frac{\beta ah - \beta k^1 Kc/k}{2Q^1}$$

Using these solutions, the times for 90% of the reaction to equilibrium to occur have been calculated for different values of β. These are shown in Table I, with the corresponding change of HCO_3^- that occurs at equilibrium, x_e, when $a = 0.027M$, $h = 4 \times 10^{-8}M$ and $c = 10^{-3}M$. The values of k, k^1 and K have been taken from Roughton (12), and at 38°C are 89, 0.13 and 3.4×10^{-4} respectively. The time of 220 secs, for $\beta = \infty$ and c constant, compares with that of 250 sec. quoted by Roughton (10, 12), when no back-reaction is involved. The value of β in the red blood cell is of the order 6×10^5 (Adair (15), German and Wyman (16)) and, allowing for the plasma buffer power to be ten times less than that within the cell (Henderson (17)), β is 6×10^4 in the plasma. The respective times for the uncatalysed reaction to be 90% complete are therefore 102 sec. and 18.2 sec. in the cell interior and plasma. If allowance is made for the more acid pH of 7.17 in the cell, the value of h is increased to 6.76×10^{-8} and the time for 90% reaction falls further to 80 sec. Thus, with carbonic anhydrase in the cell speeding this up by a factor of 13,000 (Kernohan and Roughton (18)), the respective 'in vivo' times for 90% reaction in the cell and plasma, with an open system where 'c' is kept constant, are of the order 6.1 msec. and 18.2 sec. As is discussed later, the value of β within the red blood cell alters during the respiratory cycle, so some variation of the reaction time in the cell occurs, but within the wider limits for β, from 2×10^5 to 10^6, that pertain, the velocity is so fast that this is not a decisive factor. Taking the value of K as 1.6×10^{-4} or introducing a correction for the activity coefficient of HCO_3^- being 0.6 (Roughton (10)) does not significantly alter the more critical time of 18.2 sec. for the reaction in plasma.

Further examination of Table I also reveals that the variation in time is directly related to the amount of HCO_3^- converted

TABLE I

Values of the time for 90% of the reaction to occur, t_{90}, and the total change of HCO_3^- at equilibrium, x_e, for different values of the buffer factor, β.

β	Open system		Closed system	
	t_{90}/sec	x_e mM	t_{90}/sec	x_e mM
1	3.25×10^{-4}	2.1×10^{-5}	3.25×10^{-4}	2.1×10^{-5}
10	3.25×10^{-3}	2.1×10^{-4}	3.25×10^{-3}	2.1×10^{-4}
10^2	3.25×10^{-2}	2.1×10^{-3}	3.25×10^{-2}	2.1×10^{-3}
10^3	0.325	2.14×10^{-2}	0.325	2.1×10^{-2}
10^4	3.26	0.214	2.75	0.18
6×10^4	18.2	1.20	8.95	0.59
10^5	28.3	1.87	10.9	0.71
6×10^5	102	6.80	15.0	1.00
10^6	132	8.65	15.5	1.02
10^7	206	13.6		
10^8	218	14.4		
∞	220	14.5	16.0	1.10

With $a = 0.027$ M, $h = 4 \times 10^{-8}$ M and $c = 1 \times 10^{-3}$ M initially

$$t_{90} = (1.52 \times 10^4)\, x_e$$

to CO_2 when equilibration is reached. Thus physically preventing the hydrogen ion changing, by buffering, simply maintains the forward reaction to produce more CO_2. The relation between the time in seconds, t_{90}, for 90% reaction, and x_e, in mols., for the values of 'a', 'c' and 'h' given previously, is approximately

$$t_{90} \simeq (1.52 \times 10^4)\, x_e$$

irrespective of whether the open or closed system is considered. This provides a simple means of estimating the reaction time from the initial displacement from the equilibrium, a procedure which will be utilized in the following discussion.

The reverse situation of carbon dioxide reacting with water

to form bicarbonate can be formulated in a similar manner. The kinetic equation representing this is

$$-\frac{dx}{dt} = k'(c_2) - \frac{k}{K}\left(h_1 + \frac{x}{\beta}\right)\left(a_1 + x\right) \quad (5)$$

where dx/dt represents the change of CO_2 with a_1, h_1 and c_1 the arterial values of bicarbonate, hydrogen ion and carbon dioxide concentrations respectively, and a_2, h_2 and c_2 the corresponding venous values. At equilibrium

$$\frac{k}{K}(h_2)(a_2) = k'(C_2) \quad \text{and} \quad \frac{k}{K}(h_1)(a_1) = k'(c_1)$$

Expanding equation (5) and substituting for (h_1) (a_1) gives

$$-\frac{dx}{dt} = k'(c_2) - k'(c_1) - \frac{k}{K}\left(h_1 + a_{1/\beta}\right)x - \frac{kx^2}{K\beta} \quad (6)$$

By comparison, equation (4) can be written in a form representing the change from venous to arterial values as

$$\frac{dx}{dt} = k'(c_2) - k'(c_1) - \frac{k}{K}\left(h_2 - a_{2/\beta}\right)x + \frac{kx^2}{K} \quad (7)$$

When $x = o$, the initial slope of these reactions, though opposite in direction, is identical. Further when β is large, $kx^2/K\beta$, a_2/β and a_1/β approach zero and $h_1 \simeq h_2$, so that the slopes, and hence the overall reaction, become the same. Under physiological conditions it is thus possible to regard the rates of the forward and reverse arterio-venous CO_2 exchanges as being equivalent.

CO_2 Equilibration between Two Compartments of Different Reaction Velocity

When the blood in the capillaries is exposed to the carbon dioxide tension prevailing in the tissues, CO_2 is initially removed by reacting relatively slowly in the plasma and very quickly in the red blood cell. If this situation is simplified, by considering that no buffer is present and the CO_2 is in instantaneous equilibrium between equal volumes of cell and plasma, the overall process can be represented by Fig. 3.

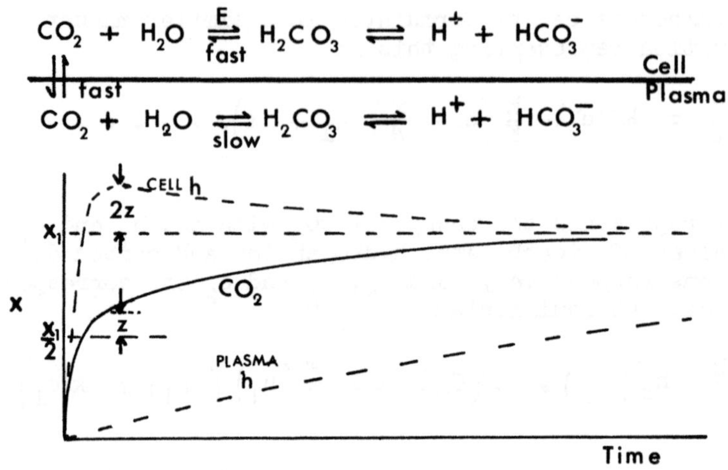

Fig. 3 The rate of reaction with two equal compartments, following a sudden increase of CO_2 concentration, when an enzyme, E, is present in one compartment and CO_2 may move freely between them. The curves represent the change of CO_2 concentration and overshoot of the cell hydrogen ion and bicarbonate ion concentrations for a closed system.

Because the CO_2 reacts 13,000 times more quickly in the cell, the plasma reaction can be initially ignored and the kinetic equation, for a closed system is

$$\frac{dx}{dt} = Ek'(c_2 - x) - \frac{Ek}{K}(h_1 + 2x)(a_1 + 2x) \quad \ldots\ldots \quad (8)$$

An initial pseudo-equilibrium is reached inside the cell, given by

$$k'(c_2 - x_2) = \frac{k}{K}(h_1 + 2x_2)(a_1 + 2x_2) \quad \ldots\ldots\ldots \quad (9)$$

where 'x_2' is the equilibrium value of 'x'. The factor of two, in the right-hand side brackets, arises because for a reduction of CO_2 concentration within the cell, an equivalent and simultaneous fall in the external CO_2 occurs. With c_2 kept constant this reduces to

$$k'(c_2) = \frac{k}{K}(h_1 + 2x_2)(a_1 + 2x_2) \quad \ldots\ldots\ldots\ldots \quad (10)$$

This is followed by the reaction of carbon dioxide in the plasma until true equilibrium is reached, with identical conditions in both compartments, so that now x has a new equilibrium value 'x_1', given by

$$k' (c_2 - x_1) = \frac{k}{K} (h_1 + x_1)(a_1 + x_1) \quad \ldots \ldots \quad (11)$$

or with the open system

$$k' (c_2) = k/K (h_1 + x_1)(a_1 + x_1) \quad \ldots \ldots \quad (12)$$

By comparing equations (9) and (11), it can be seen that, with a closed system, such as in the rapid-reaction apparatus

$$\frac{x_1}{2} < x_2 < x_1$$

The internal concentrations of hydrogen and bicarbonate ions thus transiently rise above the true final equilibrium value corresponding to x_1. With the open system however, such as occurs between the blood and tissues, where c_2 is effectively constant, x_2 must be equal to $\frac{x_1}{2}$ and the advantage of the overshoot is lost.

With CO_2 and HCO_3^- Exchanging between Fast and Slow Reacting Compartments

In this system both CO_2 and HCO_3^-/Cl^- are considered to rapidly cross the boundary separating the two compartments, as indicated in Fig. 4.

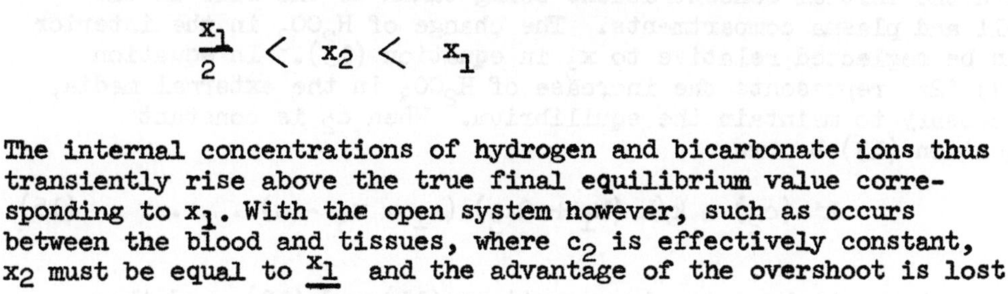

Fig. 4 A diagramatic representation of the process of equilibration between two equal compartments, following a sudden increase of the CO_2 concentration, when an enzyme, E, is present only in the cell compartment and both CO_2 and HCO_3^- may move freely between them.

If the external rate of hydration of CO_2 is slow, relative to the rate of transfer of HCO_3^- across the boundary, the reaction again proceeds rapidly via the internal pathway until the pseudo-equilibrium

$$k' (c_2 - x_3) = k/K (h_1 + 2x_3) (a_1 + x_3 - z). \; . \; . \; . \; . \; . \quad (13)$$

is reached in the cell interior. In the external plasma, instantaneous equilibration of H_2CO_3 is maintained, so that at the same time

$$K (H_2CO_3 + 2z) = (h_1 - 2z) (a_1 + x_3 - z) \; . \; . \; . \; . \; . \quad (14)$$

with the initial concentrations being taken as the same in the cell and plasma compartments. The change of H_2CO_3 in the interior can be neglected relative to x_3 in equation (13). In equation (14) '2z' represents the increase of H_2CO_3 in the external media, necessary to maintain the equilibrium. When c_2 is constant equation (13) becomes

$$k' (c_2) = k/K (h_1 + 2x_3) (a_1 + x_3 - z) \; . \; . \; . \; . \; . \quad (15)$$

It can be seen, by comparing equations (11) and (13), and then (12) and (15), that x_3 must be greater than $x_1/2$ but less than x_1. Then as the reaction progresses in the slow compartment the true equilibrium, as given by equations (11) and (12), is established.

It is thus a consequence of the permeability of the compartment boundary to both CO_2 and HCO_3^-, and a fast reaction in the cell relative to that in the plasma, that a transient increase in the cell, and decrease in the plasma, of hydrogen ions occurs. In this case the problem of the time of equilibration of the CO_2 in the plasma compartment is now complicated by the increase in bicarbonate brought about by the HCO_3^-/Cl^- shift. It is still necessary for the same amount of CO_2 to be hydrated in the external compartment in order to increase the hydrogen ion concentration from h_1 to $(h_1 + x_1)$. The kinetics of this process are now given by

$$\frac{dx}{dt} = k'(c_2) - k/K (h_1 + 2x) (a_1 + x_3 + x)$$

and it therefore takes longer, than in the previous example, for this to occur.

The Overall Equilibration of CO_2 in the Lung

The previous analysis and data will now be used to examine the process of CO_2 dehydration and oxygen uptake by the blood as it passes through the lungs. While the system is much more complicated, in that the carbon dioxide is not partitioned between two compartments of equal volume, the buffer factor varies from the plasma to the cell interior and the time constant for the Hamburger HCO_3^-/Cl^- Shift is finite, it is none-the-less possible to draw some interesting conclusions.

When the blood enters the lung capillaries the CO_2 is removed and the dissolved carbon dioxide level falls rapidly to an effectively constant value. The action of the lungs can be regarded as a 'sink' and the process corresponds to the open system discussed previously. Because of the very rapid reaction occurring in the cell, with a time for 90% completion of the order 6 msec., the internal hydrogen and bicarbonate ion concentrations rapidly fall. This in turn initiates the Hamburger Shift and bicarbonate enters from the plasma, in exchange for chloride to maintain electroneutrality. The process can be represented as:-

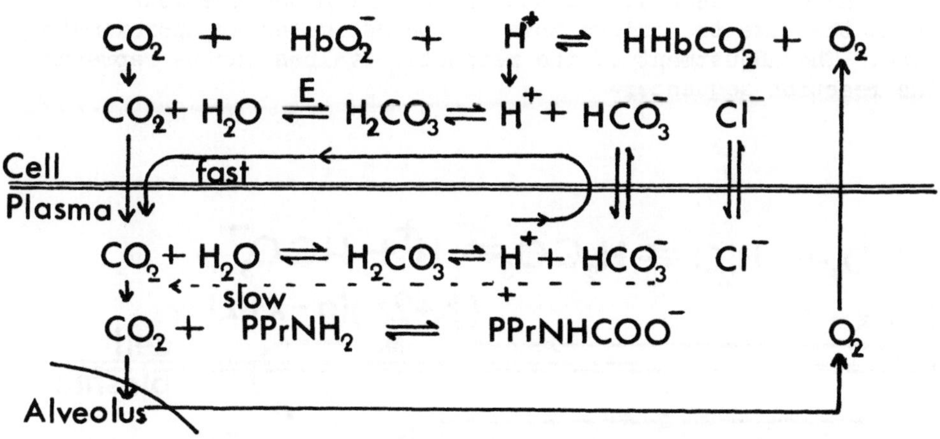

Thus the cell interior becomes more alkaline. In the plasma, apart from the removal of bicarbonate by the Hamburger Shift, there is a slower removal taking place by means of the direct reaction, taking approximately 18 sec. for 90% completion, and the possibility of hydrogen ions being removed by the carbamino-reaction with the plasma proteins. Because of the much faster rate of the HCO_3^-/Cl^- Shift, with a half-time of the order 0.1 sec. (Dirken and Mook (19), Luckner (20)(21)) at 38°C, bicarbonate is being removed faster by this process than by direct dehydration, during the 0.75 sec. the blood is exposed in the lung. This means that the direct rate of dehydration is further slowed and some H^+ is replaced by dissociation of H_2CO_3. Overall the concentration of hydrogen ion falls by less than 4% of the total change required. Some removal of plasma H^+ probably also occurs during this phase due to reaction with the plasma protein carbamino-groups. Published values (Stadie and O'Brien (22), Albritton (23)) of the difference of plasma carbamate, between venous and arterial blood at rest, vary from zero to a maximum of 0.1 ml CO_2 per 100 ml. of blood. The latter corresponds to a change of 0.045 mM of hydrogen ion concentration, compared to a difference of 0.174 mM (i.e. $0.29 \times 10^{-8} \times 6 \times 10^{4}$) necessary to change the pH from 7.37 to 7.40, taking into account the buffer factor. It could be that some of the amino-bound CO_2 is dissociated during this process, while the amount of dissolved CO_2 is reduced, and then combines again later, after the hydrogen ion concentration falls. This is unlikely to be significant however with the small pH differences observed at rest. The overall conclusion therefore is that the plasma pH is nearly the same on leaving the lungs as on entry, and the pH within the cell is correspondingly more alkaline, though the cell CO_2, H_2CO_3 and HCO_3^- are in equilibrium.

After the blood leaves the alveoli, the conditions change to that of a closed system, and the equilibration of the plasma carbonic acid proceeds at a faster rate. From the previous analysis the time to achieve this can be estimated as approximately 2.5 sec. The adjustment of the respective values can be represented by the reaction sequence:-

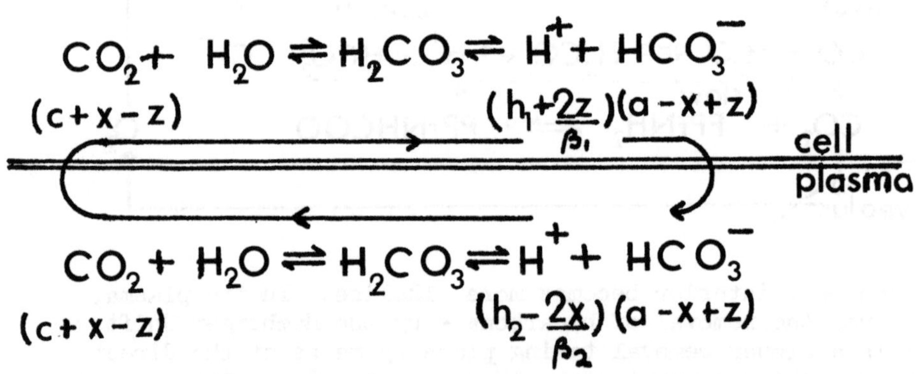

where 'c', 'h_1', and 'a' now represent the values in equilibrium in the cell upon leaving the alveoli. The relative changes are such that the total algebraic sum of the change of the carbon containing compounds is zero and electroneutrality is maintained. The two values x and z, represent the changes of concentrations that occur to reach the new equilibrium, and 'z' can be calculated from the initial equilibrium condition that

$$K'c = (h_1)(a)$$

and the final relationship

$$k'(c + x - z) = (h_1 + 2z/\beta_1)(a - x + z)$$

Hence

$$\frac{c}{(c + x - z)} = \frac{h_1}{h_1 + 2z/\beta_1} \cdot \frac{a}{(a - x + z)}$$

and since (x - z) can be neglected, relative to 'a', this reduces to

$$x = z\left(1 + \frac{2c}{h_1\,\beta_1}\right)$$

The change of dissolved CO_2, (x - z), can then be calculated given that 2x is equal to 0.174 mM. This is equivalent to a rise of the partial pressure of CO_2 by 0.15 mm., after leaving the lungs. In making these calculations it is assumed that the plasma HCO_3^- concentration is the same as that in the cell, as the relative difference, (x - z), is small, this approximation is acceptable. A rise of the arterial pCO_2, relative to the alveoli pCO_2, has been observed in some of the experiments by Jones, et al (24), with subjects at rest. The rise in pCO_2 ranged from 0.1 to 0.8 mm. It is probable that these results correspond to the conditions considered above when there is only a small difference of venous to arterial pH.

The process of oxygen uptake and carbon dioxide removal, during the passage of the blood through the lung capillaries, can be represented diagramatically by Fig. 5. Initially carbon dioxide is rapidly removed from the cell, by the action of carbonic anhydrase, thus reducing the cell hydrogen ion concentration before the oxygen can be taken up by the reduced haemoglobin. The corresponding pH change, from A to B, thus follows more closely a constant venous HbO_2 characteristic. The resulting more alkaline medium correspondingly favours the uptake of oxygen by the reduced haemoglobin. As oxygen is taken up, the oxyhaemoglobin being more acidic, reduces the excessive swing of the pH. At the same time however the Hamburger Shift ensures that all the hydrogen ion change occurs within the cell, and as a consequence the point C is not in equilibrium but still slightly to the alkaline side. A small overshoot of HbO_2, represented by the line from C to D, thus follows. After the blood leaves the alveoli, the final equilibration takes place, with the plasma pH rising and that in the cell falling, and the change from D to E completes the cycle to the new arterial pH and HbO_2 values. A similar cycle, indicated by the broken line, can be constructed for the reverse process of deoxygenation during passage through the tissues. The degree of overshoot of HbO_2, at rest, is probably not significant, the relative change of the cell H^+ ion concentration being less, by the ratio of the cell buffer factor β_1, to the plasma factor β_2, than that in the plasma.

During work, when the change from venous to arterial plasma pH is greater, the HbO_2 overshoot will be increased. Other factors however make this situation much more difficult to analyse. It is probable that the larger pH and CO_2 changes will also lead to a reversible unloading and re-uptake of CO_2 by haemoglobin and the plasma proteins. A similar cycle, indicated by PQRSD in Fig. 5, may be constructed. Whereas

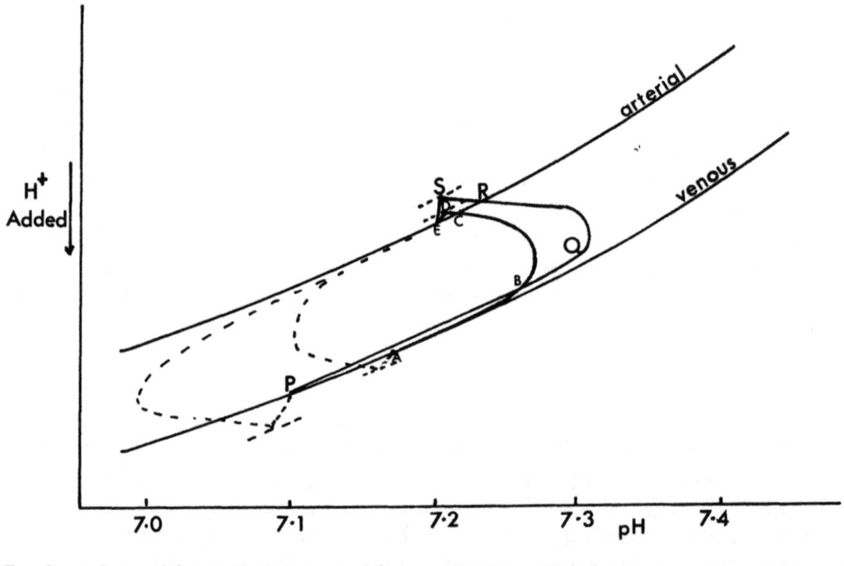

Fig. 5 A schematic representation of the pH changes occurring within the red blood cell during passage through the lung and tissue capillaries. ABCDE represents the uptake of O_2 and pH changes in the lung at rest, and PQRSE during work.

previously no change of the plasma amino-bound CO_2 was considered likely, in this case the equilibrium condition, given by Roughton (12)

$$\frac{Z}{Z - 1} = \frac{K_1 K_2 (pCO_2)}{K_2 h + h^2}$$

where Z is the proportion of carbamino-bound CO_2, and K_1 and K_2 equilibrium constants, will be significantly altered when the venous CO_2 concentration is lowered by the alveoli. If the venous pH is taken as 7.3 and the arterial pH as 7.4, an exchange of 0.6 mM H^+ is necessary in the plasma. The total carbamate present in the plasma is of the order 0.45 mM, so that even if complete conversion of this to CO_2 took place, which is not possible, equilibration of the plasma H^+ concentration would still not occur. However during the post-alveoli phase, the adjustment of the pH to its more alkaline value will mean that an almost equivalent amount of CO_2 will combine again. A similar situation arises in the erythrocyte, due to the increased alkalinity forming more HbO_2, at point S, than is present at the final equilibrium. During the stage S to D, the formation of some reduced haemoglobin means that a corresponding amount of CO_2 will be bound by haemoglobin amino-groups again. Thus the change of plasma pH in the post-alveoli flow leads to a rise of approximately 2 mm. in the plasma pCO_2, whereas the recombination by plasma proteins and haemoglobin will produce the reverse effect. The greater relative change of the plasma pH means, that of these two latter reactions, the plasma protein binding is the more important factor. Unfortunately data on the amount of carbamino-bound CO_2 carried by the plasma proteins during work does not appear to be available. A fall of up to 14 mm pCO_2 between the alveoli and artery during work, has been reported by Jones et al (24)(25) and Gurtner et al (26).

DISCUSSION

The previous analysis suggests that the blood gases, O_2 and CO_2, may not come to a complete equilibrium during passage through the capillaries of the lung alveoli and tissues (Forster (27)). This is a direct consequence of carbonic anhydrase being present only in the erythrocyte interior (Roughton (10)) and of the Hamburger HCO_3^-/Cl^- Shift. If carbonic anhydrase were present in the plasma, even in very small amounts, the rate of hydration and dehydration of the plasma CO_2 would be sufficient for equilibration to occur in the lungs. A major advantage of this is that all the hydrogen ion change, during exposure to the lungs and tissues, occurs within the cell, facilitating the gaseous exchange. Ancillary to these two factors is the buffer role of the plasma and cell interior, and the formation of carbamate by proteins. If the buffer power of the plasma were reduced, the time for 90% equilibration would be decreased and the overshoot effect minimized. On the other hand increasing the plasma buffer power and reducing the haematocrit will accentuate the CO_2, HbO_2 and pH differences.

The effect of HbO_2 being reduced to Hb, and hence increasing the effective pO_2, downstream from the alveoli, could have similar consequences to secretion of oxygen by the lung (Haldane (28)). There is an important difference however in that the change in oxygen tension is relative to that in the blood just leaving the alveoli capillary, and it does not necessarily follow that the latter is equal to that in the gas phase in the alveoli. The magnitude of this effect is probably only significant under conditions of relatively large plasma pH changes between the veins and arteries. This would be enhanced under partially saturated conditions where a change of pH has more effect on the dissociation curve. These correspond with the conditions postulated by Douglas, et al (29), namely exercise and low oxygen tension, to demonstrate a significant arterial to alveoli pO_2 increase. It may be that this has not been demonstrated on subjects living and working at high altitudes (Hurtado et al.(30)), because acclimatization considerably increases the haematocrit so that pH changes within the cell are less. The optimum conditions required to demonstrate this effect would be exercise, and a large venous-arterial pH difference, low inspired pO_2 tension, low haematocrit, high buffering capacity of the plasma and negligible haemolysis, so that there is no carbonic anhydrase present in the plasma.

It remains to explain how the 'in vitro' overshoot effect, obtained with the rapid-reaction apparatus, occurs. In this system the CO_2 is equilibrated with the Ringer-Locke solution prior to mixing with the erythrocyte suspension, so that the conditions are not comparable to those occurring in the blood during passage through the lung. The effect is produced in reverse to that 'in vivo', in that as oxygen is taken up by the haemoglobin within the cell, the carbon dioxide also enters. The carbonic anhydrase rapidly removes the dissolved CO_2 and the internal H^+ concentration rises while oxygen is being taken up. The latter reaction is also relatively rapid, so that, before the plasma bicarbonate has had chance to dissociate, a pseudo-equilibrium is established. As more plasma bicarbonate continues to dissociate, the carbon dioxide continues to diffuse through into the cell. The further decrease of the internal pH then leads to dissociation of the cell oxyhaemoglobin. This continues until the plasma CO_2 reaches equilibrium. Again this effect cannot be produced with carbonic anhydrase in the external media, so that fresh blood must be used. The plasma protein and buffer power must not be depressed by using a too dilute cell suspension, but conversely a lowered haematocrit makes for a larger effect within the cell, as has been discussed previously.

REFERENCES

(1) F.J.W. Roughton, Proc. Roy. Soc. London. Ser. B 111 (1932) 1.

(2) F.J.W. Roughton, Progr. Biophys. Biophys. Chem. 9 (1959) 55.

(3) J.A. Sirs and F.J.W. Roughton, J. Appl. Physiol. 18 (1963) 158.

(4) J.A. Sirs, Biorheology. 5 (1968) 1.

(5) E.S. Carlsen and J.H. Comroe, J. Gen. Physiol. 42 (1958) 83.

(6) J.A. Sirs, Biorheology, 3 (1966) 169.

(7) J.A. Sirs, 1st International Conference on Hemorheology. Oxford Pergamon (1968) p. 505.

(8) J.A. Sirs. Biochim. Biophys. Acta. 112 (1966) 538.

(9) J.A. Sirs. J. Physiol. 189 (1967) 461.

(10) F.J.W. Roughton. Physiol. Reviews. 15 (1935) 241.

(11) J.A. Sirs. Biochim. Biophys. Acta. 126 (1966) 19.

(12) F.J.W. Roughton. Handbook of Physiology. Respiration I. American Physiological Society. Washington (1964) p. 767.

(13) R. Brinkman, R. Margaria and F.J.W. Roughton. Phil. Trans. Roy. Soc. London. Ser. A 232 (1933) 65.

(14) C. Faurholt. J. Chim. Phys. 21 (1924) 400.

(15) G.S. Adair. J. Biol. Chem. 63 (1925) 517.

(16) B. German and J. Wyman. J. Biol. Chem. 117 (1937) 533.

(17) L.J. Henderson. Blood. Yale University Press. New Haven (1928) p. 66.

(18) J.C. Kernohan, W.W. Forrest and F.J.W. Roughton. Biochim. Biophys. Acta 67 (1963) 31.

(19) M.N.J. Dirken and H.W. Mook, J. Physiol. 73 (1931) 349.

(20) H. Luckner. Pflug. Arch. ges. Physiol. 241 (1939) 753.

(21) H. Luckner. Pflug. Arch. ges. Physiol. 250 (1948) 303.

(22) W.C. Stadie and H. O'Brien. J. Biol. Chem. 117 (1937) 439.

(23) E.C. Albritton (Ed). Standard Values in Blood. W. B. Saunders Company. Philadelphia and London (1952) p. 120.

(24) N.L. Jones, E.J. Moran Campbell, R.H.T. Edwards and W.G. Wilkoff. J. Appl. Physiol. 27 (1969) 356.

(25) N. L. Jones, E.J. Moran Campbell, G.J.R. McHardy, B.E. Higgs and M. Clode. Clin. Sci. 32 (1967) 311.

(26) G.H. Gurtner, S.H. Song and L.E. Farhi. Physiologist. 10 (1967) 190.

(27) R.E. Forster. Physiol. Reviews. 37 (1957) 391.

(28) J.S. Haldane. Respiration. Yale University Press. New Haven. (1922) p. 220.

(29) C.G. Douglas, J.S. Haldane, Y. Henderson and E.C. Schneider. Philos. Trans. Roy. Soc. London. Ser. B 203 (1913) 185.

(30) A. Hurtado, T. Velasquez, C. Reynafarje, R. Lozano, R. Chavez, H.A. Salazar, B. Reynafarje, C. Sanchez and J. Munoz. U.S.A.F. School of Aviation Medicine (1956) Rep. No. 56 - 1.

PHYSICOCHEMICAL PROPERTIES OF BLOOD CONSTITUENTS DERIVED FROM PORPHYRINS

G. Blauer

The Hebrew University, Jerusalem, Israel

BLOOD PIGMENTS*

In the respiration of mammals and of many other species the oxygen transporting hemoglobin of erythrocytes plays a major role. Mammalian hemoglobins, having molecular weights of about 65,000, normally consist of four subunits of two identical pairs. Oxygen is bound reversibly by a non-protein moiety which is attached in a specific manner to each subunit of the protein. This non-protein moiety is the prosthetic group iron-protoporphyrin IX (Formula I) which consists of four pyrrole derivatives linked by four methene bridges into a rigid conjugated ring structure. The eight side chains consist of two propionic acid residues, two vinyl groups, and four methyl groups arranged in a specific order. The central ferrous ion is coordinated by the four pyrrole nitrogens and, most likely, by an imidazole nitrogen of a histidyl residue, while the sixth ligand in the octahedral arrangement is oxygen in the case of oxyhemoglobin.

Upon destruction of the erythrocyte, whose life span is limited, the heme of hemoglobin undergoes several chemical changes. These include oxidation of the α-methene bridge, removal of the iron and reduction of the methene bridge to a methylene group at position γ. The composition of the side chains and their order are the same as in iron-protoporphyrin IX. The resulting bile pigment, bilirubin IXα (Formula II), is transported to the liver in the form of a complex with serum albumin (see, for example, refs. 1-6).

*Abbreviations used: Heme, iron-protoporphyrin IX; ORD, optical rotatory dispersion; CD, circular dichroism.

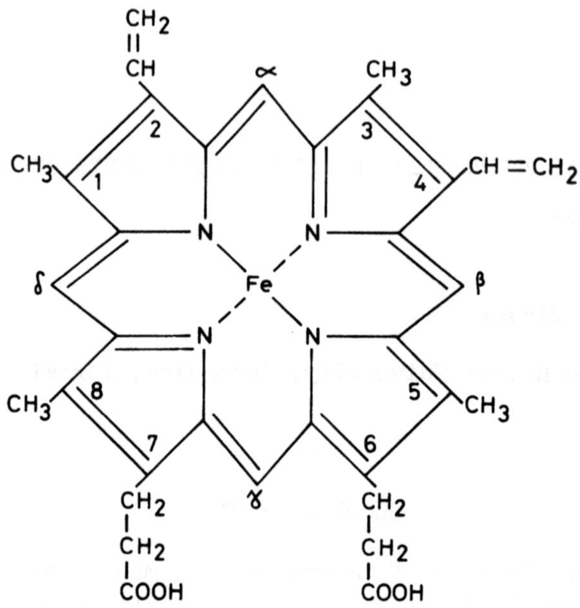

Formula I. Iron-protoporphyrin IX. The total charge on the molecule depends on the valence of the Fe ion, on the charges of the coordinated groups and on the charges of the propionic acid residues.

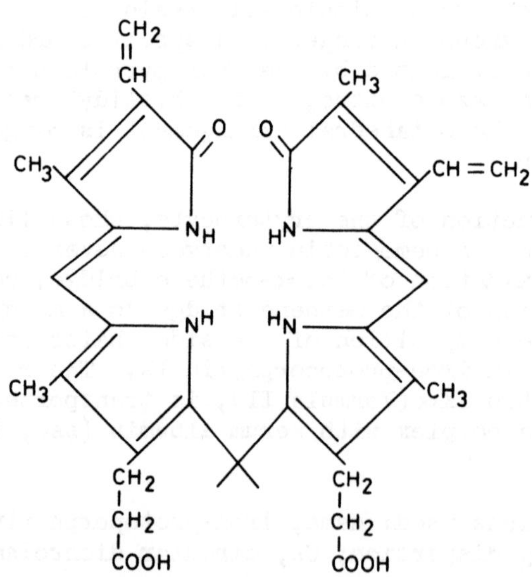

Formula II. Bilirubin IX∝.

Some selected properties of both heme and bilirubin, which are not necessarily included in the above or other reviews, are surveyed below with the aim of elucidating some structural and functional properties of these blood constituents.

STRUCTURE-FUNCTION RELATIONSHIPS

There are a number of proteins, besides hemoglobin, which contain iron-protoporphyrin IX or a modified form of it. One such protein is myoglobin, found in animal muscle, whose function is oxygen storage. Catalases and peroxidases, peroxide decomposing enzymes which are widespread in animal and higher plant tissues, respectively (7, 8) both also contain heme as their active group. The same prosthetic group is also found in the electron carriers of the cytochrome b type and, in modified form, in other hemoproteins of the respiratory chain of animal tissues, such as cytochrome c and cytochrome c oxidase; the function of the latter is the reduction of molecular oxygen (9-12).

Three-dimensional structures for heme proteins, obtained by X-ray analysis, are known for crystalline sperm whale myoglobin and horse oxyhemoglobin (2, 13-15), and, to a lesser degree of resolution, for crystalline horse heart ferricytochrome c (16). For many hemoproteins, even the nature of the coordinating groups at the porphyrin iron is still uncertain (see refs. cited in this section).

Notwithstanding the structural information available, the structure-function relationships in solution remain one of the most interesting but largely unsolved problems not only for hemoproteins but also for the general case of enzyme-substrate interactions. Despite considerable advances in this field in recent years, such as our understanding of cooperative effects (17), many basic structure-function relationships on a molecular level remain obscure (see ref. 18). In the case of hemoproteins, we have seen that one basic type of prosthetic group is responsible for a variety of functions in various proteins. This poses a fundamental question: which factors determine the particular function which this prosthetic group will fulfill in a given protein? The bond types at the central iron and the associated spectral and magnetic properties are certainly important, but they are clearly insufficient criteria for determining function. Environmental and steric factors must also be decisive in determining the various mechanisms of action.

MODELS AND THEIR INFERENCES

a) Prosthetic Group Interactions

One way to approach the study of the molecular mechanism and function of a hemoprotein is to investigate the chemical properties of the detached prosthetic group and its interactions with other molecules. In working with such simple chemical models, a kind of uncertainty principle pertaining to structure-function relations may be applied and considered to be non-trivial: As we investigate the intact structure of a functional unit in a biological system, we remain close to physiological conditions, but we usually cannot differentiate details of mechanism because of the complexity of the system. Conversely, if we detach the active unit from its native environment, we may gain in chemical and physical understanding of this unit (e.g., single proteins, prosthetic groups, etc.), but we may loose some essential interactions with the whole structure. Despite these limitations, the use of models has contributed to our understanding of many systems (see, for example, refs. 1, 19, 3). Thus, ferroprotoporphyrin IX in solution is readily oxidized by molecular oxygen. However, when its diethyl ester is coordinated at the ferrous iron by 1-(2-phenylethyl)-imidazole and embedded in a film of polystyrene, via a mixed CO-complex, oxygen combines reversibly at the sixth coordination position of the iron, which remains in the bivalent state even if the film is kept in contact with air-saturated water for days (20, 21). This system constitutes a model for the reversible oxygenation of hemoglobin or myoglobin. It has been suggested (20) that non-polar side chains in the vicinity of the heme create a medium of low dielectric constant in which the reaction leading to electron transfer between ferrous ion and oxygen should be slow. The back dissociation of "oxyheme" to ferroheme and molecular oxygen should be unhindered. Indeed, in myoglobin the heme is located in the interior of the protein and surrounded almost entirely by non-polar residues (22, see also ref. 15). One would expect, however, that, besides the medium, other local factors would also be important for the full function of the hemoprotein.

We shall now turn to other experiments which again suggest that prosthetic group interactions other than those usually considered (e.g. bond type at the iron) may be of importance in native systems. Let us first consider two basic types of light-absorption spectra for ferric heme which were distinguished many years ago (23, 1) in relation to older concepts: Ferric compounds with essentially ionic bonds appear brown or green in solution and are characterized by an absorption band in the red region (e.g., ferriheme, ferrihemoglobin, peroxidase, catalase). On the other hand, ferric compounds with essentially covalent bonds (d^2sp^3 in octahedral complex) appear red in solution, with one single or composite absorption band between 500 and 600 mμ (e.g., ferrihemoglobin cyanide, ferricytochrome c). In the terms of the newer ligand-field theory (24), the cyanide anion, nitrogen- and some sulfur-containing ligands (the latter two are likely in cytochrome c; see ref. 16 and refs. cited therein) are strong-field ligands which increase the orbital separation energy. Thus, low-spin complexes will be formed and, as a result of spin-pairing, the Fe^{3+} complex will show a magnetic moment of about 2 molar Bohr magnetons.* This follows from the known relation between the magnetic moment $\mu_{eff.}$ and the spin-only value

$$\mu_{eff.} = \sqrt{4S\,(S + 1)}$$

where S is the effective spin (S = 1/2 for one unpaired electron in Fe^{3+}). The magnetic moment follows from the experimental paramagnetic susceptibility χ according to:

$$\mu_{eff.} = 2.84\sqrt{\chi(T + \theta)}$$

where T is the absolute temperature and θ is the Curie-Weiss constant (assumed to be zero in our case; see, for example, 25-28). χ_{Fe} will be about 2000 x 10^{-6} electromagnetic cgs units* at room temperature for ferrihemochromes containing two strong-field ligands at coordination positions 5 and 6 of the ferriheme. This value includes some orbital moment contribution (see ref. 25, and refs. cited therein; 29). On the other hand, χ_{Fe} of ferriheme in dilute aqueous NaOH solution will be about 14000 x 10^{-6} cgsu at 20° in the absence of strong-field ligands (29).

Returning now the heme models, one can indeed reproduce a low-spin hemochrome of the cytochrome c type by mixing ferriheme with a synthetic polyamino acid such as poly-α, L-lysine in aqueous solution at pH 11 or poly-L-histidine near neutral pH (30-32). At pH 11, two non-protonated ε-amino groups of the helical polylysine are considered to provide the strong-field ligands to give a "red complex" of $\mu_{eff.}$ = 2.2 Bohr magnetons at 20°. At pH 8-9, however, a different spectral type ("green complex") is produced which is normally not observed in hemoproteins [absorption bands at 360, 385-390 (40-50), 570 (5), 590-600 mμ; millimolar extinction coefficient* per cm of maxima in parentheses

---

*Molar paramagnetic susceptibilities, magnetic moments and extinction coefficients are calculated per gram atom of heme iron.

(30, 33, 31). Magnetically, this green complex was found to be of the low-spin type (32). This result was unexpected. A similar complex could be obtained with other positively charged polyelectrolytes (34); it could also be observed upon the addition of simple electrolytes to high-spin ferriheme in aqueous alkaline solutions (33) (see Fig. 1). Upon extrapolation of the plot of the paramagnetic susceptibility against the reciprocal of sodium-chloride concentration (Fig. 2), a low-spin value of $\mu_{eff.} = 2.4$ molar Bohr magnetons can be evaluated for the complex at infinite salt concentration (29). Since the chloride anion should not

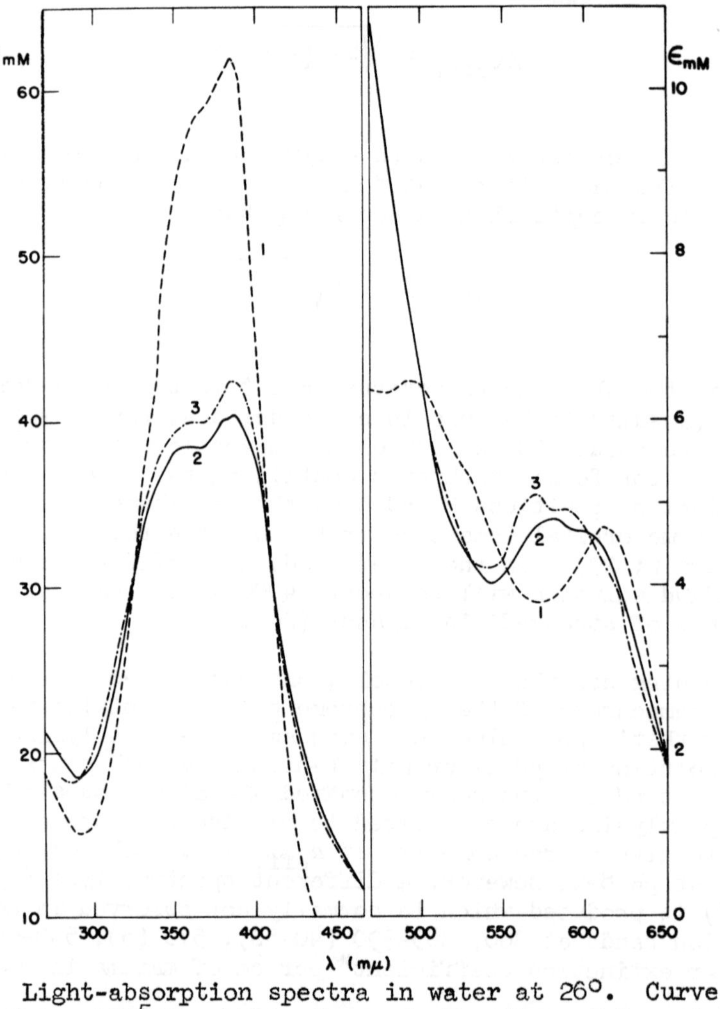

Figure 1. Light-absorption spectra in water at 26°. Curve 1: Ferriheme 5.1×10^{-5} M; pH 11. Curve 2: Ferriheme 5.1×10^{-5} M; NaCl 1.9 M; pH 11. Curve 3: Ferriheme 5.1×10^{-5} M; poly-α, L-lysine 5.0×10^{-4} residue M; pH 8.9 (from ref. 33).

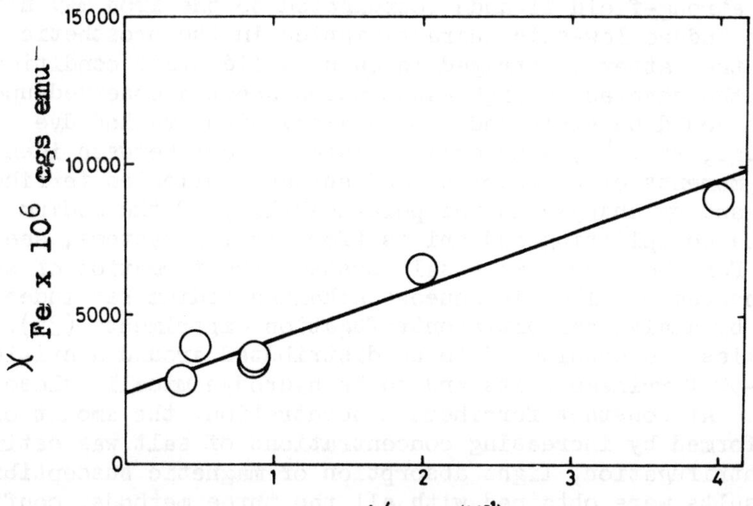

Figure 2. Molar paramagnetic susceptibility at 20° of 0.2 mM ferriheme vs. the reciprocal of NaCl concentration, Aqueous solutions at pH 11 (from ref. 29).

produce strong-ligand fields (ref. 24, p. 46), and since the protonated polybases mentioned above would probably not coordinate to the heme iron, a suitable explanation must be sought for the low-spin characteristics observed for these "green complexes." It has been suggested that side-to-side aggregation of ferriheme molecules occurs (33), with concomitant changes in the energy separation of the molecular orbitals through π-π or other interactions (29, 35).

Recently, the magnetic moment of extensively dried tetraphenylporphine-aquoiron (III) was found to be 1.74 Bohr magnetons at room temperature (36). It has been claimed that iron-oxo-iron bridges form a dimer in this case and that all magnetic moments observed for hematins in solution can be accounted for by mixtures of monomers (high spin) and dimers (low spin) (36). This is at variance with our measurements (32, 29) on dilute ferriheme in aqueous alkaline medium (pH 11-12) for which a moment of about 5.7 Bohr magnetons was obtained at 20°. Under these conditions, ferriheme is present mainly in the form of dimers or oligomers (37, 38, 35, 39).

Thus, strong-field ligands coordinated to the iron may not be required to induce low-spin characteristics in the prosthetic group when the latter is arrayed in quasi-solid state conditions. Similarly, the changes in light-absorption spectra observed upon aggregation could be explained - by analogy with stacked dye systems (e.g., ref. 40) - by exciton interactions between identical transition moments of neighboring and suitably oriented ferriheme molecules, and by changes in the polarizability of the medium which may cause splitting and shifts (for similar systems, see ref. 41). For the ferriheme - salt system, the formation of sodium chloride-induced micelles in aqueous alkaline medium has indeed been shown by analytical ultracentrifugation experiments (35). These micelles are considered to be distributed around a critical size of 45-50 ferriheme units and to be hydrodynamically close to spheres. At constant ferriheme concentration, the amount of aggregate formed by increasing concentrations of salt was estimated by ultracentrifugation, light absorption or magnetic susceptibility. Similar results were obtained with all the three methods, confirming the assumption that the interaction influences the physicochemical properties of the prosthetic group. In hemoproteins the heme is usually not in direct contact with other heme molecules. However, in myoglobin, for example, there are several aromatic rings arranged parallel, or nearly parallel, to the pyrrole rings or vinyl groups of the heme and π-π electron interactions should be significant in this case (22). More data and model experiments would be required to assess the influence of juxtaposition and π-interactions in hemoproteins (see refs. 42, 43, and refs. cited therein).

These interactions between the heme and aromatic or heterocyclic residues may not only influence spectral or magnetic properties but may also facilitate electron flow. A simple model of a cytochrome c oxidase system, based on the above green low-spin complex between poly-L-lysine or poly-4-vinylpiperidine and ferriheme at pH 8-9, illustrates this point (44). These soluble complexes catalyze the oxidation of cysteine by molecular oxygen according to the overall equation

$$4RS^- + O_2 + 4H^+ \rightarrow 2RS-SR + 2H_2O$$

The initial rate of oxidation in the presence of protonated polybases is about 15 times the rate in the absence of polymer (Fig. 3). It can be shown that the increased rate in the presence of polymer is due to an higher overall entropy of activation rather than to a difference in overall energy of activation (in the range of 9-40°). Apparently, stacked aggregates of ferriheme are formed in this system. It appears that a greater-than-one electron transfer from bound cysteine anions to oxygen is facilitated

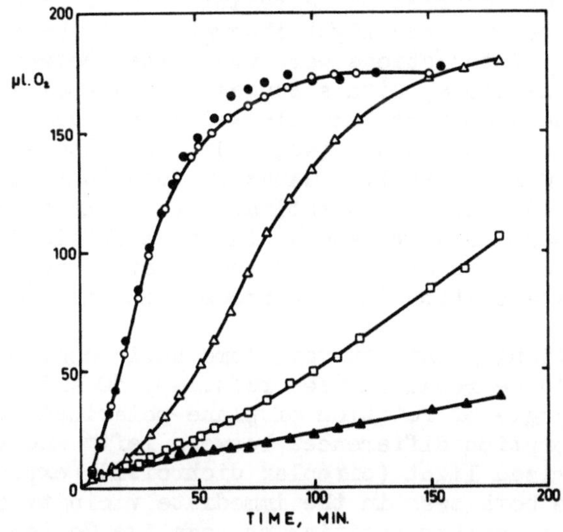

Figure 3. Typical rate curves for the oxidation in water at pH 8.2-8.4 of L-cysteine (initial 1.7 x 10^{-2} M) by various catalysts involving ferriheme (5 x 10^{-6} M) as initial component. Amounts of molecular oxygen (at NTP) consumed as a function of time. Temp., 31.2 ± 0.1°; o, 1.7 x 10^{-4} residue M poly-α, L-lysine ($\overline{DP}_w$ = 700); ●, 1.6 x 10^{-4} residue M poly-4-vinylpiperidine ($\overline{DP}_w$ = 3000); Δ, 0.8 M pyridine; □, ferriheme alone; ▲, control; (from ref. 44).

in these arrays of the prosthetic groups on the polybase. These aggregates may function as semiconductors. An efficient synthetic catalyst has been described which also acts like cytochrome c oxidase and consists of insoluble complexes of heme linked between the iron atoms by 4,4'-dipyridyl (45). When the connecting ligand was 1,2-di-(4-pyridyl)-ethane, there was little catalytic reduction of oxygen. In therefore appears that the conjugated double-bond system which links the heme units together is important in this catalysis (45).

Other recent model systems for interacting heme moieties consisted of the heme undecapeptide and the ferriheme octapeptide of cytochrome c (46, 47) which were obtained by peptic, or by peptic and subsequent tryptic digestion of cytochrome c, respectively. Depending on the conditions in solution, these peptides can be obtained at various degrees of aggregation (48). From the results of ORD, CD (see below), and light absorption of the aggregates, exciton resonance interactions were postulated between Soret transitions of the hemes, with a suggested head-to-tail alignment for the heme undecapeptides, and stacked or oblique conformations for the ferriheme octapeptides (46, 47). From the splitting energies obtained from resolved Gaussian functions, the magnitude of the heme-heme distances was estimated to be less than about 7 Å in the octapeptide aggregates. The data obtained for the above model systems are also considered relevant to the problem of heme-heme juxtaposition in cytochrome c oxidase (46).

Before continuing this survey, some basic concepts regarding ORD and CD should be recalled (see refs. 49, 50 and refs. cited therein). The angle of rotation of plane-polarized light and the related absorption differences between left- and right-circularly polarized light (circular dichroism), expressed as ellipticity, are both seen in the immediate vicinity of an isolated absorption band as either positive or negative Cotton effects (Fig. 4). The integrated dichroism band gives the rotational strength, which depends upon the charge displacements associated with an electronic transition which is induced by the electromagnetic field of light. It can be shown that the rotational strength R_i is given by the scalar product of the corresponding electric ($\vec{p}$) and magnetic (m) dipole moments

$$R_i = \mathrm{Im}\,(\psi_0|\vec{p}|\psi_i)\cdot(\psi_i|\vec{m}|\psi_0)$$

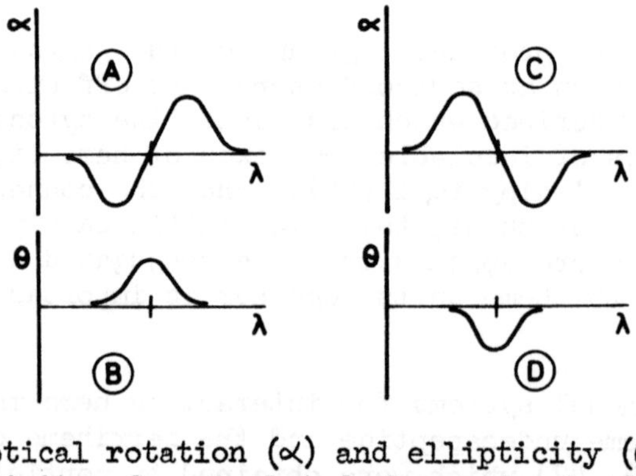

Figure 4. Optical rotation (∝) and ellipticity (θ) in the immediate vicinity of an isolated absorption band (idealized). A and B, positive Cotton effect; C and D, negative Cotton effect (from ref. 49).

For a symmetrical molecule, this product will vanish and no optical activity will be observed (50).

For evaluation of the helical (asymmetric) structure of proteins or polypeptides, the Moffitt equation (51, 52) has proved useful in wavelength regions remote from the Cotton effect

$$[m']_\lambda = \frac{a_0\lambda_0^2}{\lambda^2 - \lambda_0^2} + \frac{b_0\lambda_0^4}{(\lambda^2 - \lambda_0^2)^2}$$

where $[m']_\lambda = (M_0/100)\cdot[3/(n_\lambda^2 + 2)]\cdot[\alpha]_\lambda$ is the effective reduced residue rotation; M_0, mean mol.wt. of residues; n_λ, refractive index of solvent; $[\alpha]_\lambda$, specific rotation at wavelength λ; a_0, b_0, λ_0 are constants.

The rotation at the peptide Cotton effect in the far ultraviolet range has also been used as a measure of helix content (for a discussion on these methods and their validity, see refs. 53, 54).

b) Protein-Prosthetic Group Interaction and Stability

Some related aspects of protein conformation and stability may now be considered. Evidently, any conformational change in the protein caused by the prosthetic group or by its ligands may also have kinetic and thermodynamic consequences for the function of the hemoprotein. In several cases, the prosthetic group has been reversibly detached from the protein. By using slightly modified iron-porphyrins, artificial hemoproteins have been prepared and their properties have been studied (for a detailed discussion of hemoglobin and myoglobin, see ref. 5 and refs. cited therein; peroxidases, e.g., refs. 55, 56). It has been observed that the degree of binding of various hemes by globin (human) decreases in the order proto-, meso-, deutero- heme (57, 58). The same order was found in the formation of ferriheme micelles in the presence of sodium chloride in aqueous medium, where progressively higher concentrations of salt were required, in that order, to obtain equal spectral effects under otherwise identical conditions (59). Since deuteroheme lacks substituents in positions 2 and 4 of the porphyrin ring and mesoheme has ethyl groups, while protoheme carries vinyl groups instead, the above differences in stability have been attributed to hydrophobic interactions involving the side chains (58). Apparently, π-bonding by the vinyl groups also has some influence (see above). It has also been observed that upon removal of the ferriheme from sperm-whale myoglobin (60, 61), hemoglobin (62; 63, and refs. cited therein) or horse-radish peroxidase (64), a significant decrease in α-helical content of the protein in solution could be evaluated by using ORD or CD. In most cases, the decrease is about 10% in the maximum helical content, neglecting possible contributions to the rotation from heme and aromatic residues. These changes were reversible. For inducing this effect on the secondary structure of human hemoglobin, mesoporphyrin IX and its Cu and Mn complexes did not differ from ferriprotoporphyrin IX (65); nor did protoporphyrin IX differ (65; for sperm-whale myoglobin, see ref. 66). An analogous effect can be shown by evaluating the ORD spectrum of a synthetic complex between ferriheme and poly-α, L-ornithine, which is a ferrihemochrome at pH 11 (see above). In dilute aqueous solution, poly-L-ornithine of an average degree of polymerization of 100 has an average content of 15-20% right-handed α-helix at pH 11.4 and 25°, as estimated by ORD, using either the Moffitt parameter b_0 or the rotation at 233 mμ (67, 68). When ferriheme was added to this polyamino acid at the above conditions, an increase of up to about 30% in the apparent helix content (at a mole ratio of ornithyl residues: ferriheme of 35:1) could be calculated. Thus the binding of the prosthetic group either to the apoprotein or to the polyamino acid appears to stabilize the secondary structure in each case (68). The stabilization in the case of myoglobin is apparently due largely to hydrophobic interactions - about 90 van der Waals contacts of the porphyrin with neighboring atoms have been observed (22, 69). In the synthetic complex it is the cross-linking of ferriheme between molecules of polyornithine which is considered to provide the stabilization of the secondary structure (70, 32, 67).

Recently, some ligand-induced conformational changes in native human hemoglobin in solution, as also measured by changes in ORD in the far UV region, have been observed (71, 72). Molecular oxygen or carbon monoxide increased the apparent helix content slightly (about 8% decrease in optical rotation at 233 mμ). The isolated α- or β-chains or sperm-whale myoglobin did not show this effect but modified hemoglobins with no heme-heme interaction showed the full change. The possibility of aromatic side-chain contributions to the Cotton effect at 233 mμ could not be excluded (72).

The present survey of some physicochemical properties of heme has been confined to a few selected topics. It is believed that these have demonstrated the influence of environment and of subtle changes in or near the prosthetic group on some of its properties and also on stereochemical conditions for certain reactions. Many other important relationships of this type cannot be elaborated here, e.g., the influence of the position of the central iron with respect to the porphyrin plane (73-77, 42) or the role of the propionic acid groups. The latter are considered to have structural functions only (22; ref. 5, and refs. cited therein; 8). Important results from optical, magnetic resonance and theoretical spectroscopy have also not been included (see, for example, refs. 77-83 and refs. cited therein; 28).

In the last experiments cited above, ORD or CD have been used as criteria for conformational changes either in the polypeptide component of hemoproteins or in their models. We shall now see that these methods can also yield valuable information about the conformation of a bile pigment attached to a protein.

THE BILIRUBIN-SERUM ALBUMIN COMPLEX

This physiologically important complex is also present in blood plasma. As shown above, bilirubin has no rigid ring structure like its parent protoporphyrin, and its methylene bridge allows the molecule to assume a variety of conformations.

An induced ("extrinsic") Cotton effect can be observed upon binding of an optically inactive dye such as an acridine orange, to helical poly-L-glutamic acid (84, 85) where the helical polyamino acid provides the asymmetric environment. Similarly, ferriheme bound to helical poly-L-lysine (86) or poly-L-ornithine (68) will show Cotton effects in the Soret absorption bands of the heme with molar amplitudes of the order of 10^5 per heme.

As would be expected, bilirubin (2.5×10^{-5}M) in aqueous medium does not show measurable optical activity in the visible and UV regions. However, when it is bound to bovine serum albumin at pH 5, a large Cotton effect is measured in the region of the visible absorption bands of the complex (87, 88). The amplitude per mole bilirubin (protein rotations were deducted), on the long wavelength side, was of the order of 1.5×10 degrees$\cdot$cm$^2\cdot$decimole^{-1}. In this case, the shape of the complex Cotton effect also suggests the existence of dipole-dipole interactions between the transition moments of the chromophores of bilirubin (couplet, see ref. 89). Two associated dichroic bands of opposite sign were indeed measured. Under the conditions of the above system, a mole ratio of 1:1 between bovine serum albumin and bilirubin was determined by titrations involving both light absorption and optical rotation. The apparent association constant was estimated to be $(5.7 \pm 1.5) \times 10^6$ M^{-1} at pH 5 and 25°C. The rotation effects observed are interpreted by the formation of a dissymmetric conformation between the transition moments of the juxtaposed dipyrrylmethene chromophores appears to be possible (Fig. 5a). However, extrinsic effects of the asymmetric protein environment could also contribute to the

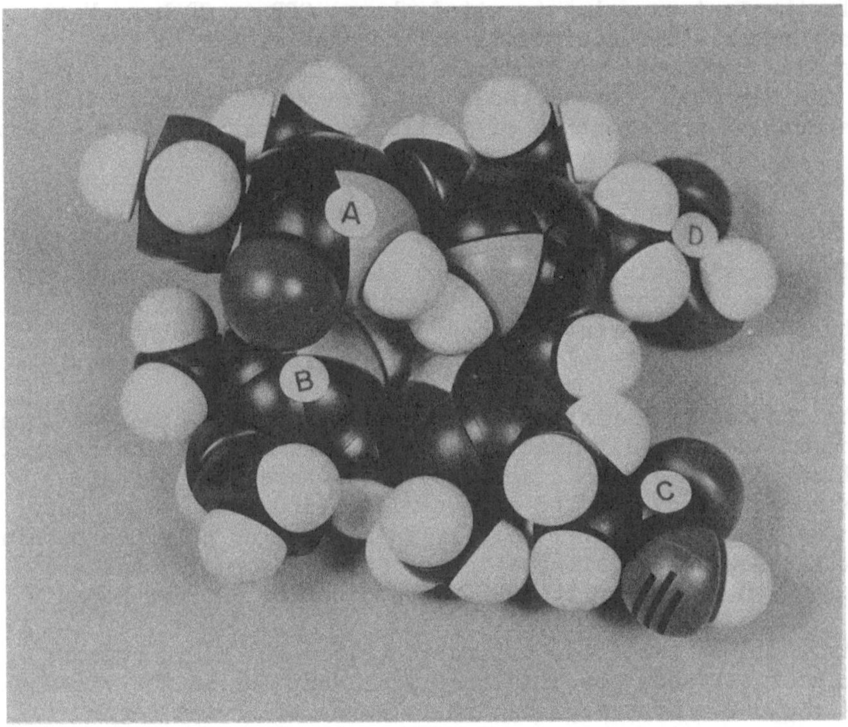

Figure 5 (a). Molecular models of possible conformations of bilirubin. Proposed model of bilirubin bound to BSA at pH 5. Note opposite arrangement of the two terminal pyrrole rings A and B.

observed rotations (87, 88).

Thus, ORD (and CD) measurements can yield important information on the mode of binding of blood pigments to proteins and other biopolymers, or of other small molecules to macromolecules in general.

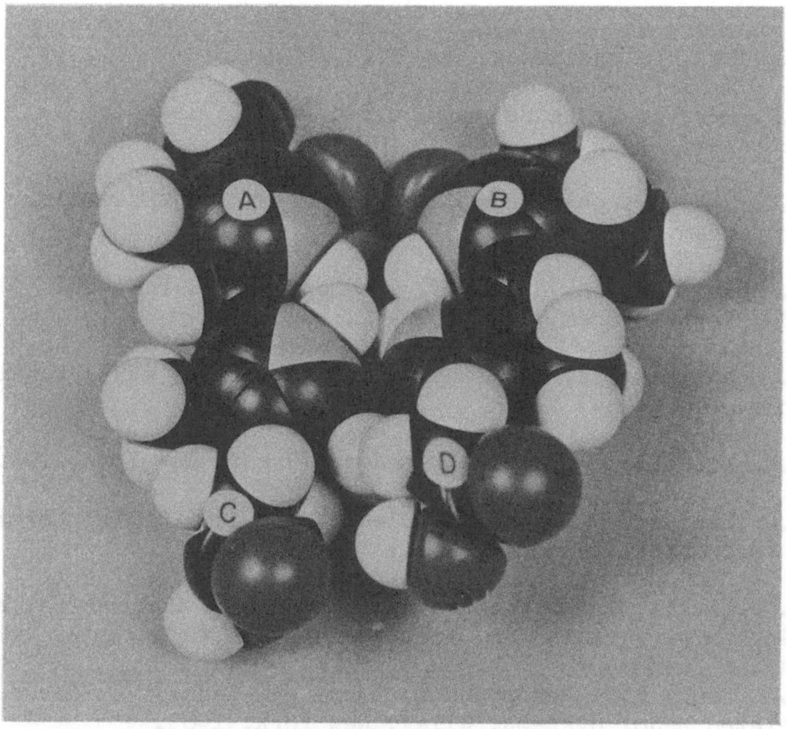

Figure 5 (b). Molecular models of possible conformations of bilirubin. Porphyrin-like conformation of bilirubin (from ref. 88).

CONCLUSIONS

Various physicochemical studies of prosthetic groups detached from their apoproteins yield interesting information *per se* which may also be valuable in interpretation of the properties *in vivo* of hemoproteins from blood or other sources. The measurement of optical rotatory properties of asymmetric systems involving hemoproteins and their models or of bile pigment-protein complexes is very useful in the evaluation of molecular conformation of both proteins and attached molecules.

REFERENCES

1. R. Lemberg and J.W. Legge, Haematin Compounds and Bile Pigments, Interscience, New York, 1949.

2. M.F. Perutz, Proteins and Nucleic Acids, Elsevier, Amsterdam, 1962.

3. J.E. Falk, Porphyrins and Metalloporphyrins, Elsevier, Amsterdam, 1964.

4. G. Braunitzer, K. Hilse, V. Rudloff and N. Hilschmann, Adv. Protein Chem., 19 (1964) 1.

5. A. Rossi-Fanelli, E. Antonini and A. Caputo, Adv. Protein Chem., 19 (1964) 74.

6. T.K. With, Bile Pigments, Academic Press, New York, 1968.

7. B.C. Saunders, A.G. Holmes-Siedle, and B.P. Stark, Peroxidase, Butterworths, London, 1964.

8. A. S. Brill, in M. Florkin and E.H. Stotz, Comprehensive Biochemistry, Elsevier, Amsterdam, 1966, Vol. 14, p. 447.

9. T.E. King, H.S. Mason and M. Morrison, (eds.), Oxidases and Related Redox Systems, Wiley, New York, 1965.

10. E. Margoliash and A. Schejter, Adv. Protein Chem., 21 (1966) 113.

11. M.R. Lemberg, Enzymologia, 32 (1967) 18.

12. M.R. Lemberg, Physiol. Revs., 49 (1969) 48.

13. J.C. Kendrew, H.C. Watson, B.E. Standberg, R.E. Dickerson, D.C. Phillips and V.C. Shore, Nature, 190 (1961) 666.

14. A.F. Cullis, H. Muirhead, M.F. Perutz, M.G. Rossmann and A.C.T. North, Proc. Roy. Soc. A265 (1962) 161.

15. M.F. Perutz, H. Muirhead, J.M. Cox and L.C.G. Goaman, Nature, 219 (1968) 131.

16. R.E. Dickerson, M.L. Kopka, J. Weinzierl, J. Varnum, D. Eisenberg and E. Margoliash, J. Biol. Chem., 242 (1967) 3015.

17. J. Wyman, Quart. Rev. Biophys., 1 (1968) 35.

18. M.F. Perutz, Eur. J. Biochem., 8 (1969) 455.

19. O. Warburg, Heavy Metal Prosthetic Groups and Enzyme Action, Oxford Univ. Press, London, 1949.

20. J.H.Wang, J. Am. Chem. Soc., 80 (1958) 3168.

20a. J.H. Wang, A. Nakahara and E.B. Fleischer, J. Am. Chem. Soc., 80 (1958) 1109.

21. J.H.Wang, in Oxygenases, ed. O. Hayaishi, Academic Press, New York, 1962, p. 506.

22. J.C. Kendrew, Brookhaven Symp. Biol., 15 (1962) 216.

23. H. Theorell, Arkiv Kemi, 16A (1942) No. 3.

24. L.E. Orgel, An Introduction to Transition-Metal Chemistry. Ligand-Field Theory. 2nd ed., Methuen, London, 1966.

25. A. Ehrenberg, Svensk Kem. Tidskr., 74 (1962) 53.

26. M. Kotani, in Adv. Chem. Phys., Interscience, 7 (1964) 159.

27. G. Schoffa, in Adv. Chem. Phys., Interscience, 7 (1964) 182

28. M. Kotani, Adv. Quant. Chem., 4 (1968) 227.

29. G. Blauer and A. Ehrenberg, Biochim. Biophys. Acta, 112 (1966) 496.

30. G. Blauer, Nature, 189 (1961) 396.

31. M. Tohjo and K. Shibata, Arch. Biochem. Biophys., 103 (1963) 401.

32. G. Blauer and A. Ehrenberg, Acta Chem. Scand., 17 (1963) 8.

33. G. Blauer and H. Rottenberg, Acta Chem. Scand., 17 (1963) S216.

34. G. Blauer, Biochim. Biophys. Acta, 79 (1964) 547.

35. G. Blauer and B. Zvilichovsky, Arch. Biochem. Biophys., 127 (1968) 749.

36. I.A. Cohen, J. Am. Chem. Soc., 91 (1969) 1980.

37. J. Shack and W.M. Clark, J. Biol. Chem., 171 (1947) 143.

38. Y. Inada and K. Shibata, Biochem. Biophys. Res. Commun., 9 (1962) 323.

39. W.B. Elliott and J.C. Hellriegel, Jr., Abstr. 158th Am. Chem. Soc. Natl. Meeting, New York, 1969, No. 235.

40. D.F. Bradley, Trans. N.Y. Acad. Sci., 24 (1961) 64.

41. I. Tinoco, Jr., A.Halpern and W.T. Simpson, in M.A. Stahmann, Polyamino Acids, Polypeptides and Proteins, Univ. of Wisconsin Press, 1962, p. 147.

42. W.S. Caughey, Ann. Rev. Biochem., 36 (1967) 611.

43. J.R. Cann, Biochem., 6 (1967) 3435.

44. G. Blauer and B. Zvilichovsky, Biochim. Biophys. Acta, 110 (1965) 215.

45. J.H.Wang and W.S. Brinigar, Proc. Natl. Acad. Sci. U.S., 46 (1960) 958.

46. D.W. Urry, J. Am. Chem. Soc., 89 (1967) 4190.

47. D.W. Urry and J. W. Pettegrew, J. Am. Chem. Soc., 89 (1967) 5276.

48. A. Ehrenberg and H. Theorell, Acta Chem. Scand., 9 (1955) 1193.

49. A. Moscowitz, in C. Djerassi, Optical Rotatory Dispersion, McGraw Hill, New York, 1960, p. 150.

50. H. Eyring, H-C. Liu and D. Caldwell, Chem. Rev., 68 (1968) 525.

51. W. Moffitt, J. Chem. Phys., 25 (1956) 467.

52. W. Moffitt and J.T. Yang, Proc. Natl. Acad. Sci. U.S., 42 (1956) 596.

53. J.T. Yang, in G.N. Ramachandran, Conformation of Biopolymers, Academic Press, London, 1967, p. 157.

54. W.B. Gratzer and D.A. Cowburn, Nature, 222 (1969) 426.

55. H. Theorell and A.C. Maehly, Acta Chem. Scand., 4(1950) 422.

56. K.G. Paul, Acta Chem. Scand., 13 (1959) 1239.

57. E. Antonini, M. Brunori, A. Caputo, E. Chiancone, A. Rossi-Fanelli and J. Wyman, Biochim. Biophys. Acta, 79 (1964) 284.

58. Q.H.Gibson and E. Antonini, in B. Chance, R.W. Estabrook and T.Yonetani, Hemes and Hemoproteins, Academic Press, New York, 1966, p. 67.

59. G. Blauer and B. Zvilichovsky (1969), in the course of publication.

60. E. Breslow, S. Beychok, K.D. Hardman and F.R.N. Gurd, J. Biol. Chem., 240 (1965) 304.

61. S.C. Harrison and E.R. Blout, J. Biol. Chem., 240 (1965) 299.

62. T. Asakura, S. Minakami, Y. Yoneyama and H. Yoshikawa, J. Biochem. Tokyo, 56 (1964) 594.

63. K. Javaherian and S. Beychok, J. Mol. Biol. 37 (1968) 1.

64. W.D. Ellis and H.B. Dunford, Can. J. Biochem., 46 (1968) 1231.

65. T.L. Fabry, C. Simo and K. Javaherian, Biochim. Biophys. Acta, 160 (1968) 118.

66. E. Breslow, R. Koehler and A.W. Girotti, J. Biol. Chem., 242 (1967) 4149.

67. G. Blauer and Z.B. Alfassi, Biochim. Biophys. Acta, 113 (1967) 206.

68. G. Blauer and B. Zvilichovsky, Abstr. 3rd Intl. Biophys. Congr., Cambridge, Mass., 1969, p. 35.

69. H.C. Watson, in B. Chance, R.W. Estabrook and T. Yonetani, Hemes and Hemoproteins, Academic Press, New York, 1966, p.63.

70. J.A. Schellman, Compt. Rend. Trav. Lab. Carlsberg, Sér. Chim., 29 (1955) 230.

71. M. Brunori, J. Engel and T.M. Schuster, J. Biol. Chem., 242 (1967) 773.

72. G. Hänisch, J. Engel, M. Brunori and H. Fasold, Eur. J. Biochem., 9 (1969) 335.

73. J.C. Kendrew, Science, 139 (1963) 1259.

74. J.S. Griffith, Biopol. Symp., 1 (1964) 35.

75. J.L. Hoard, M.J. Hamor, T.A. Hamor and W.S. Caughey, J. Am. Chem. Soc., 87 (1965) 2312.

76. J.L. Hoard, in B. Chance, R.W. Estabrook and T. Yonetani, Hemes and Hemoproteins, Academic Press, New York, 1966, p. 9.

77. J.R.P. Williams, in B. Chance, R.W. Estabrook and T. Yonetani, Hemes and Hemoproteins, Academic Press, New York, 1966, pp. 557, 585.

78. J.E. Bennett and D.J.E. Ingram, Nature, 177 (1956) 275.

79. A. Ehrenberg, Z. Naturwiss. Med. Grundlagenforsch., 2 (1965) 203.

80. G. Lang and W. Marshall, J. Mol. Biol. 18 (1966) 385.

81. M. Zerner and M. Gouterman, Theoret. Chim. Acta 4 (1966) 44.

82. A. Ehrenberg, B.G. Malmström and T. Vänngård, Magnetic Resonance in Biological Systems, Pergamon, Oxford, 1967.

83. A.H. Corwin, A.B. Chivvis, R.W. Poor, D.G. Whitten and E.W. Baker, J. Am. Chem. Soc., 90 (1968) 6577.

84. E.R. Blout, Biopol. Symp., 1 (1964) 397.

85. L. Stryer and E.R. Blout, J. Am.Chem.Soc., 83 (1961) 1411.

86. L. Stryer, Biochim. Biophys. Acta, 54 (1961) 395.

87. G. Blauer and T.E. King, Biochem. Biophys. Res. Commun., 31 (1968) 678.

88. G. Blauer and T.E. King, J. Biol.Chem., in press.

89. J.A. Schellman, Accounts Chem. Res., 1 (1968) 144.

EQUILIBRIUM AND KINETIC ASPECTS OF THE REACTION OF HEMOGLOBIN, ISOLATED CHAINS AND INTERMEDIATES WITH OXYGEN

Eraldo Antonini

Institute of Biological Chemistry of the

Universities of Rome, Italy

INTRODUCTION

One of the major recent developments in understanding hemoglobin function is represented by the preparation and study of the isolated subunits of the protein, i.e. the α and β chains, and of compounds which contain both α and β chains, but in which only one type of chain may undergo reaction with ligand. The latter ones may be regarded as artificially produced stable intermediates of the reaction of hemoglobin with ligands.

The purpose of the present paper is to review and to discuss the main equilibrium and kinetic features of the reaction with oxygen of hemoglobin in relation to those of isolated hemoglobin chains and artificial intermediates. The differences between the behaviour of these compounds and hemoglobin may be related to reversible changes in the tridimensional structure, which occur when one type of chain is separated from the other or when the ligand occupies some of the sites in the protein.

HEMOGLOBIN

Oxygen Equilibrium

Characteristically, the oxygen equilibrium of purified hemoglobin solutions is dominated by interaction effects of great physiological significance (1) (2) (3).

The shape of the equilibrium curve versus pO_2 is sigmoid and this implies functional interactions or cooperativity, in the O_2 binding process. It is still useful to describe the shape of the equilibrium curve in terms of the exponent, n, of the Hill empirical equation $\frac{y}{1-y} = K(pO_2)^n$, where y is $\frac{HbO_2}{(Hb)+(HbO_2)}$. This n value for purified, dilute, hemoglobin solutions is close to 2.8-3, similar to that found for whole blood.

It is possible to calculate from the lower and higher portions of the equilibrium curve, a value for the total interaction energy in the binding process (2); this turns out to be about 3 kcal/mole. However it should be pointed out that this is a minimum figure based on the implicit assumption that the system is homogeneous in respect to the properties of the hemoglobin molecules and that the hemes in every one molecule are functionally equivalent. Even with these reservations it is clear that the interactions energy, of such great physiological importance, is only a small fraction of the total free energy change in the oxygen binding process.

In case of mammalian hemoglobins, the shape of the O_2 equilibrium curve remains the same or very nearly so, under the different conditions which affect the affinity of the protein for the ligand; this might imply that the interaction energy is independent of the total free energy change in the binding of the ligand. The same situation applies to the equilibrium of hemoglobin with different ligands; the shape of the equilibrium curve is the same or nearly so in spite of large differences in affinity.

The O_2 affinity of hemoglobin changes widely and reversibly as a result of changes in the environmental conditions; over the range where the protein is stable, at constant temperature, the combined effects of pH and salt concentration may produce a change in $p\frac{1}{2}$ of about 2 orders of magnitude, where p is the pressure of oxygen in equilibrium with the hemoglobin.

The effect of solvent components on the O_2 affinity of hemoglobin implies reciprocal interactions between the O_2 binding sites and sites binding the solvent components. These, heterotropic, interactions can be treated in terms of linkage relations which can be derived from standard chemical thermodynamics (2) by the use of the "binding potential" recently introduced by Wyman (4).

It is now generally admitted that the homo and heterotropic interactions in hemoglobin are mediated by conformational transitions associated with ligand binding. The conformational changes can be easily detected in solution and are being elucidated by high resolution X-ray diffraction analysis of crystalline oxy and deoxy hemoglobin.

Effect of pH - The familiar Bohr effect is the effect of pH on the O_2 affinity of hemoglobin; in view of the linkage relations it implies that the proton binding capacity of hemoglobin depends on the degree of saturation with ligands.

The Bohr effect in human hemoglobin over the pH range where the protein is stable and at different temperatures (5) is shown in Fig. 1. It may be noticed that the Bohr effect measured by the change in O_2 affinity in the presence of 0.1-0.4 M acetate, phosphate or borate buffers agrees well with that measured by differential titrations in 0.2 M NaCl. This indicates that over this salt concentration range, the Bohr effect is independent of the type of ions present in the solution.

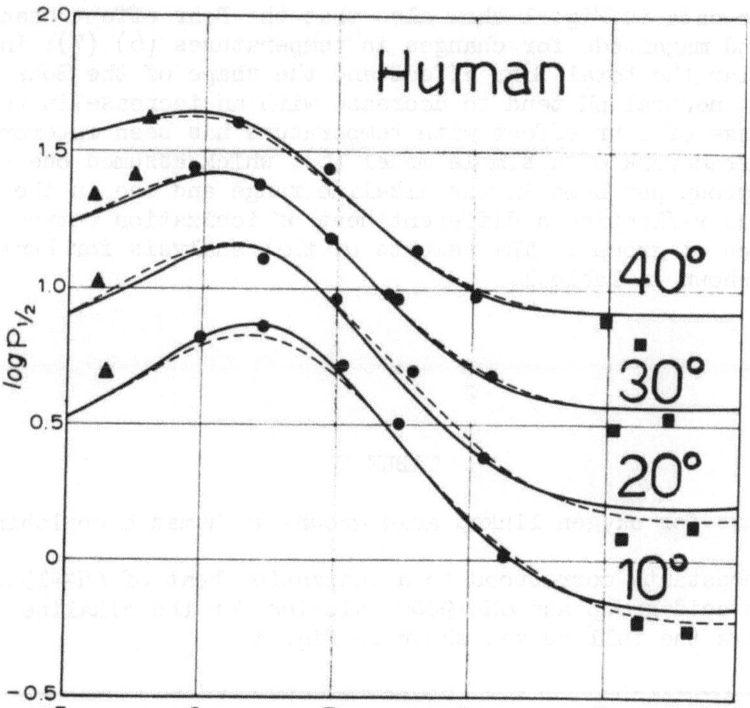

Fig. 1 Bohr effect in human hemoglobin at four temperature. Points show experimental values of log $p_{\frac{1}{2}}$ read from oxygen equilibrium curves. ●, in 0.15 M phosphate buffer; ▲, in 0.4 M acetate buffer; ■, in 0.05 M borate buffer. Dashed lines correspond to a graphical integration of titration data and involve an arbitrary constant of integration. Solid lines are calculated from the constant given in Table I. (From 5).

Several features of the Bohr effect depicted in Fig. 1 are **worthy** of comment since they may have physiological significance. The O_2 affinity of hemoglobin is at minimum near pH 6.5 and increases on both sides of pH; on the alkaline side for the "alkaline" or "normal" Bohr effect, on the acid side for the "acid" or "reverse" Bohr effect.

At high temperatures, near 40°, the acid and alkaline Bohr effect appear to overlap even at physiological pH values; thus, the acid Bohr effect appears to contribute, even if slightly, to the physiological regulation of oxygen affinity. It may be interesting to note that in human fetal hemoglobin the acid Bohr effect is much lower than in the adult protein, while the alkaline Bohr effect is the same in the two hemoglobins (6).

The data in Fig. 1 show also that the Bohr effect changes in shape and magnitude for changes in temperatures (6) (7); in particular the total Bohr effect and the shape of the Bohr effect curve at neutral pH tend to decrease with an increase in temperature. The change of Bohr effect with temperature has been interpreted in the framework of a simple model (6), which assumed one oxygen linked group per heme in the alkaline range and one in the acid range, as reflecting a different heat of ionization between the two types of groups. The results of this analysis for human hemoglobin shown in Table 1.

TABLE I

Constants for oxygen linked acid groups in human hemoglobin (5)

These constants correspond to a ionization heat of $\Delta H=-1500$ calories for the acid group and $\Delta H=+9000$ calories for the alkaline group. They give the full curves shown in Fig. 1.

Temperature	$pK_1(Hb)$	$pK_1(HbO_2)$	$pK_2(Hb)$	$pK_2(HbO_2)$
10°	5.42	6.22	8.08	6.68
20°	5.46	6.26	7.85	6.45
30°	5.495	6.295	7.63	6.23
40°	5.54	6.34	7.42	6.02

A stereochemical interpretation of the Bohr effect has been recently attempted by Perutz et al., (8). It is hoped that it will culminate in a detailed description of the molecular events leading to the changes in pK of the relevant ionizing groups. The tentative conclusion at present is that the imidazole of the C-terminal histidine in the β chains and the αNH_3^+ of the N-terminal α chains are among the groups contributing the Bohr effect protons.

Effect of organic phosphates and inorganic ions. The effect of organic phosphates, notably of 2, 3 diphosphatoglycerate (9) (10), is treated together with the effect of inorganic ions since the various types of ions differ from one another only quantitatively and one may substitute for the other in producing the same change in affinity although at a different concentration (11).

Formally these effects are analogous to the Bohr effect and imply a different affinity for ions between oxy and deoxy hemoglobin. The effect of several salts on the O_2 affinity of hemoglobin at pH 7 and 20°C is shown in Figs.2 and 3 (12).

Effect of CO_2. Carbon dioxide has a specific effect, apart from its properties as an acid, on the O_2 affinity of hemoglobin. This has been unequivocally shown in experiments (13) where the total CO_2 content of oxy and deoxy hemoglobin was measured at constant pH, after equilibration with the same pCO_2. This effect is due to the formation of carbamino compounds which take place preferentially with deoxy, rather than oxy hemoglobin.

Recent work with chemically modified hemoglobins has indicated that the oxygen linked carbamino formation takes place at the N-terminal amino groups of the α chains (14).

Kinetics of Hemoglobin Reactions

In the kinetics of hemoglobin reactions it is important to consider not only the rates for uptake and release of the iron ligand but also the rates of any ligand linked process, such as the Bohr effect, the effect of organic phosphates, of ions, of CO_2, of ligands linked association-dissociation processes.

Kinetics of the reaction of hemoglobin with O_2. The kinetics of the reaction of hemoglobin with O_2 represents, experimentally, a harder problem than that of other ligand reactions, and this explains why even its phenomenological aspects are still incompletely known (15).

Recent work has shown, just like the equilibria, the kinetics of the hemoglobin reactions show the same features for all the

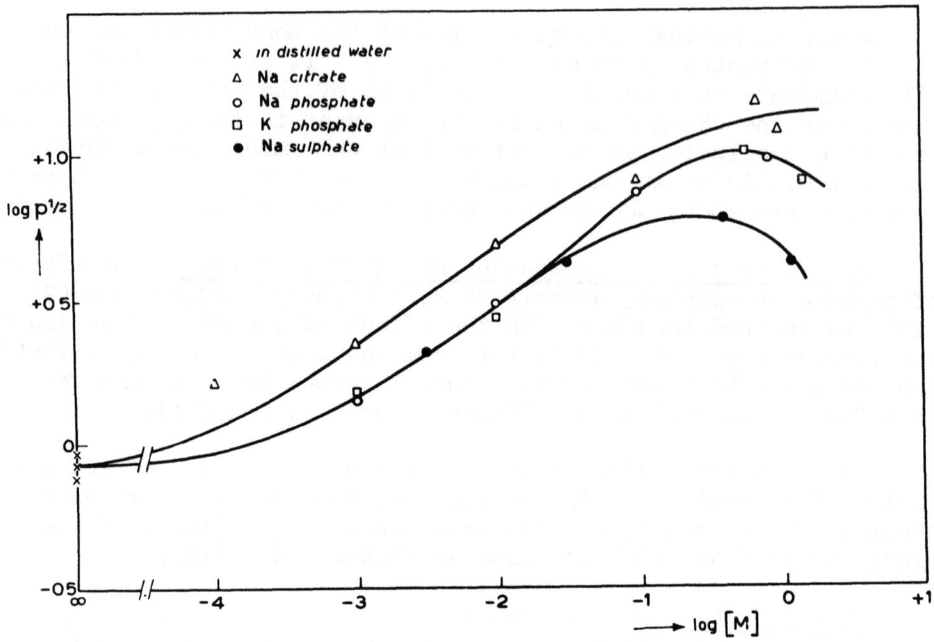

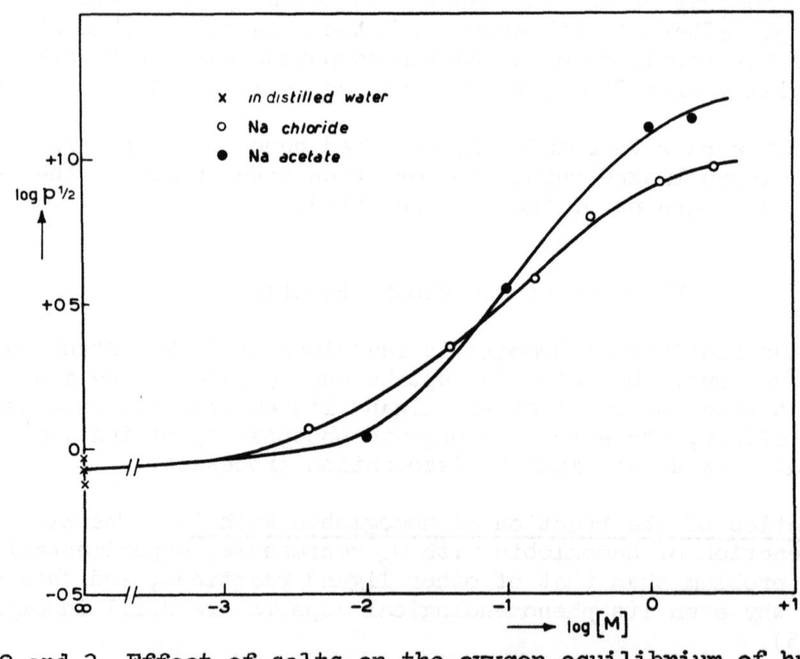

Figs. 2 and 3 Effect of salts on the oxygen equilibrium of human hemoglobin. Hemoglobin concentration $\simeq 3\times10^{-4}$M in heme pH 7 (in the solutions of oxyhemoglobin) and 20°C. (From 12).

ligands in spite of large differences among them in the absolute values of the rates of the "on" and "off" processes (16).

When the ligand concentration is high, so that the contribution of the "off" rates can be neglected, the kinetics of combination of deoxy hemoglobin with oxygen shows the same characteristics as found earlier for the combination with CO (17). The shape of the progress curve is autocatalytic, the value of the second order rate increasing slightly as the reaction proceeds. Under these conditions, as shown in Fig. 4, the rate reckoned at any given degree of saturation with ligand is proportional to the oxygen concentration.

On the other hand at low oxygen concentration the shape of the progress curve changes and the approach to equilibrium is diphasic (15). In T-jump experiments two or three relaxation times are evident (18); at least one of them shows a clear dependence on reagent concentration.

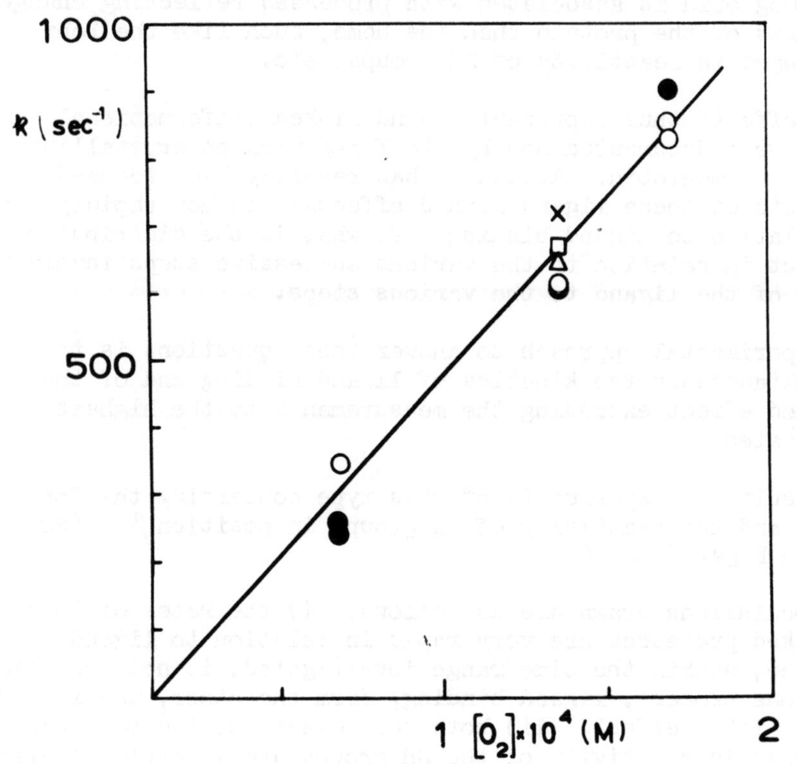

Fig. 4 Dependence of the pseudo first order rate constant for the reaction of hemoglobin with oxygen on ligand concentration. The values of k have been computed for values of $\bar{Y} \geq 0.5$. Solvent = 0.05 M phosphate buffer, pH 7; temperature = 22° - $23^{o}C$. (From 17).

The dissociation of oxygen in experiments in which oxyhemoglobin is mixed with dithionite, corresponds closely to a first order process (15). However, there are indications that in the process leading from fully oxy to fully deoxy hemoglobin after the first ligand molecule has dissociated, dissociation of the others follows at much higher rate (19) (20). This might explain the shape of the kinetics curves and their concentration dependence in the rapid mixing experiments at low oxygen concentration and in the relaxation experiments.

Thus, kinetically, interaction effects are present both on the combination and dissociation side being reflected by increases in the rates along the chains of reactions corresponding to the combination of the ligand with deoxyhemoglobin and to its dissociation from the fully saturated protein.

Kinetics of ligand linked processes. Binding of the ligand to deoxy hemoglobin is associated with processes reflecting changes in other parts of the protein than the heme, such like the Bohr effect, changes in reactivity of SH groups, etc.

These effects thus represent ligand linked conformational transitions well documented now by the X-ray work on crystalline oxy and deoxy hemoglobin. Attention has recently been focused on two aspects of these ligand linked effects: 1) how rapidly they occur in relation to ligand binding. 2) what is the distribution of the effect in relation to the various successive steps involving the binding of the ligand to the various steps.

The experimental approach to answer these questions is to follow simultaneously the kinetics of ligand binding and of the ligand linked effect extending the measurements to the highest attainable rates.

The results of experiments of this type concerning the Bohr effect (21) and the reactivity of SH groups in position β_{93} (22) are shown in **Figs.** 5 and 6.

The conclusions drawn are as follows: i) the rates of these ligands linked processes are very rapid in relation to ligand binding since, within the time range investigated, is not possible to resolve one process, ligand binding, from the other, the appearance of the linked effect. ii) both the release of the Bohr protons and the change in reactivity of the SH groups are linearly related, in the kinetic experiments, to the amount of ligand bound: this in turn indicates that these ligand linked effects reflect intrachain conformational changes brought out by the presence of the ligand on that particular chain and are not concerted within the whole molecule.

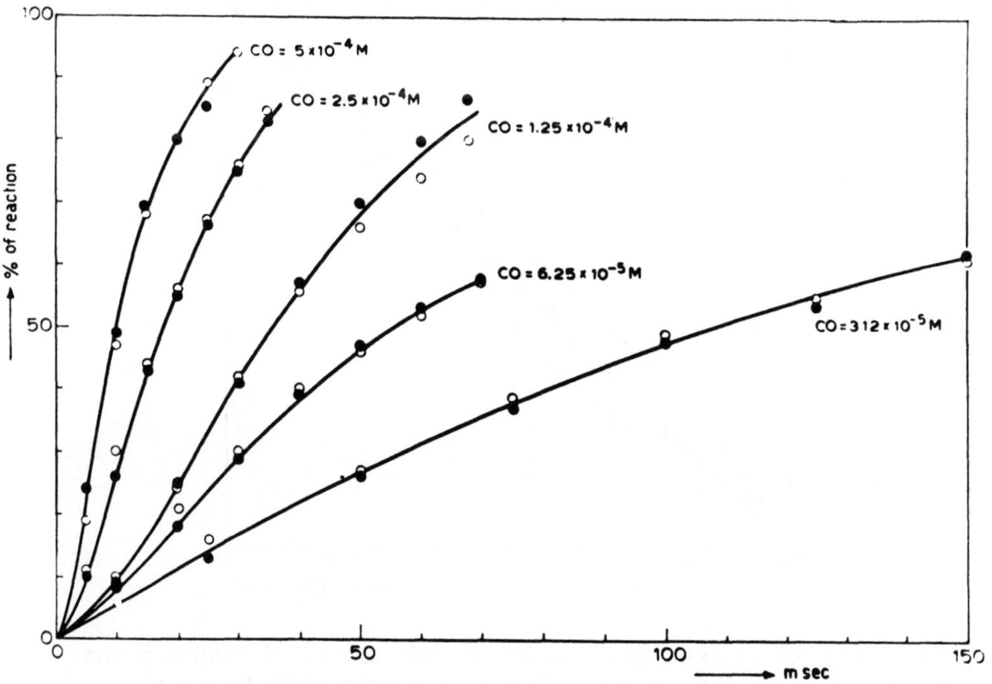

Fig. 5 Time course of the optical density changes at 482 m (O) and 563 m (●) in the reaction of human hemoglobin with CO. Hemoglobin concentration, 3×10^{-5}M, phenol red concentration, 1.1×10^{-5}M; pH (Hb) 7.15; ΔpH 0.150; temperature, 20°C.

This conclusion has been entirely substantiated by experiments on artificial intermediates.

Kinetic effects of ligand linked association-dissociation processes. The hemoglobin tetramer, $\alpha_2\beta_2$, may reversibly dissociate into subunits of lower molecular weight, $\alpha\beta$ dimers and single chain molecules and the association-dissociation equilibria may be ligand linked (1); thus effects due to association-dissociation processes are expected to show up when kinetic experiments are performed under conditions which favor dissociation.

This indeed has been verified by experiments: on sudden addition or removal of the ligand, with dilute hemoglobin solutions, ligand uptake or release is followed by "slow" changes in the properties of the system; the size of the changes depends on protein concentration (Fig. 7) (23).

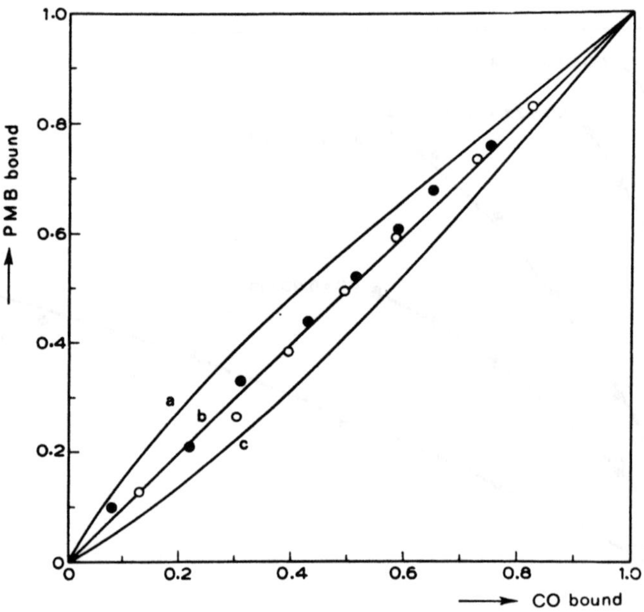

Fig. 6 Relation between the uptake of PMB and CO throughout the course of the reaction process when deoxyhemoglobin is **mixed** simultaneously with both reagents. O and ●, experimental points obtained at two different PMB concentrations; i.e. 0,5x10^{-5}M, PMB; ●, 2.5x10^{-4}M PMB. The CO concentration was 5x10^{-5}M; Hb, 1.5x10^{-5}M (in heme); at pH 7, in 0.1 M potassium phosphate, at 20°C. The experimental points, calculated from the optical density changes at 255 and 455 nm, are compared with different possibilities within a model based on the ($\alpha\beta$) dimer. Complete kinetic equivalence in the reaction with CO of the two different chains in the dimer is also assumed. The lines represent analogue solutions at the following schemes: a, the change in reactivity of the -SH groups is fully concerted within the dimer, and binding of a ligand to either α or β chains results in a change in reactivity of the β_{93} -SH groups; b, the change in reactivity of the -SH groups occurs upon binding of a ligand to the β chains only, independently of the state, liganded or not, of the α chains; c, the change in reactivity of the -SH groups is concerted within the dimer, but occurs only when both chains have reacted with the ligand. (from 22).

These effects were originally interpreted as reflecting association-dissociation processes involving single chain molecules

(23); however, more recently it has been realized that they may involve dissociation into two types of dimers, $\alpha_1\beta_2$ and $\alpha_1\beta_1$, and ligand linked dismutation of one type of dimer into the other (24, 25).

ISOLATED α AND β CHAINS

Haemoglobin can be reversibly dissociated into its constituent polypeptide chains. The α and β chains may be isolated under conditions which promote dissociation such as reactions of the masked

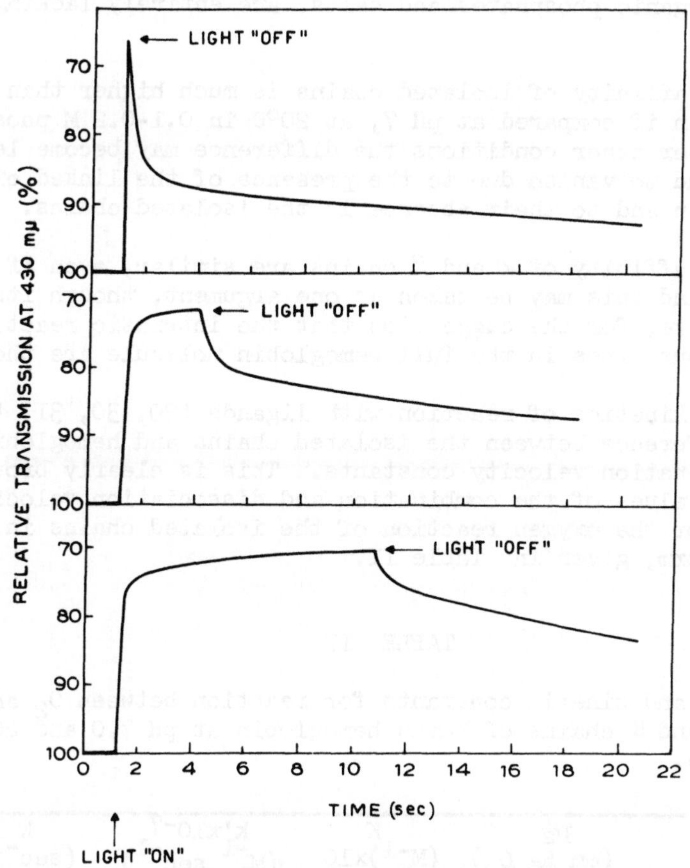

Fig. 7 Tracing of oscilloscope records showing changes occurring on illumination of dilute HbCO solutions. Arrows indicate beginning-end of illumination. The record in the top panel is obtained with a brief photolytic flash (300 joules; duration to 10% peak~300μ sec). Hb 6.2×10^{-7}M; 0.1 M phosphate buffer, pH 7.0 temperature 21°C. Degree of saturation of hemoglobin with CO in the dark, 0.92; light path 10 cm. (From 23).

SH groups with the mercurial para-mercurybenzoate PMB (26). The PMB containing chains (α^{PMB} and β^{PMB}) once isolated can be treated with thiols to regenerate their SH groups; the SH chains so obtained held rapidly native hemoglobin when mixed in stoichiometric amounts (26) (27) (28).

The oxygen equilibrium of isolated chains is characterized by the complete abolition of the homo - and hetero-tropic interaction typically present in hemoglobin. The equilibrium curves are hyperbolic, with a value of n in the Hill equation of one; the Bohr effect and other ligand linked effects, such as the effect of organic phosphates and salts, are entirely lacking (27, 28, 29, 30).

The O_2 affinity of isolated chains is much higher than that of hemoglobin if compared at pH 7, at 20°C in 0.1-0.2 M phosphate buffer. Under other conditions the difference may become less and even tend to vanish due to the presence of the linked effects in hemoglobin and to their absence in the isolated chains.

The O_2 affinity of α and β chains are similar, even if not identical, and this may be taken as one argument, though itself not conclusive, for the suggestion that the intrinsic reactivities of the various sites in the full hemoglobin molecule are the same.

In the kinetics of reaction with ligands (29, 30, 31) the largest difference between the isolated chains and hemoglobin is in the combination velocity constants. This is clearly brought out by the values of the combination and dissociation velocity constants for the oxygen reaction of the isolated chains in their PMB or SH form, given in Table II.

TABLE II

Equilibrium and kinetic constants for reaction between O_2 and isolated α and β chains of human hemoglobin at pH 7.0 and 20°C. (18, 29, 30).

	$p_{\frac{1}{2}}$ (mm Hg O_2)	K (M^{-1})x10	k'x10^{-7} (M^{-1} sec^{-1})	k (sec^{-1})
α -PMB	0.45	12.5	5.0	31
α -SH	0.46	12	4.9	28
β -PMB	2.65	2.1	7.1	156
β -SH	0.40	14	6.8	16

The functional behaviour of the isolated hemoglobin chains may be correlated, instead of by their state of polymerization (the α chains are monomeric, but the β tetrameric) by the absence of conformational changes associated with ligand binding (29, 32).

INTERMEDIATES

Isolation of hemoglobin chains has offered the long desired opportunity to prepare and study artificial intermediates of hemoglobin, i.e. compounds in which only part of the sides in each molecule may undergo reaction with ligand (32, 33). These intermediates, which may be represented as $(\alpha \times \beta)_1$ or $(\alpha \times \beta)_2$, are prepared by combination of α or β chains bound with ligand in a stable form with their deoxygenated partners. The ligand must be such as not to redistribute during the time of the experiments and therefore NO, ferric or ferric cyanide derivatives of chains are used to obtain stable intermediates (32, 34, 35).

The O_2 equilibrium of artificial ferric-CN-intermediates is shown in Figs. 8 and 9 (34). It is evident that, while the homotropic interactions are largely abolished, the heterotropic interactions, notably the Bohr effect, is maintained. The very low value of n (1 - 1.2 values) in the oxygen equilibrium of intermediates indicates that in hemoglobin homotropic interactions occur mainly within $\alpha\beta$ units. The two types of intermediates behave alike except for the reverse Bohr effect. At pH 7 and 20°C in phosphate buffer their O_2 affinity is higher than that of hemoglobin but lower than that of isolated chains; on the other hand the two types of intermediates have similar oxygen affinities.

The most important kinetic finding is that the rates of combination with ligand of both types of intermediates are only slightly higher (2-4 fold) than those of hemoglobin (32, 34); thus, the rates of combination with ligand of intermediates are much lower than those of isolated chains (Fig. 10).

Contrary to what occurs with the isolated chains, binding of the ligand to the intermediates is associated with conformational changes occurring in the chains which undergo the reaction with ligand (34, 36, 37). From here the conclusion, substantiated by result with hemoglobin, that to a large extent, ligand linked conformational changes reflect intrachain changes and occur on binding of the ligand to any given chain, independent of the state, ligated or not of their partners.

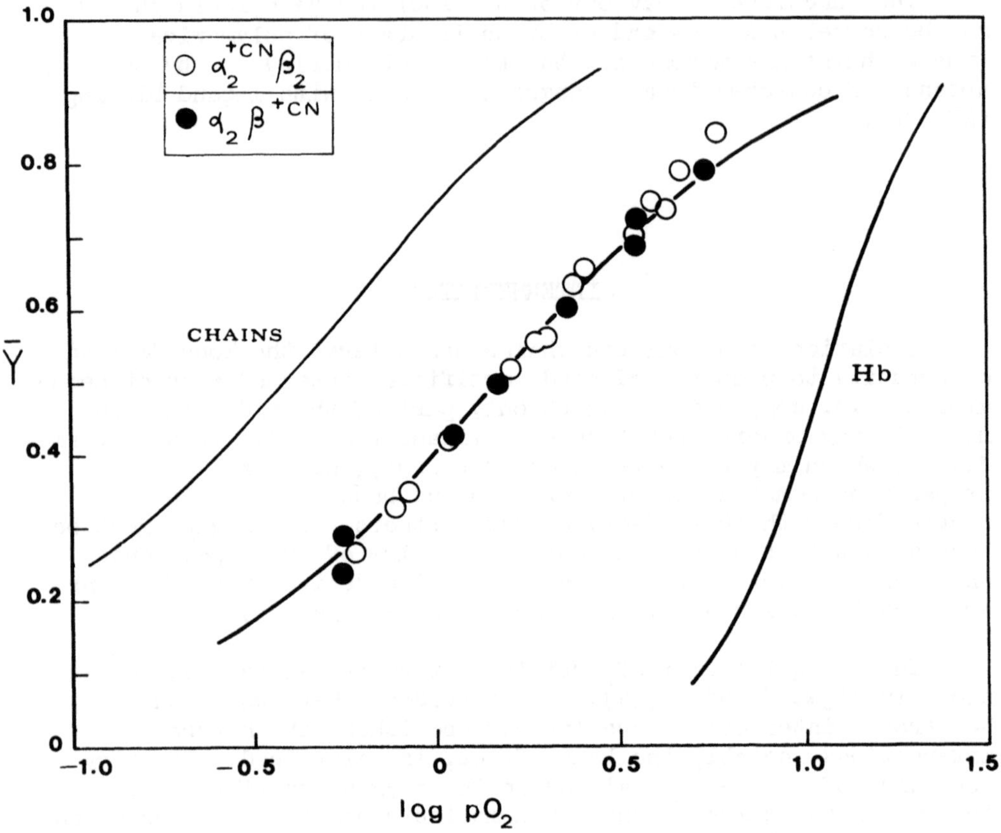

Fig. 8 Oxygen equilibrium curves for normal hemoglobin and artificial CN intermediates. From left to right, results corresponding to chains, to $(\alpha_2\beta_2^{+CN})$ and $(\alpha_2^{+CN}\beta_2)$ and normal Hb, respectively. Smooth curves are theoretical for n = 1 for the chains and intermediates, and n = 2.8 for normal Hb. Different points for each intermediate are from experiments performed at different pH values, and normalized for pH 7. (from 34).

Concluding Remarks

Comparison of the properties of hemoglobin, isolated chains and intermediates led directly to important conclusions regarding structure function relationships in hemoglobin.

These may be summarized as follows: Hemoglobin function requires the presence of the two different kinds of polypeptide chains. Only when both kinds of chains are present in the molecule do ligand linked conformational transitions occur and functional interactions operate between different sites in the molecule. The cooperativity of the O_2 binding process in hemoglobin is largely due to interactions between the hemes in $\alpha\beta$ units.

Heterotropic interactions, such as the Bohr effect, reflect intrachain conformation changes which occur, sequentially, as the ligand progressively binds to the hemes in the various chains.

The α and β chains are approximately equivalent in respect to hemoglobin function; they make the same contribution to the alkaline Bohr effect but a different one to the acid Bohr effect and to the carbamino formation.

Kinetically, homotropic interactions are reflected both on the combination and dissociation velocity constants; on the other hand, chain-chain interactions, as distinct from site-site interactions, produce very large effects in the combination velocity constants.

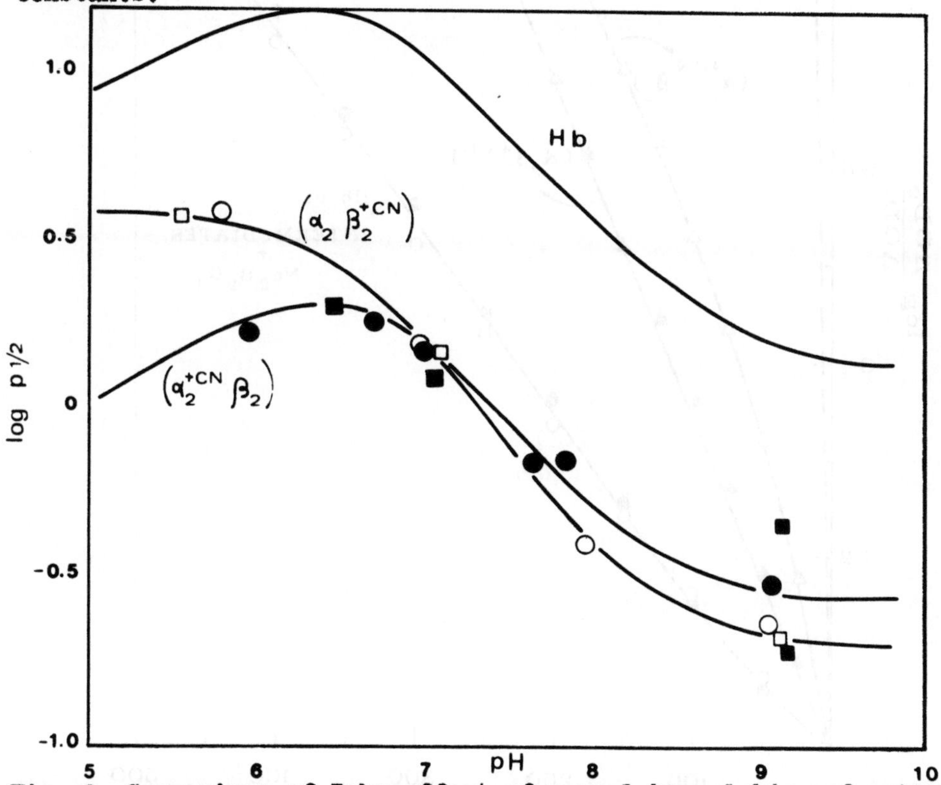

Fig. 9 Comparison of Bohr effect of normal hemoglobin and artificial CN intermediates. Conditions: protein concentration ~ 3 mg/ml 0.2 M potassium phosphate and 0.05 M sodium borate buffers, 20°C. (From 34).

Intramolecular conformational changes in hemoglobin are very rapid in comparison with ligand binding and hence they do not appear as distinct steps in the kinetics of the reactions of hemoglobin with ligands. In contrast, ligand linked association-dissociation processes, involving protein subunits clearly show up, kinetically, when the reactions with ligands are studied under conditions which favour dissociation of the protein.

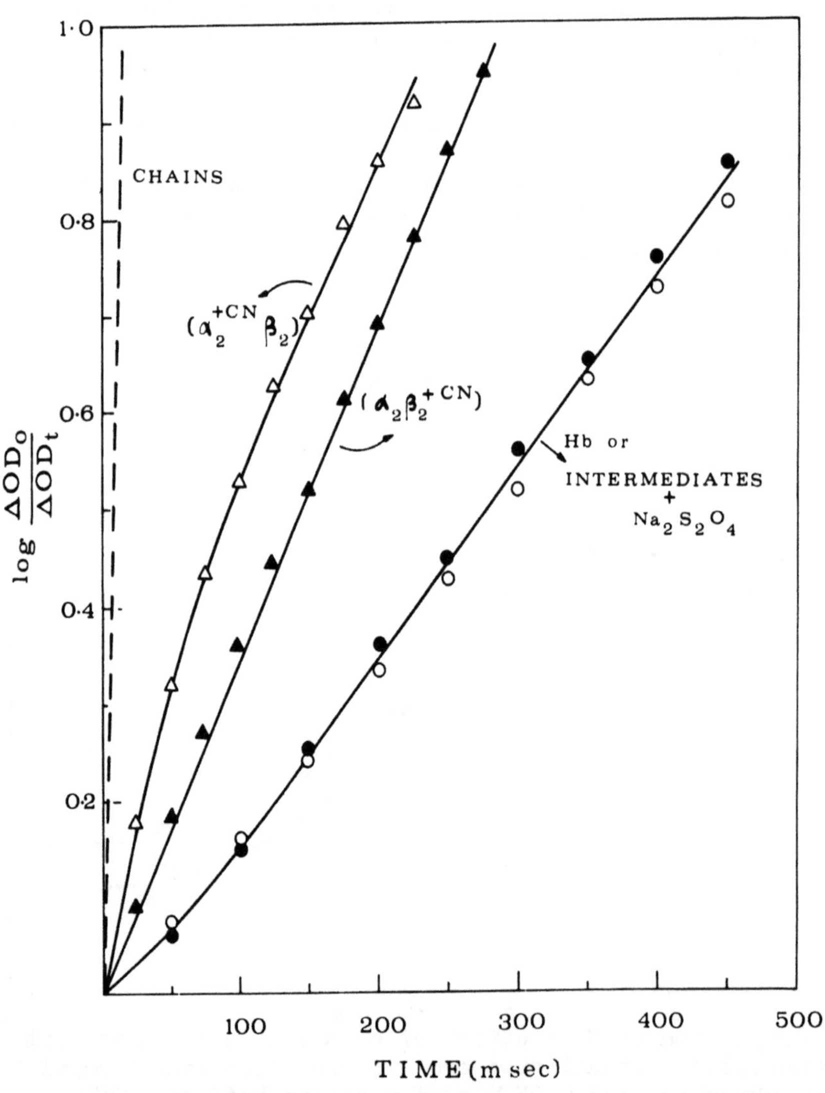

Fig. 10 Combination with CO of deoxyhemoglobin and deoxy artificial CN intermediates. Protein concentration≃ 3x10^{-6}M (in heme); CO concentration = 2.5x10^{-5}M; observation at λ = 435 nm; pH 7, 0.1 M potassium phosphate buffer and 20°C. (From 34).

REFERENCES

1) ROSSI FANELLI A., ANTONINI E. and CAPUTO A., Advan. Protein Chem., 19, 74 (1964).
2) WYMAN J., Advan. Protein Chem., 19, 123 (1964)
3) ANTONINI E., Physiol. Rev., 45, 123 (1965)
4) WYMAN J., J. Mol. Biol., 11, 631 (1965)
5) ANTONINI E., WYMAN J., BRUNORI M., FRONTICELLI C., BUCCI E.,and ROSSI FANELLI A., J. Biol. Chem., 240, 1096 (1965)
6) ANTONINI E., WYMAN J., BRUNORI M., FRONTICELLI C., BUCCI E., REICHLIN M. and ROSSI FANELLI A., Arch. Biochem. Biophys,, 108, 569 (1964)
7) CHIPPERFIELD J.R., ROSSI BERNARDI L. and ROUGHTON F.J.W., J.Biol. Chem. 242, 777 (1967)
8) PERUTZ M.F., MUIRHEAD M., MAZZARELLA L., GOWTHER R.A., GREEN J. and KILMARTIN J.W., Nature, 222, 1240 (1969)
9) BENESCH R. and BENESCH R.E., Biochem. Biophys. Res. Commun., 26, 162 (1967)
10) CHANUTIN A. and CURNISH R.R., Arch. Biochem. Biophys., 121, 96 (1967)
11) AMICONI G., ANTONINI E., BRUNORI M. and WYMAN J., work in progress.
12) ANTONINI E., Symposium on Intracellular Regulation of Hemoglobin Affinity to Oxygen, Gimo (Sweden) January 24-26 (1969)
13) ROSSI BERNARDI L. and ROUGHTON F.J.W., J. Physiol., 189, 1 (1967)
14) KILMARTIN J.W. and ROSSI BERNARDI L., Nature, 222, 1249 (1969)
15) GIBSON Q.H., Progr. Biophys. Biophys. Chem., 9, 1 (1959)
16) ANTONINI E., Science, 158, 1417 (1967)
17) BERGER R.L., ANTONINI E., BRUNORI M., WYMAN J. and ROSSI FANELLI A., J. Biol. Chem., 242, 4841 (1967)
18) BRUNORI M. and SCHUSTER T.M., J. Biol. Chem., 244, 4063 (1969)
19) ANTONINI E., and BRUNORI M., Atti del IV Convegno Nazionale della Societa Italiana di Biofisica e Biologia Molecolare, Pavia, 5-6 Maggio 1967.
20) GIBSON Q.H., Personal communication
21) ANTONINI E., SCHUSTER T.M., BRUNORI M. and WYMAN J., J. Biol. Chem., 240, PC2262 (1965)
22) ANTONINI E. and BRUNORI M., J. Biol. Chem., 244, 3909 (1969)
23) ANTONINI E., BRUNORI M. and ANDERSON S., J. Biol. Chem., 243, 1816 (1968)
24) ANTONINI E., BRUNORI M., CHIANCONE E. and WYMAN J., Johnson Foundation Symposium (1969) in press.
25) EDELSTEIN S. and GIBSON Q.H., Johnson Foundation Symposium (1969) in press.
26) BUCCI E. and FRONTICELLI C., J. Biol. Chem., 240, PC551 (1965)
27) DE RENZO E.C., IOPPOLO C., AMICONI G., ANTONINI E. and WYMAN J., J. Biol. Chem., 242, 4850 (1967)

28) TYUMA I., BENESCH R.E. and BENESCH R., Biochemistry, 5, 2957 (1966)

29) ANTONINI E., BUCCI E., FRONTICELLI C., WYMAN J., and ROSSI FANELLI A., J. Mol. Biol., 12, 375 (1965)

30) BRUNORI M., NOBLE R.W., ANTONINI E., and WYMAN J.; J. Biol. Chem., 241, 5238 (1966)

31) NOBLE R.W., GIBSON Q.H., BRUNORI M., ANTONITI E. and WYMAN J., J. Biol. Chem., 244, 3905 (1969).

32) BRUNORI M., ENGEL J., and SCHUSTER T.M., J. Biol. Chem., 242, 773, (1967)

33) ANTONINI E., BRUNORI M., WYMAN J., and NOBLE R.W., J. Biol., 241, 3236 (1966)

34) BRUNORI M., AMICONI G., ANTONINI E., WYMAN J. and WINTERHALTER K.H., J. Mol. Biol., (1969) submitted.

35) BANERJEE R. and CASSOLY R., J. Mol. Biol., 42, 251 (1969)

36) SHULMAN R.G., OGAWA S., WUTHRICH K., YAMANE T., PEISACH J. and BLUMBERG W.E., Science, 165, 251 (1969)

37) OGAWA S., Mc CONNELL H.M., and HORWITZ A., Proc. Nat. Acad. Sci. U.S., 61, 401 (1968).

A MATHEMATICAL ANALYSIS PREDICTING CEREBRAL TISSUE REOXYGENATION TIME AS A FUNCTION OF THE RATE OF CHANGE OF EFFECTIVE CEREBRAL BLOOD FLOW

D. D. Reneau, Dept. of Chem. Engr.
La. Poly. Inst., Ruston, Louisiana
H. I. Bicher, Dept. of Anatomy
Med. Univ. of S.C., Charleston, S.C.
D. F. Bruley, Dept. of Chem. Engr.
Clemson Univ., Clemson, S.C.
M. H. Knisely, Dept. of Anatomy
Med. Univ. of S.C., Charleston, S.C.

INTRODUCTION

In vivo experimental investigations in the cortex of cat brains with the new, ultramicro oxygen electrode (2), (3) have produced surprising results that need to be correlated with adequate convection-diffusion theory for proper analysis and future predictions. Transient changes (Figure 1) forced on arterial oxygen partial pressure (oxygen tension) of healthy cats produced tissue responses similar to that shown in Figures 2 and 3. On returning to normal conditions, oxygen tension was observed to "overshoot" normal values before leveling out at the previously recorded normal level. Identical experiments performed on the same animals when under conditions of intravascular red cell aggregation (sludge), hemorrhage, or hemorrhage combined with aggregation produced tissue responses similar to that shown in Figure 4. In these latter cases of pathologic physiology the "reoxygenation time" was greatly extended, and no overshoot was recorded.

The purpose of this paper is to analyze the above phenomena by means of mathematics in order to demonstrate a manner by which precise control of normal or abnormal cerebral oxygenation can be maintained or relinquished through the integrated response of system variables. The experimental responses shown in Figure 1, 2, 3, and 4 were found in cat brains, but subsequent analysis in this paper will simulate the responses to be expected in the human brain. Physical and chemical parameters of the human brain are used (11).

Although second order in appearance, it will be shown that the responses given in Figures 2 and 3 can be obtained from a first order diffusion system provided that proper regulation of the "effective flow" for cerebral oxygenation is maintained. A clear distinction will be made between the normally used term "cerebral blood flow rate", and the term "effective flow" for cerebral oxygenation.

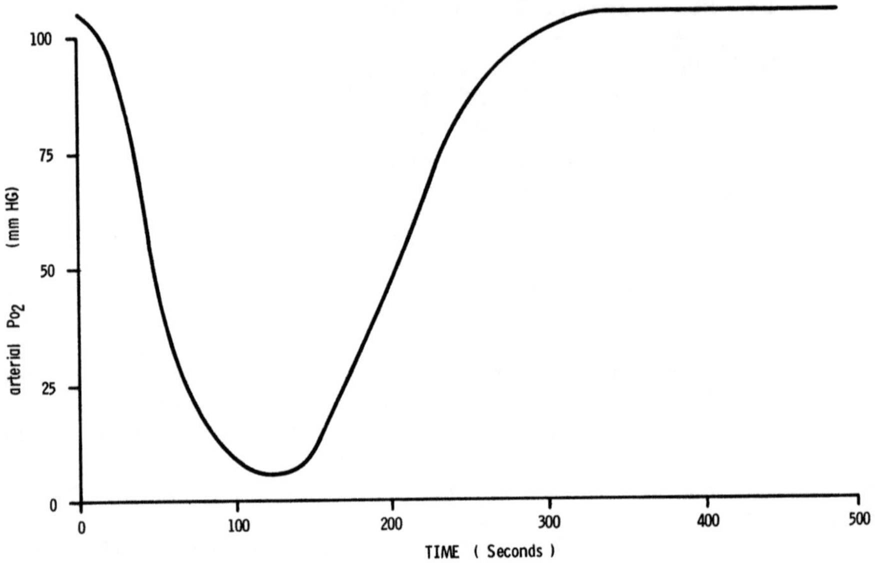

Figure 1. Arterial Oxygen Tension Changes as a Function of Time.

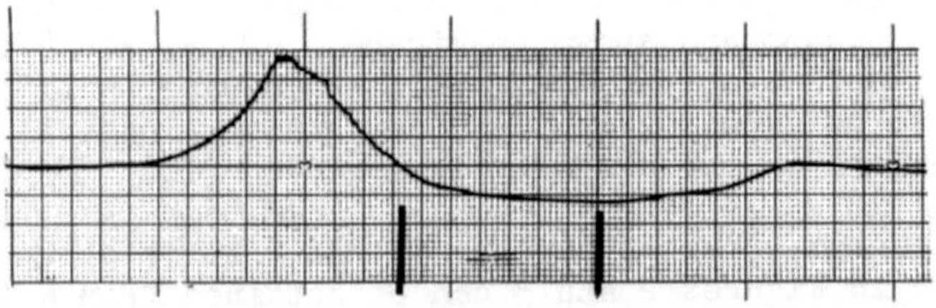

Figure 2. Transient Oxygen Response in Cerebral Tissue Following Deoxygenation and Reoxygenation of Arterial Blood. (Read from right to left - abscissa represents time). Note the Overshoot.

MATHEMATICAL ANALYSIS

Adequate oxygenation of cerebral tissue is dependent on the synchronous functions of several physiological, biochemical, and anatomical factors. Major factors affecting oxygen tension in tissue are rates of flow of blood through capillaries, oxygen tension of capillary blood, oxygen content of capillary blood, kinetics of the rate of release of oxygen from oxyhemoglobin, tissue metabolism, capillary dimensions, and the physical dimensions of tissues supplied by capillaries.

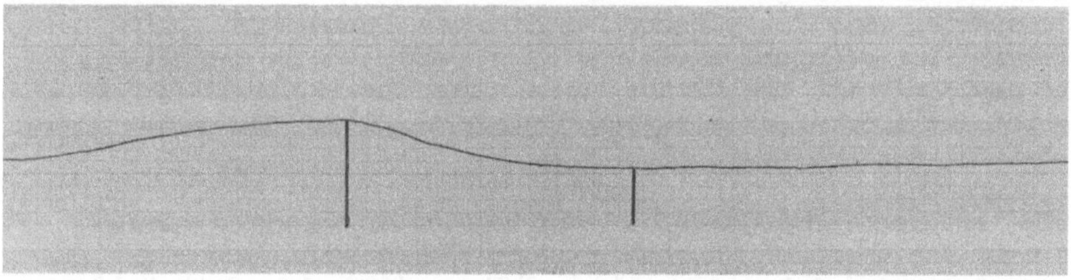

Figure 3. Transient Oxygen Response in Cerebral Tissue Following Deoxygenation and Reoxygenation of Arterial Blood. (Read from right to left - abscissa represents time). Note the Overshoot.

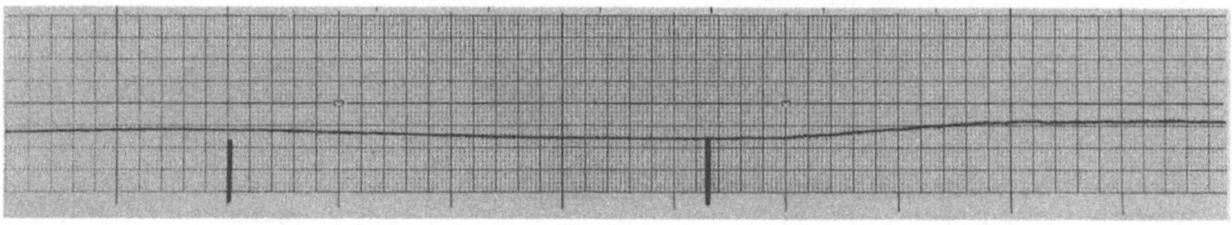

Figure 4. Transient Oxygen Response in Cerebral Tissue Following Deoxygenation and Reoxygenation of Arterial Blood When Under Conditions of Intravascular Red Cell Aggregation (Sludge). (Read from right to left - abscissa represents time). Note the Absence of Overshoot.

Mathematical analyses incorporating all or part of the above listed factors were conducted in three phases - distributed parameter, lumped parameter, and distributed-lumped parameter. In the following material each analysis will be developed, and its significance to the present experimental study discussed.

Mathematical Analysis A: Distributed Parameter

In conducting a conventional non-statistical mathematical analysis of a physiological transport problem it is important to select a basic anatomical unit which can be used as a representative sample of the whole. Such a unit is a capillary and the surrounding tissue that is supplied only by that capillary. Figure 5 is an ideal anatomical representation of a capillary-tissue system which was originally devised by August Krogh (10) as representative of striated muscle. Krogh found that in striated muscle the capillaries were dispersed with conspicuous regularity. Consequently, if concurrent flow of equal magnitude is present in each capillary, a single capillary and the tissue surrounding the capillary can be extracted and treated as a representative sample of the whole. (For details see Reneau et al. (13)).

The apparent arrangement of capillaries in cerebral tissue might make one question the concept of a homogeneous geometry. However, Opitz and Schneider (11) and Thews (21) showed that if the capillary length per volume of tissue or the number of capillaries per unit area was known for a given tissue section, an effective capillary radius for the section could be calculated with appropriately derived equations. This effective radius represents the radial region each capillary would need to supply if parallel homogeneity were present and actually accounts for all tissue present in the particular section being studied.

Reneau, et al. (13) showed how a standardized capillary length corresponding to the effective radius could be determined provided

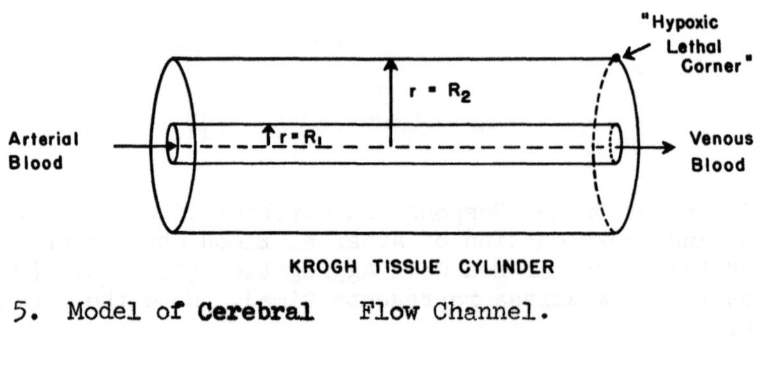

Figure 5. Model of **Cerebral** Flow Channel.

experimental values for arterial and venous concentrations of substances were known or measured in the tissue under investigation. The representation has been used quite successfully for oxygen diffusion studies in cerebral cortex (Reneau, et al. (13, 14), Reneau and Knisely (17), Knisely, et al. (9), Bruley and Knisely (5), and equally as well in preliminary glucose studies, Reneau and Knisely (18).

In addition to the above information, Figure 5 allows one to study that basic unit which is most sensitive to a decrease or increase in oxygen supply. Accordingly, the idealized capillary-tissue arrangement shown in Figure 5 was selected for a distributed parameter study of the dynamics of oxygen supply to cerebral tissue.

Mathematical Model. Based on known phenomena and certain assumptions a mathematical model has been developed which describes the change in oxygen partial pressure in capillary blood and tissue as a function of time, position, flow rate, pH, oxygen capacity, metabolic rate, and various constants such as diffusion coefficients and solubility. Now in an improved form the model consists of Equations (1), (2), (3) and (4) - one each for the capillary and tissue and two for the interface between tissue and blood. In essence the equations describe in mathematical terms the mechanisms by which molecular oxygen is transported from red cells while the red cells are transported in blood flowing longitudinally through capillaries into moving plasma, thence radially out to and through the capillary wall into the surrounding tissues for tissue cell respiration. For this particular simulation, one of the more basic assumptions is that the time required for the release of oxygen from hemoglobin is negligible compared to the time required for oxygen to diffuse through the red cell, through plasma and into tissue. If this is true, mass transfer is the controlling factor and the oxygen dissociation curve (Figure 6) can be used to describe the relation between combined oxygen and free oxygen at all longitudinal positions within the capillary. As shown below, the equations are coupled, non-linear, partial differential equations with three independent variables. Additional details concerning the model and its application are given in previous publications (13, 14, 15, 16 and 17).

Mathematical Model

Capillary:

$$\left(1+\frac{Nkmp^{n-1}}{c_1(1+kp)^n}\right)\frac{\partial p}{\partial t}=D_1\left(\frac{\partial^2 p}{\partial r^2}+\frac{1}{r}\frac{\partial p}{\partial r}\right)+D_1\frac{\partial^2 p}{\partial x^2}-V_x\frac{\partial p}{\partial x}-\left(\frac{V_x Nkmp^{n-1}}{c_1(1+kp^n)^2}\right)\frac{\partial p}{\partial x} \quad \textbf{(1)}$$

Interface:

$$P_i\Big|_{BLOOD} = P_i\Big|_{TISSUE} \tag{2}$$

$$D_1 C_1 \frac{\partial P}{\partial r}\Big|_{\substack{r=R_1 \\ BLOOD}} = D_2 C_2 \frac{\partial P}{\partial r}\Big|_{\substack{r=R_2 \\ TISSUE}} \tag{3}$$

Tissue:

$$\frac{\partial P}{\partial t} = D_2 \left[\frac{\partial^2 P}{\partial r^2} + \frac{1}{r} \frac{\partial P}{\partial r} \right] + D_2 \frac{\partial^2 P}{\partial x^2} - \frac{A}{C_2} \tag{4}$$

Boundary conditions imposed on the model are as follows:

I.C.	$P(x,r,o)$	Solution from steady-state equations
B.C.1	$P(o,r,t)$	Given step change $t > o$, $o \leq r < R_1$
B.C.2	$\frac{\partial P}{\partial x} = 0$	$x=0$ $R_1 \leq r \leq R_2$
B.C.3	$\frac{\partial P}{\partial r} = 0$	$r=0$
B.C.4	$\frac{\partial P}{\partial r} = 0$	$r=R_2$
B.C.5	$\frac{\partial P}{\partial x} = M$	$x=L$

Note that the two interface equations indicate that the oxygen partial pressure profile is continuous across the blood-tissue interface, and oxygen is transported across the interface according to Ficks' First Law.

Solution of Model. Numerical solution techniques for the above model have been developed and programmed on IBM 360 and 7094 digital computers for the following four steady-state situations.

1. Steady-state neglecting axial gradients (13).
2. Steady-state including axial gradients (13).
3. Steady-state (neglecting axial gradients) for conditions in which capillary oxygen tension becomes insufficient to supply completely all tissue positions (15) (16).
4. Steady-state neglecting axial gradients for the condition of countercurrent capillary flow (17).

Dynamic analysis of cerebral oxygenation required solving the model for the unsteady-state, and solution techniques presented the greatest difficulty. Neglecting the second-term (accounts for

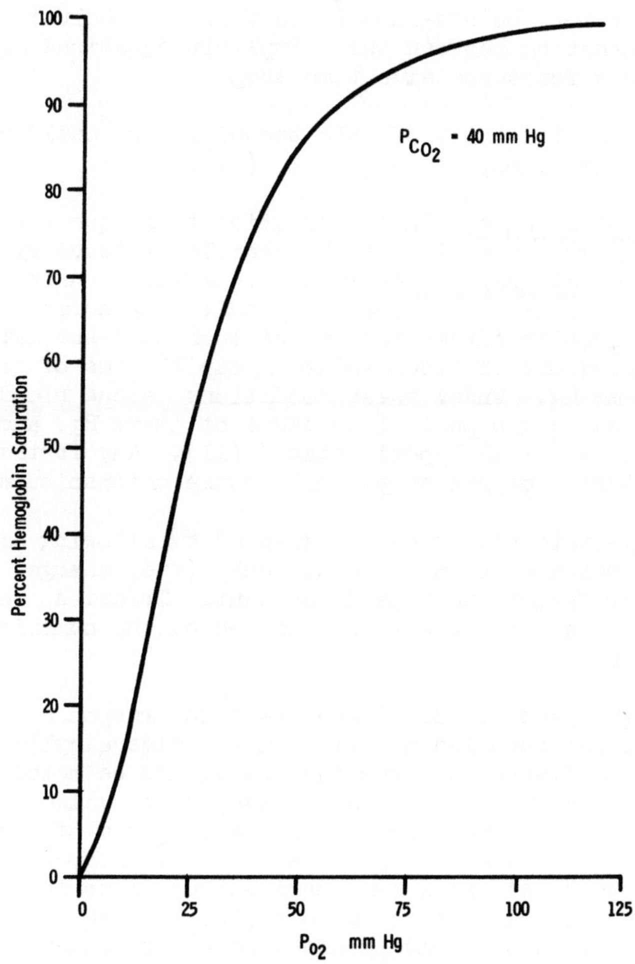

Figure 6. Oxygen Dissociation Curve.

accumulation of oxyhemoglobin) on the left-hand side of Equation 1, an implicit solution technique based on numerical analysis was developed to solve the equations. The numerical method developed is a combination of the original Peaceman-Rachford technique (12) and the modified Peaceman-Rachford technique (1). First, the x-derivatives of the tissue and capillary equations are implicitly differenced and the r-derivatives explicitly differenced. The resulting equations are of a form easily solved by the method of Thomas (4). Second, the r-derivatives of each equation are implicitly differenced and the x-derivatives explicitly differenced. These equations are also solvable by the method of Thomas. The tissue and capillary equations are linked by the interface equations which are adapted to the implicit scheme by alternately applying them in two different ways. A special equation is used to generate an approximate value for the non-linear coefficient. Solution then proceeds by alternating between the x-implicit equations and r-implicit equations for successive time steps.

The mathematical details of this technique are outlined in a forthcoming publication, Reneau et al. (14).

Discussion of Results. Once a solution technique became available, the dynamic behavior of the capillary-tissue system was explored for cerebral gray matter in the following manner. The case selected for transient analysis was that of a normal person existing under alveolar oxygen conditions such that the arterial oxygen partial pressure of blood entering capillaries of gray matter was 28.5 mm Hg. Under these conditions venous blood leaving capillaries has an oxygen partial pressure of 19 mm Hg. and the person is in the "arterial hypoxic state" (11). Any further reduction in arterial oxygen supply will cause unconsciousness.

An unsteady-state situation was created by allowing arterial oxygen partial pressure to increase suddenly (step change) to 38.5 mm Hg. All other factors were held constant. Transient recovery from the hypoxic condition was then obtained by the numerical solution technique.

Figure 7 is a plot of the tissue transient response of oxygen tension at the least supplied position in the tissue cylinder, lethal corner (see Figure 5). Two factors should be noted. One, the oxygen tension approaches the new steady-state value at an exponential rate in a manner similar to the response of first order systems. No overshoot is present, but this response is in agreement with the first order time derivatives given in Equations 1 and 4. Two, the overall response is rapid. A time constant for the lethal corner is approximately one second. In other words, 63.2 percent of the total change has occurred at the lethal corner one second after the step change was applied to the capillary entrance.

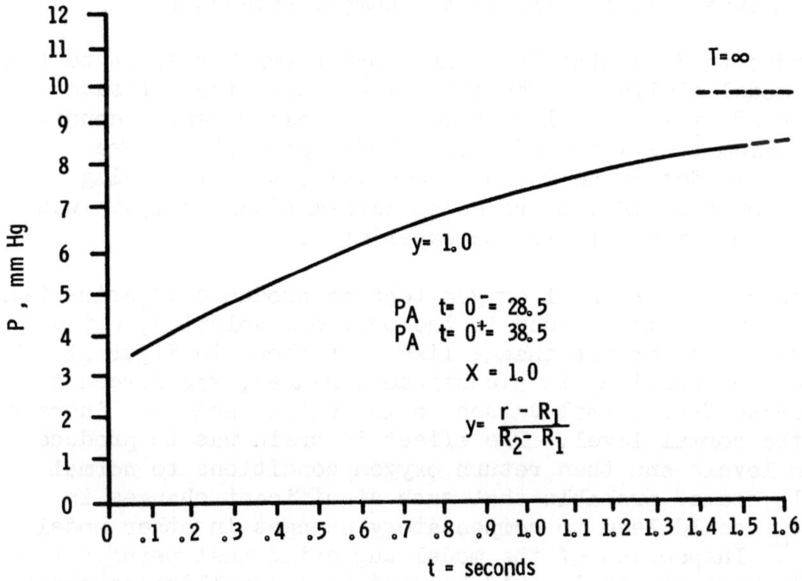

Figure 7. Oxygen Partial Pressure Time Responses of the Tissue Cylinder at the Lethal Corner.

It was stated previously that this analysis omitted the second term of Equation (1). Recent solutions including all terms of the model (unpublished results) give time responses of the same order of magnitude.

A response with overshoot found in the dynamic investigation of cat brains and similar to that of second order systems (Figures 2 and 3) was not found in the above study. Further investigation was necessary including a time variation of input oxygen tension (Figure 1) instead of a step change and a time variation of another parameter in order to approximate experimental conditions. However, computer time requirements for the simple case analyzed above and the relative tedious nature of the solution technique eliminated numerous studies with the distributed parameter model. Nevertheless, the acquired information which indicated a very fast tissue response to a change in blood oxygenation led to the development of a satisfactory approximate method to investigate the dynamics of oxygen response in cerebral tissue.

Mathematical Analysis B: Lumped Parameter

In order to determine if a first order (with respect to time) diffusion model similar to Equations 1-4 could give a tissue response found experimentally, a review of experimental conditions and model parameters was conducted. Model parameters were investigated to determine whether some value that was being maintained constant in the preceding mathematical analysis was in fact varying with time in the experiment.

The change in arterial oxygen tension produced experimentally was not a step change as used in the previous solution, but a continuously time variant change like that shown in Figure 1. The arterial oxygen tension, in a continuous manner, was forced to first decrease from normal values to low values and then increase again to the normal level. The effect in brain was to produce low oxygen levels and then return oxygen conditions to normal values. It seemed probable that such significant changes in arterial P_{O_2} could lead to compensatory changes in other model parameters.[2] Inspection of the model suggested that perhaps the most likely variable that could be used in controlling cerebral oxygenation would be the flow rate term.

Discussions on the regulation of cerebral blood flow can be found in the publications of Kety (6), Kety, et al.(7), Opitz and Schneider (11), Schmidt (19), and Sokoloff and Kety (20). In general, the rate of flow of blood to the brain is conspicuously constant regardless of the state of mental or physical excitation. In most physiological conditions where blood pressure is adequate (greater than 60 to 70 mm Hg) intrinsic mechanisms regulate cerebral circulation. Of the intrinsic mechanisms, oxygen and especially carbon dioxide predominate.* It has been shown that a decrease in carbon dioxide can reduce cerebral flow rate to a minimum of 50 percent of normal or an increase in carbon dioxide can increase flow rate to a maximum of twice the normal flow rate. Conversely, large increases in oxygen above normal values slightly decreases flow rates, but large decreases in oxygen tension can significantly increase cerebral blood flow rate. Figure 8 is a plot of cerebral flow versus arterial P_{O_2} compiled in the works of Opitz and Schneider (11). The extrapolated region is the present authors' approximation.

*In addition to oxygen and carbon dioxide, cerebral flow is known to be at least a function of blood pressure, blood acidity, blood glucose level, and certain chemical agents. There is some evidence that under certain conditions neurohumoral agents are released in the brain and act as vasodilators or vasoconstrictors to redistribute the internal flow pattern. However, quantitative data for local flow changes is incomplete, and in this paper changes in overall cerebral flow rates are considered.

Consequently it can be concluded that in the experiment changes in cerebral flow rate accompanied changes in arterial P_{O_2}. Occasional P_{CO_2} measurements indicated relatively constant values.

Mathematical Approach. Thus a mathematical model was needed which allowed rapid investigation of numerous conditions in which arterial P_{O_2} was a function of time and cerebral flow rate was a function of arterial P_{O_2}. The rapid response of tissue to changes in arterial P_{O_2} (one second or less) found in the distributed parameter study suggested that perhaps a section of brain (blood and tissue) could be treated as a thoroughly mixed **homogeneous** system as shown in Figure 9 (A). Oxygen is considered soluble in all brain tissue. The dynamic response of this system would be analogous to the electrode response. (The output concentration from the system would be approximately equal to the system venous concentration and changes in it would be analogous to changes in tissue.)

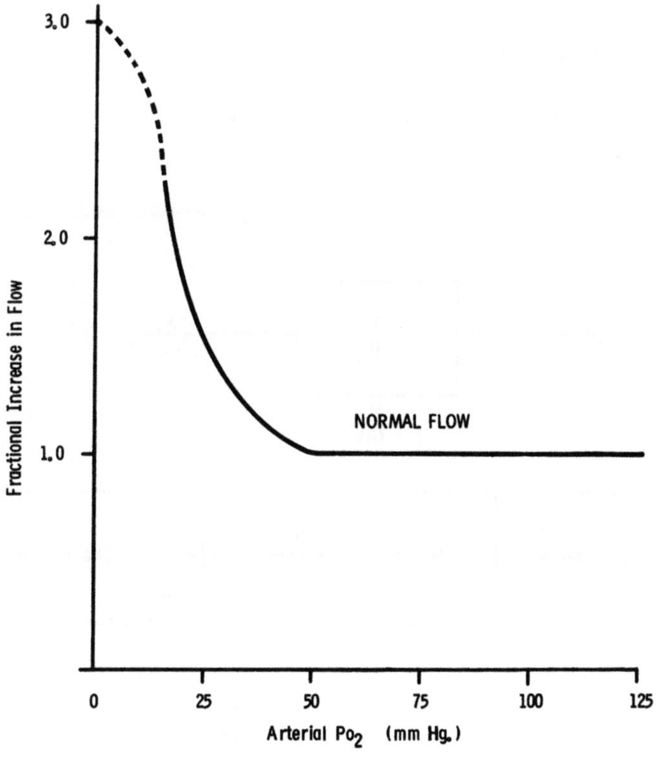

Figure 8. Cerebral Flow Rate as a Function of Arterial Oxygen Tension. (Opitz and Schneider.)

The next difficulty encountered was that oxygen is present in blood in two forms - physically dissolved in plasma and chemically combined with hemoglobin. Oxygen tension is a result of the physically dissolved oxygen, but in normal conditions over 99 percent of oxygen in blood is present in the form of oxyhemoglobin. Because of its insignificant amount, the physically dissolved oxygen was neglected and the system modeled from a combined oxygen viewpoint. The relationship of P_{O_2} to combined oxygen is given by the oxygen dissociation curve shown in Figure 6. Calculation could be conducted on the basis of combined oxygen, and then the corresponding P_{O_2} determined from Figure 6.

Performing a material balance around the system in Figure 9 (A) and rearranging, yielded the following relations:

$$\frac{VdC_o}{dt} = QC_i - QC_o - R$$

where

C = concentration of combined oxygen
V = volume of system
R = metabolism rate of system
Q = cerebral blood flow rate to system

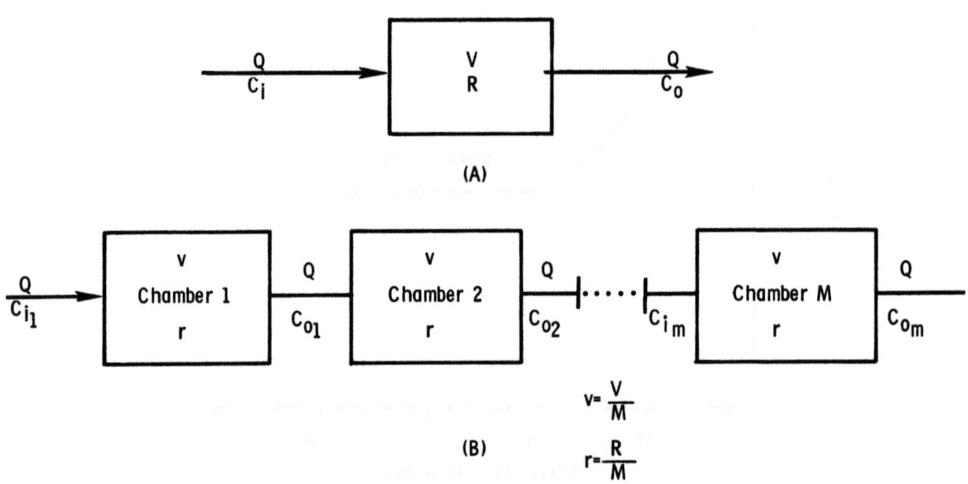

Figure 9. Diagramatic Representation of Lumped Parameter (A) and Distributed - Lumped Parameter Systems (B).

But

$$C = N\lambda$$

where

N = oxygen capacity of blood
λ = fractional saturation of hemoglobin with oxygen

Thus, substituting,

$$VN \frac{d\lambda_o}{dt} = QN\lambda_i - QN\lambda_o - R$$

Note that the first term, $QN\lambda_i$, on the left hand side of the above equation, represents the rate of flow of oxygen to a section of brain. The rate of flow of oxygen to the system is dependent on three factors: (1) the cerebral blood flow rate, (2) the oxygen capacity of blood (directly related to red cell count and/or hemoglobin content), and (3) the degree of saturation of hemoglobin with oxygen.

Normally the cerebral blood flow rate is considered a measure of the rate of supply of substances to tissues. A distinction is made here because changes in hematocrit, as found following hemorrhage, or changes in hemoglobin saturation at a given P_{O_2}, as found in carbon monoxide poisoning, will produce a decrease in the availability of oxygen to brain even if the cerebral blood flow rate is maintained constant. For a constant flow rate and a constant hemoglobin saturation, a 25 percent reduction in hematocrit, as found by Bicher, Reneau and Knisely (3) following hemorrhage in cats, would reduce the rate of flow of oxygen by approximately 25 percent. An increase in cerebral blood flow rate of 25 percent would be needed to even maintain normal conditions of oxygen availability.

Accordingly, we define the term $E_F = QN\lambda_i$, as "effective flow" for oxygenation and point out the fact that E_F controls the rate of supply of oxygen to the system - not cerebral blood flow rate. It should also be pointed out that E_F is to be used only for determining the rate of oxygen supply and not the supply of nutrients, drugs, etc.

Now, letting

$$Q = Q_N F$$

where

Q_N = normal cerebral blood flow rate to system
F = fractional change of normal flow rate (F = 1.0 for normal flow)

Substituting for Q, rearranging, and listing the additional relations for initial conditions and flow rate changes yields the following mathematical model:

$$\frac{d\lambda_o}{dt} = \gamma_N F\,(\lambda_i - \lambda_o) - \frac{R}{VN} \tag{5}$$

λ_i = f (t) Obtained from Figure 1 by use of Figure 6. **(6)**

F = f (λ_i)Obtained from Figure 8 by use of Figure 6. **(7)**

where

$\gamma_N = \frac{Q_N}{V}$

λ_i = fractional saturation of hemoglobin with oxygen in blood entering the system

λ_o = fractional saturation of hemoglobin with oxygen in blood leaving the system

The basis for analysis in this technique was taken as 100 cm^3 of human brain. Corresponding values for equation components were taken as those previously collected from the literature (11) and are presented below:

V = 100 cm^3 of brain
R = 3.3 $cm^3 O_2$ (STP)/100 cm^3 brain/min.
N = 0.204 $cm^3 O_2$ (STP)/ cm^3 blood
Q_N = 54 cm^3 blood/100 cm^3 tissue/min.

Solution of Mathematical Model. Because of the nonlinearity and constant time variation of Equations (6) and (7), an analytical solution of Equation (5) was not considered feasible. Instead, Equation (5) was numerically differenced with an implicit forward difference formula as follows:

$$\frac{\lambda_o(t+\Delta t) - \lambda_o(t)}{\Delta t} = \gamma_N F\left(\lambda_i(t+\Delta t) - \lambda_o(t+\Delta t\right) - \frac{R}{VN}$$

Rearranging and solving for λ_o $(t+\Delta t)$, yielded

$$\lambda_o(t+\Delta t) = \frac{\Upsilon_N F \Delta t \, \lambda_i(t+\Delta t) - \frac{R}{VN}\Delta t + \lambda_o(t)}{(1.0 + \Upsilon_N F \Delta t)} \qquad (8)$$

Equation (8) along with Figures 1, 6, and 8 were programmed on an IBM 360 digital computer and rapid solution of transient conditions were obtained. Standard procedures of numerical analysis were used at all times to check for stability and convergence.

Mathematical Analysis C: Distributed-Lumped Parameter

The lumped parameter analysis allows investigation of only one transient oxygen tension response for the system and is analogous to venous conditions. All radial and axial concentrations are "lumped" into one conglomerate. In order to more closely approach the capillary-tissue system shown in Figure 5 and in order to check the qualitative and quantitative accuracy of the conclusions that will be drawn in the previous mathematical analysis, the homogeneous system shown in Figure 9 (A) was divided into a series of continuous homogeneous systems as given in Figure 9 (B). Since the output of each system was taken as the input to the next system, the resulting system of simultaneous equations is distributed longitudinally and only lumped radially. Equations describing the dynamic behavior of this distributed-lumped parameter system are given below and allow investigation of oxygen tension response at various longitudinal positions from arterial to venous conditions.

$$\frac{d\lambda_{o1}}{dt} = \Upsilon_N F(\lambda_{i1} - \lambda_{o1}) - \frac{r}{VN} \qquad (1)$$

$$\frac{d\lambda_{o2}}{dt} = \Upsilon_N F(\lambda_{i2} - \lambda_{o2}) - \frac{r}{VN} \qquad (2)$$

$$\vdots \qquad\qquad \vdots \qquad\qquad \vdots$$

(9)

$$d\lambda_{oM} = \Upsilon_N F(\lambda_{iM} - \lambda_{oM}) - \frac{r}{VN} \qquad (M)$$

$$\lambda_{i1} = f(t) \qquad (M+1)$$

$$F = f(\lambda_{i1}) \qquad (M+2)$$

DISCUSSION OF RESULTS

In a discussion of the results from the distributed parameter analysis, it was pointed out that oxygen diffusion time between capillary blood and tissue was short. It needs to be emphasized that this analysis considered only a capillary and the blood contained within the capillary. The transient response analyzed was for a step change placed on arterial blood entering the capillary. In the lumped and distributed-lumped analyses, sections of tissue, $V = 100\ cm^3$, were investigated and overall flow rates into and out of these systems were used. Consequently, blood is present in the system in arteries of varying size, arterioles, and venules as well as capillaries. The "time constant" for tissue response in this particular lumped system will be logically longer than those for the capillary distributed parameter system since much larger volumes of residual blood are present in the lumped systems than in the distributed system.

The actual degree of mixing of blood between different arteries and between different arterioles would appear to be a function of the anastomoses in the section of tissue under consideration and the ability of the body to adjust blood supply from one minute section to another. There may be total mixing, partial mixing or no mixing, but the present state of knowledge concerning this relationship in brain is incomplete. Without the above information, the actual magnitude of "time constants" in the lumped and distributed-lumped analyses should be accepted with a degree of caution. On the other hand, equilibration between oxygen in capillary blood and oxygen in surrounding tissue would be of the same order of magnitude in the lumped analysis as in the distributed analysis.

Lumped Parameter Analysis: As a first step in the theoretical investigation, the arterial P_{O_2} was varied with time as shown in the experimentally measured curve of Figure 1. Flow rate to the brain was correlated with the arterial P_{O_2} as indicated in Figure 8, and as can be seen, blood flow rate was maintained constant until arterial oxygen tension dropped below 50 mm Hg and then increased to a maximum of 3 times normal when P_{O_2} dropped to approximately 0 mm Hg. As arterial P_{O_2} began increasing, flow rate decreased from the maximum increase to normal flow at a P_{O_2} of 50 mm Hg and remained constant until normal conditions were reached.

The oxygen tension response of the system for this approach is shown as the curve labelled $\gamma_D = 0$ in Figure 10. The P_{O_2} decreased from the normal steady-state value of 35 mm Hg to a low of 8 mm Hg in 150 seconds (deoxygenation phase) and then increased to 31 mm Hg during the next 300 seconds. No overshoot was found, and the pre-existing oxygen level of 35 mm Hg was approached at a relatively slow rate.

Obviously this type of response was not similar to that found in normal, healthy cats' brains (Figures 2 and 3). In a search for the proper integration of system variables that would indicate the type of response found experimentally, the rate of increase in cerebral flow rate was analyzed. For instance, a change in arterial oxygen tension from 20 mm Hg to 50 mm Hg indicates that flow rate will decrease from 80 percent above normal to normal (Figure 8). But does this change in flow rate occur at the same instant that arterial P_{O_2} changes or at the same rate of change as arterial oxygen tension? It seems likely that a time lag of some degree is necessary to effect the capillary contraction, blood pressure, etc., changes that must occur to reduce the flow rate. To test this hypothesis, flow rate was not allowed to increase instantaneously with the indicated arterial P_{O_2} as was done previously, but the response was delayed a given number of seconds. Flow rate response was forced to lag arterial P_{O_2} by a factor, τ_D.

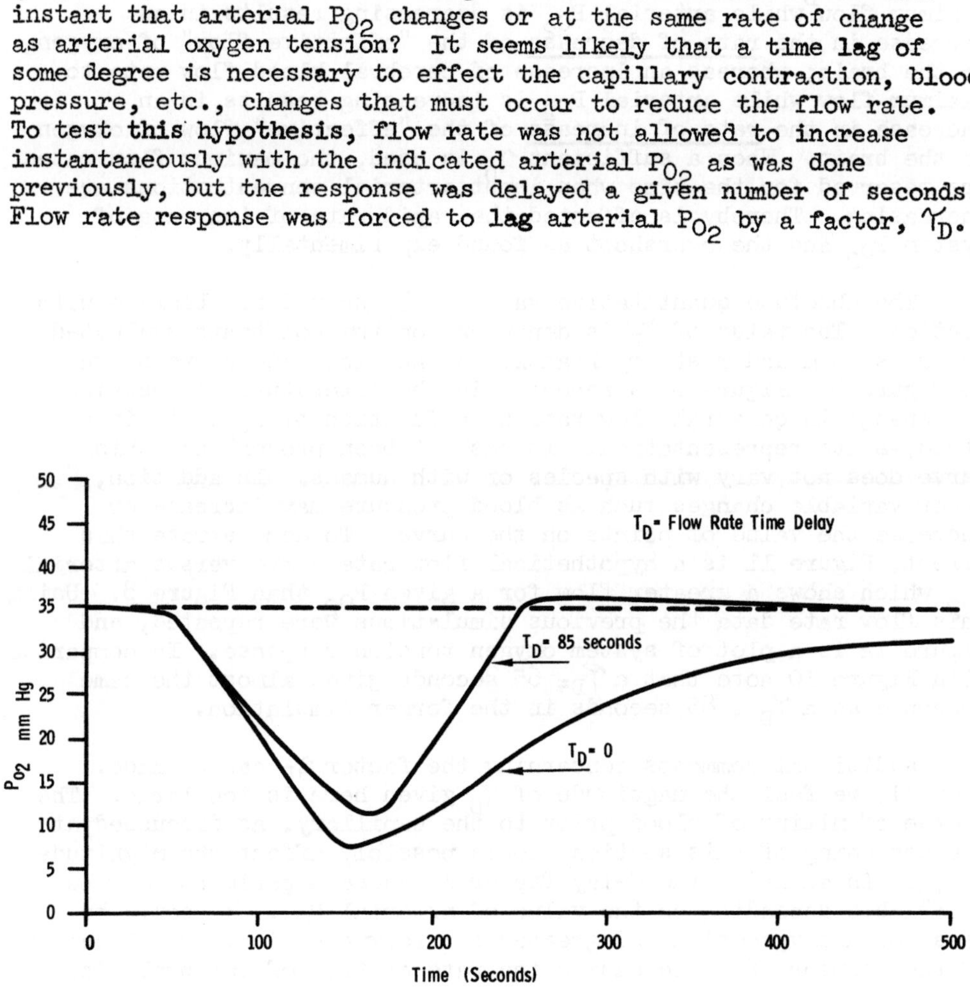

Figure 10. Lumped Parameter System: Oxygen Tension Response Curves as a Function of Time. Parameters are Flow Rate Time Delay.

The upper curve on Figure 10 shows the system response to a flow rate time delay of 85 seconds. The deoxygenation phase is approximately 165 seconds, and the reoxygenation time is 80 seconds. The system P_{O_2} is not only returned to a normal value but a noticeable overshoot is obtained with P_{O_2} reducing to the normal steady-state value as time increases. In the deoxygenation phase an oxygen tension low of 11 mm Hg was reached as compared to a low of 8 mm Hg in the simulation where $\tau_D = 0$. This is expected since maximum flow rates at low arterial concentrations are experienced when $\tau_D = 0$ and the system is "washed out" faster than when $\tau_D = 85$. The increase of cerebral blood flow rate to maximum flow while arterial P_{O_2} is decreasing results in an increase in the rate of decrease of the "effective flow" of oxygen to the brain; whereas an increase of cerebral blood flow rate to maximum flow while arterial P_{O_2} is increasing results in an increase in the rate of increase of the "effective" flow of oxygen to the brain. When a sufficient τ_D is used, the maximum flow rates are reserved for the time when input arterial concentrations are increasing. Thereby is effected the rapid rate of increase of system P_{O_2} and the overshoot as found experimentally.

The absolute quantitative value of τ_D should be observed with caution. The value of τ_D is dependent on the constants evaluated in the system and most importantly on the flow rate curve shown in Figure 8. Figure 8 is reported in the literature to describe the change in cerebral flow rate as a function of P_{O_2}. It is a steady-state representation. It has not been proven that this curve does not vary with species or with humans. In addition, other variable changes such as blood pressure may increase or decrease the value of points on the curve. To demonstrate this effect, Figure 11 is a hypothetical flow rate curve versus arterial P_{O_2} which shows a greater flow for a given P_{O_2} than Figure 8. Using this flow rate data the previous simulations were repeated, and Figure 12 is a plot of system oxygen tension response. In comparison with Figure 10 note that a $\tau_D = 65$ seconds gives almost the same response as a $\tau_D = 85$ seconds in the former simulation.

Additional comments concerning the factor τ_D can be made. Overall, we feel the magnitude of τ_D given here is too large. The degree of mixing of blood prior to the capillary, as discussed at the beginning of this section, could possibly affect the magnitude of τ_D. In addition the delay factor in these experiments may be a variable depending on the value of arterial P_{O_2}. τ_D could be less for deoxygenation and greater for reoxygenation. The location of the "sensor" for increasing the rate of flow of cerebral blood may be near the venous end of capillaries and not near arterial blood. If so, the value of τ_D would be decreased. To answer these questions, additional experimental work is necessary.

The important conclusion is that the proposed time delay in the rate of change of cerebral blood flow can (1) minimize the lowest system P_{O_2}, (2) produce a rapid rate of recovery, and (3) produce the overshoot phenomena found experimentally.

Distributed-Lumped Parameter Analysis. In order to investigate possible longitudinal differences in oxygen response and to approach a distributed parameter analysis, the system of Figure 9 (A) was divided into 10 systems connected in series as shown in

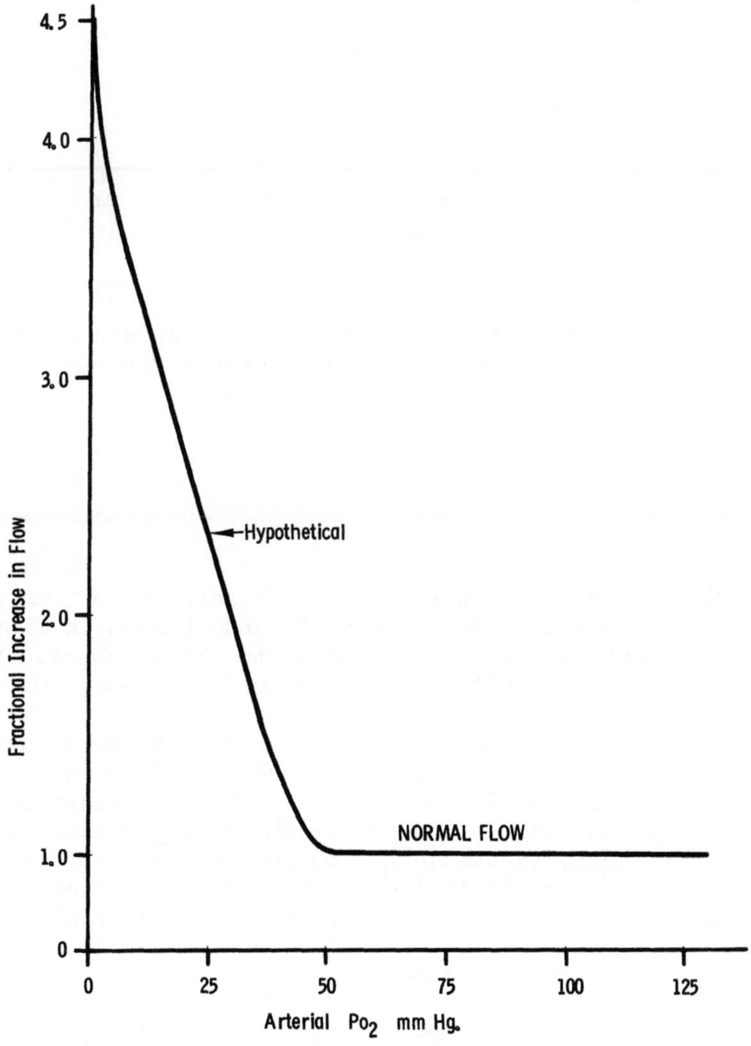

Figure 11. Cerebral Flow Rate as a Function of Arterial Oxygen Tension. Hypothetical Curve.

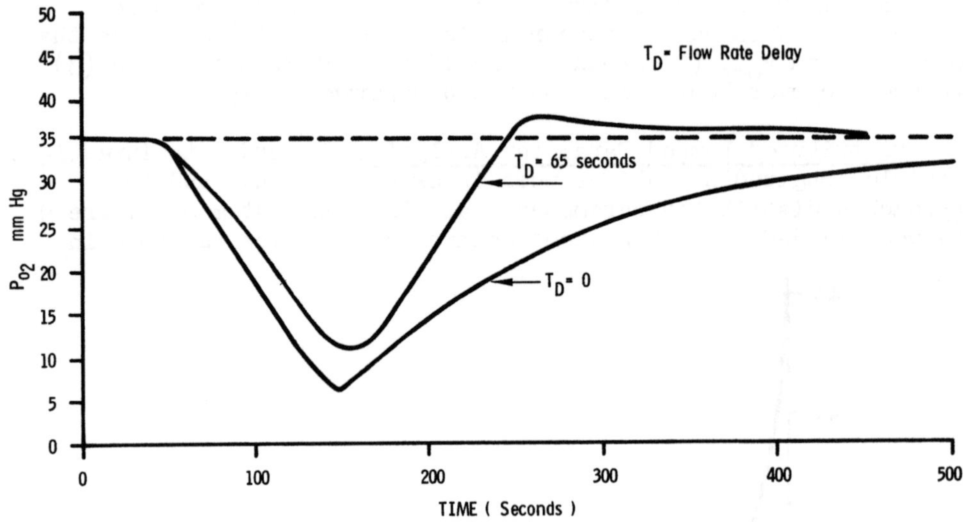

Figure 12. Lumped Parameter: Oxygen Tension Response Curves as a Function of Time. Hypothetical Large Increase Flow Rate Curve was used. Parameters are Flow Rate Time Delay.

Figure 9 (B). Changes in arterial P_{O_2} with respect to time (Figure 1) were used as input data for chamber 1, and changes in flow rate with respect to arterial P_{O_2} were applied to all chambers. Figure 13 is a plot of the oxygen tension response for Chamber 10.

When the flow rate time delay, τ_D, is equal to zero, oxygen tension drops from the normal level to zero in 160 seconds and remains at this condition for 40 seconds. The reoxygenation time is approximately 250 seconds and no overshoot is present. When a flow rate time delay is used, τ_D = 85 seconds, oxygen tension in the system drops to 5 mm Hg in 180 seconds and reoxygenation is accomplished within 60 seconds. An overshoot phase is obtained within 60 seconds, and thereafter oxygen tension peaks and reduces to normal levels as time increases.

A comparison between the results of the lumped and distributed parameter systems yields the same conclusions. The delay in flow rate is essential in minimizing the lowest system P_{O_2}, producing a rapid rate of reoxygenation, and in producing an overshoot phase.

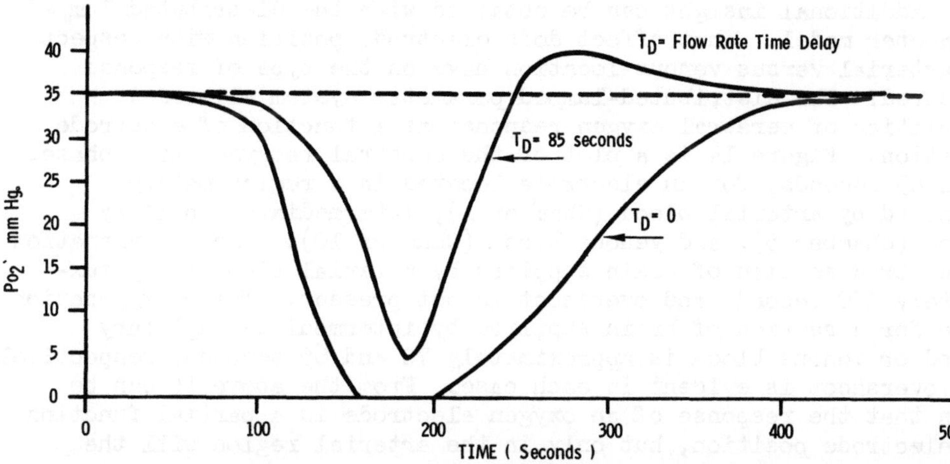

Figure 13. Distributed-Lumped Parameter System: Oxygen Tension Response Curves as a Function of Time (chamber 10).

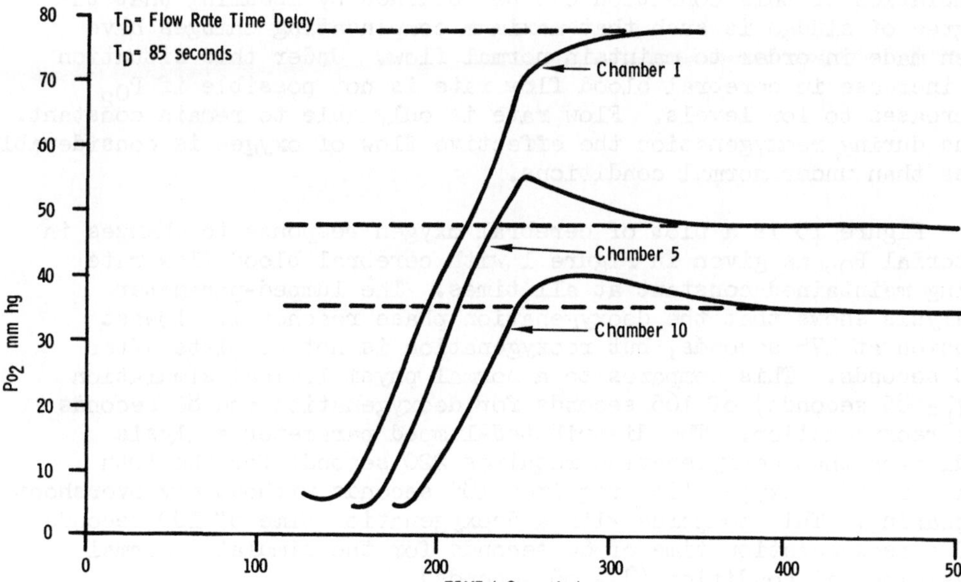

Figure 14. Distributed-Lumped Parameter System: Cerebral Oxygen Response as a Function of Electrode Location.

Additional insight can be obtained with the distributed-lumped parameter model. What effect does electrode position with respect to arterial versus venous location have on the type of response measured? The distributed-lumped parameter system allowed a simulation of cerebral oxygen response as a function of electrode location. Figure 14 is a plot of the cerebral reoxygenation phase, τ_D = 85 seconds, for an electrode located in a region mainly supplied by arterial blood (Chamber 1), intermediate capillary blood (Chamber 5), and venous blood (Chamber 10). The reoxygenation time for a section of brain supplied by arterial blood is approximately 150 seconds and overshoot is not present. The reoxygenation time for a section of brain supplied by intermediate capillary blood or venous blood is approximately 70 and 65 seconds, respectively, and overshoot is evident in each case. From the above it can be seen that the response of an oxygen electrode is a partial function of electrode position, but only in the arterial region will the oxygen response curve have a different shape.

Cerebral Oxygenation Under Conditions of Intravascular Red Cell Aggregation. After red cell aggregation, the shape of experimentally measured cerebral oxygenation response curves were different than those obtained from the same cat under normal conditions. It is well known that red cell aggregation reduces flow in the microcirculation and in severe cases can even stop capillary flow by plugging capillaries, Knisely (8). A theoretical simulation of this condition can be obtained by assuming that the degree of sludge is such that maximum compensating changes have been made in order to maintain normal flow. Under this situation an increase in cerebral blood flow rate is not possible if P_{O_2} decreases to low levels. Flow rate is only able to remain constant. Thus during reoxygenation the effective flow of oxygen is considerably less than under normal conditions.

Figure 15 is a plot of cerebral oxygen response to changes in arterial P_{O_2} as given in Figure 1 with cerebral blood flow rate being maintained constant at all times. The lumped-parameter analysis shows that the deoxygenation phase reaches its lowest tension at 175 seconds, but reoxygenation is not complete after 325 seconds. This compares to a normal physiological simulation (τ_D= 85 seconds) of 165 seconds for deoxygenation and 80 seconds for reoxygenation. The distributed-lumped parameter analysis indicates that deoxygenation requires 220 seconds for the 10th Chamber and reoxygenation requires 185 seconds without any overshoot appearing. This compares with a deoxygenation time of 180 seconds and a reoxygenation time of 60 seconds for the simulated normal physiological condition (τ_D= 85 seconds).

The implication is clear. Red cell aggregation, which is found in over 160 diseases (Knisely (8)) and which acts to diminish

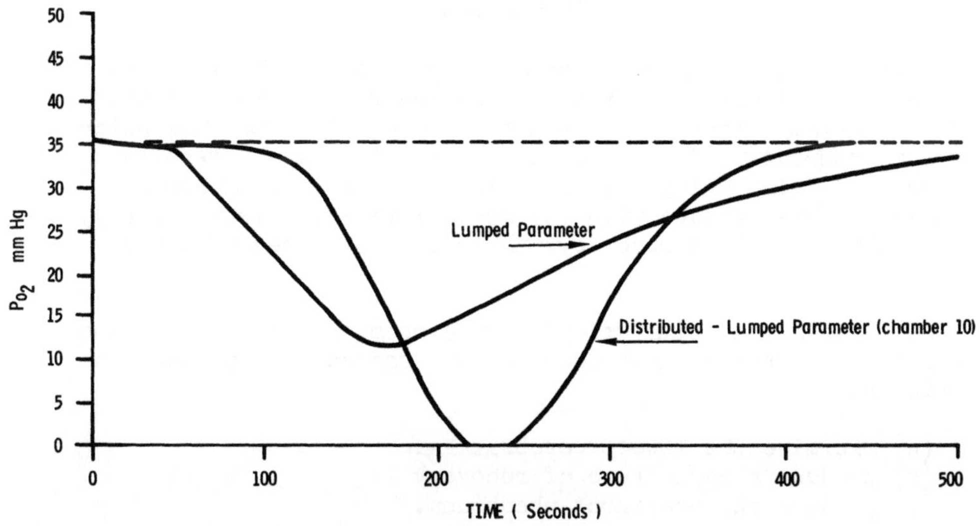

Figure 15. Tissue Oxygen Response as a Function of Time When Flow is Maintained Constant at Normal Effective Flow.

the rate of increase and magnitude of blood flow to brain as compensation for decreased blood oxygenation, seriously decreases the ability of the brain to reoxygenate.

Cerebral Oxygenation Under Conditions of Hemorrhage or Hemorrhage Combined with Sludge. The same discussion given under the above section concerning the effects of sludge can partially explain the experimentally measured large increases in reoxygenation times found after hemorrhage or hemorrhage combined with sludging. However another detrimental factor is present. After hemorrhage, an average 25 percent reduction in hematocrit was observed (created by rapid replacement of water into the blood stream). This lowered red cell concentration would indicate that even if the normal blood flow rate to the brain was able to remain constant, the effective flow of oxygen to cerebral tissue was reduced. An approximate increase in blood flow rate of 25 percent above normal would be necessary to maintain a normal effective flow of oxygen. Consequently, the extended reoxygenation times are to be expected.

CONCLUSIONS

Mathematical analyses of oxygen supply to cerebral tissue combined with experimental results has led to the proposal that there is a time delay in the change of cerebral blood flow which occurs to compensate for decreases in arterial oxygen tension. The proposed "time delay" can explain the overshoot phenomena, found in cerebral tissue with the new, ultramicro electrode (3), as a result of a first order system instead of a second order system.

Summarizing, a "time delay" in the change of cerebral blood flow resulting from a large decrease and consequent increase in arterial P_{O_2} can:

(a) minimize the lowest cerebral P_{O_2}
(b) produce a rapid rate of recovery
(c) produce the overshoot phenomena.

It can also be concluded that the theoretical and/or experimental results indicate that,

(1) The shape as well as magnitude of oxygen tension response from microelectrodes is a function of electrode location.
(2) Intravascular red cell aggregation, hemorrhage, or hemorrhage combined with aggregation appears to adversely affect the ability of the brain to regulate cerebral oxygenation.

In addition it was pointed out that a difference exists between cerebral blood flow rate and the effective flow of oxygen to the brain. An increase in blood flow rate does not always indicate an increase in the rate of flow of oxygen. Effective flow is a function of blood oxygen capacity, the degree of hemoglobin saturation with oxygen, and the rate of flow of blood.

This work was supported in part by United States Public Health Service Grants NB-08228, NB-06957, HE-04176, HE-05340, HE-07114, and FR-5420.

BIBLIOGRAPHY

1. Ananthakrishnan, V., W.N. Gill, and A.J. Barduhn: Laminar Dispersion in Capillaries, A.I.Ch.E. Journal, 11, No. 6, 1063-1072 (1965).

2. Bicher, H.I. and M.H. Knisely: Brain Tissue Reoxygenation Time Demonstrated with the New Untramicro Oxygen Microelectrode, J. App. Physiol. (In press).

3. Bicher, H.I., D.D. Reneau, and M.H. Knisely: The Effect of Intravascular Red Cell Aggregation (Sludge) and Hemorrhage on Blood and Tissue Oxygenation as Determined with a New, Ultramicro Oxygen Electrode, Proceedings of the ISBO (In press).

4. Bruce, G.H., D.W. Peaceman, H.H. Rackford, and J.D. Rice: Trans., A.I.M.E., 198, 79 (1953).

5. Bruley, D.F. and M.H. Knisely: Hybrid Simulation-Oxygen Transport in the Microcirculation, Chem. Engr. Symp. Ser. (In Press).

6. Kety, S.S.: The Cerebral Circulation in "Handbook of Physiology, Neutrophysiology III", ed. by W.F. Hamilton and P. Dow, Chapter 71, 1751-1760, Williams and Wilkins, Baltimore, Md., 1960.

7. Kety, S.S., and C.F. Schmidt: The Effects of Altered Arterial Tensions of Carbon Dioxide and Oxygen on Cerebral Blood Flow and Cerebral Oxygen Consumption of Normal Young Men, J. Clin. Invest., 27, 484-492 (1948).

8. Knisely, M.H.: Intravascular Erythrocyte Aggregation (Blood Sludge), in "Handbook of Physiology, Circulation III," ed. by W.F. Hamilton and P. Dow, Chapter 63, pp. 2249-2292. Williams and Wilkins, Baltimore, 1965.

9. Knisely, M.H., D.D. Reneau and D.F. Bruley: The Development and Use of Equations for Predicting the Limits on the Rates of Oxygen Supply to the Cells of Living Tissues and Organs, Angiology (In Press).

10. Krogh, A.: "The Anatomy and Physiology of Capillaries," Yale University Press, New Haven, Conn., 1922, 1 ed.

11. Opitz, E. and M. Schneider: The Oxygen Supply of the Brain and the Mechanism of Deficiency Effects, Ergebnisse der Physiologie, biologischem Chemie, und experimentallen Pharmakologie, 46, 126-260 (1950).

12. Peaceman, D.W., and H.H. Rachford, Jr.; The Numerical Solution of Parabolic and Elliptic Differential Equations, J. Soc. Indust. Appl. Math., 3, No. 1, 28-41 (1955).

13. Reneau, D.D., D.F. Bruley and M.H. Knisely, A Mathematical Simulation of Oxygen Release, Diffusion and Consumption in the Capillaries and Tissue of the Human Brain, in Chemical Engineering in Medicine and Biology, pp. 135-241, Plenum Press (1967).

14. Reneau, D.D., D.F. Bruley, and M.H. Knisely: A Digital Simulation of Oxygen Transport in Capillary-Tissue Systems (Cerebral Gray Matter), A.I.Ch.E. Journal (In Press).

15. Reneau, D.D., D.F. Bruley and M.H. Knisely: A Computer Simulation Showing Multiple Minute Hypoxic Areas in Cerebral Tissue as a Result of Sludged Blood, 5th Europ. Conf. Microcirculation, Gothenburg, 1968, Bibl. Anat., 10, Karger, Basel/New York (1969).

16. Reneau, D.D., D.F. Bruley and M.H. Knisely: A Computer Simulation for Prediction of Oxygen Limitations in Cerebral Gray Matter, JAAMI (In Press).

17. Reneau, D.D., and M.H. Knisely: A Mathematical Simulation of Oxygen Transport in the Human Brain Under Conditions of Countercurrent Capillary Blood Flow, 62nd Annual Meeting A.I.Ch.E., Gas Exchange in Biological Systems, Symposium, Paper 66C, Washington, D.C. (1969).

18. Reneau, D.D. and M.H. Knisely: A Mathematical Simulation of Glucose Diffusion and Consumption in Brain, Proceedings of the 21st Annual Conference on Engineering in Medicine and Biology, 10, p. 29.2, 1968.

19. Schmidt, C.F., Central Nervous System Circulation, Fluids and Barriers-Introduction, in "Handbook of Physiology, Neurophysiology III", ed. by W.F. Hamilton and P. Dow, Chapter 70, 1745-1750, Williams and Wilkins, Baltimore, Md., 1960.

20. Sokoloff, L. and S.S. Kety: Regulation of Cerebral Circulation, Physiol. Rev., 40, 38-49 (1960).

21. Thews, G.: Oxygen Diffusion in the Brain. A Contribution to the Question of the Oxygen Supply of the Organs, Pflugers Archiv, 271, 197-226 (1960).

BRAIN TISSUE OXYGENATION AS DETERMINED WITH A NEW ULTRAMICRO OXYGEN ELECTRODE

H.I. Bicher, Department of Anatomy, Medical University of South Carolina, Charleston, South Carolina

D.D. Reneau, Department of Chemical Engineering, Louisiana Polytechnic Institute, Ruston, Louisiana

M.H. Knisely, Department of Anatomy, Medical University of South Carolina, Charleston, South Carolina

SUMMARY

A new, 2 micron tip, oxygen microelectrode and a constantly circulated Beckman oxygen gas analyzer were used to measure tissue and blood PO_2 in anesthetized, curarized cats under positive pressure breathing. As a parameter for the ability of the circulation to oxygenate tissue, the "Reoxygenation Time" (defined as the time required to reach the previous PO_2 level after a short period of anoxic anoxia) was determined on blood and brain cortex.

First, it was found that hemorrhage (from 15 to 25 cc/kg) alone or hemorrhage combined with sludging of the blood (by the I.V. administration of high molecular weight Dextran, 1 g/kg) markedly diminished PO_2 levels in blood and tissue.

Further, the reoxygenation time was significantly affected by these procedures. Sludging markedly prolonged the reoxygenation time, an effect counteracted by the use of an anti-adhesive drug breaking up the red cell aggregates. Bleeding prolonged the reoxygenation time up to four times that found in the same animals previous to the bleeding.

INTRODUCTION

The constant delivery of the oxygen needed for cell survival and work in the tissues throughout the body is a vital function of blood circulation. This function is dependent on cardiovascular factors (cardiac output, blood pressure, blood flow, among others)

and hematological factors mainly related to the number and physical behavior of the red cells.

The normal circulation is characterized by rapid flow of blood in the cardiovascular system. All formed elements of blood are separate from one another. Red cells do not stick together, nor do white cells or platelets. White cells and platelets do not stick to normal healthy endothelium. The cells circulate rapidly (12,39) and pass freely along small arterioles, capillaries and venules.

The intravascular agglutination of the formed elements of blood which occurs under certain pathological conditions results in peripheral circulation disturbances that may lead to blood flow stagnation and the plugging of small vessels with cell aggregates(39).

Knisely et al. (39, 42, 43) and Bloch (13) observed long ago the pathological significance of intravascular red cell aggregation (sludge). They described the presence of large erythrocyte masses in the microcirculation of the bulbar conjunctiva of patients suffering from different thrombotic processes, such as myocardial infarction, diabetic retinopathy, and occlusive peripheral vascular disease. Borgstrom et al. (15) demonstrated in the rabbit a direct etiological relation between experimentally induced sludge and venous thrombosis. In previous work (8), we suggested that intravascular red cell aggregation may be a factor in the induction of anoxic myocardial damage. Recently, Reneau et al. (61) developed a mathematical equation which predicted the development of multiple minute areas of anoxia in brain tissue under this condition.

In previous publications we have described a group of synthetic, membrane active drugs that are able to counteract intravascular blood cell aggregation (5, 6, 7, 9). These were called the "anti-adhesive" drugs.

The obvious symptomatology of tissue oxygen deprivation during acute hemorrhage is common in clinical practice. A relation between this phenomenon and a reduction of the oxygen transport ability of the circulating blood can probably be established through diminished red cell count and hematocrit values, as well as during reduced peripheral circulation.

The purpose of this paper is to evaluate the effects of blood sludging, its counteraction by an anti-adhesive drug, and acute hemorrhage on the ability of the circulating blood to carry and deliver enough oxygen to supply brain tissue. In order to measure the tissue PO_2 changes under this experimental situation, a new ultramicro oxygen electrode has been developed. This electrode is easy to produce and calibrate and has the small size and high resolution necessary for oxygen tissue studies. With all of the necessary coatings on it, the external diameter of the electrode is 2μ or less.

EDITORS NOTE

For these Proceedings the original paper has been edited so that only the discussion of the new oxygen electrode has been presented. However the original bibliography is presented here in its entirety.

METHODS

A. Recording of Tissue PO_2 with the Oxygen

Ultramicro Electrode:

1. Construction:

The oxygen cathode basically consists of a very finely drawn (one micron or less in diameter) glass pipette which has a thin film over-coating of platinum. The use of precisely controlled vacuum deposition techniques[1] assures a good bond between platinum and glass and a high degree of reproducibility in the mechanical and electrical properties of the microelectrode.

Each basic pipette (probe) is then coated with two layers of an oxygen impervious resin, Seran F 310[2], leaving an exposed length of tip of approximately 2 microns in diameter. This basic probe could be used either as an "open tip" or an "internal reference" oxygen microelectrode.

(a). "open Tip" oxygen ultra microelectrode.

[1]The basic microelectrodes already pulled and coated with platinum and oxygen impervious Seran F 310 Resin as described in this paper were supplied by Transydine General Company, Ann Arbor, Michigan.

[2]Seran F 310 supplied by Dow Chemicals, Midland, Michigan.

The basic probe can be used, as such, to record relative changes in tissue PO_2, acting as an open electrode. According to Davies and Brink (21) and Cater (17), this type of electrode is useful for continuous measurements and responds very rapidly to changes in tissue oxygen tension. Also, as described by Cater and Silver (18), electrodes of this size are insensitive to "stirring" (rapid movements of fluid near the tip area), consume but small amounts of oxygen themselves, and cause little tissue damage in the area in which they are inserted. Their main disadvantage is that they are subject to "poisoning," particularly by the electrophoretic deposition of protein on the active platinum surface.

To overcome this "poisoning," the tip of our microelectrode was coated with a water carrying, highly adhesive oxygen pervious material, Rhoplex AC35 (Rohm Haas Co., Philadelphia, Pennsylvania). This coating was applied by dipping the tip of the microelectrode in a Rhoplex AC35 emulsion, and then drying the electrode in air for 30 minutes. In some experiments an additional coating with a water repellent substance, Formvar[1], was added. The Formvar coating was not as extensive as the Rhoplex AC35, so that electrical contact can be established between the emulsion and the surrounding tissue, as described by Silver (66). In recent experiments, a coating of collodion proved to be as effective as the Rhoplex-Formvar combination.

(b). "internal Reference" oxygen ultra microelectrode.

In order to measure absolute PO_2 values in tissue, the oxygen cathode should be connected with the reference silver-silver chloride electrode through a medium of fixed electric resistance. Clark (19) and Silver (66) described a micro and an ultra micro oxygen electrode that are subject to absolute calibration; in both cases, both the oxygen cathode and the Ag-AgCl reference anode were separated from the tissue by an oxygen pervious electrical insulating membrane. In order to reproduce this feature, the electrode construction method was modified as follows (see figs. 1a and 1b).

The basic probe (glass pipette, covered by Pt. and coated with Seran F 310) was filled with a 2M solution of KCl, using a standard filling procedure (5 minutes boiling in absolute alcohol under vacuum, then 3 successive immersions in 2M KCl solutions, 2 of 10 minutes each and the last for 24 hours). To prevent peeling of the Seran F 310 during boiling, a coating of "Dekophane"[2] was

[1]Supplied by E. C. Fulham, Schenectedy, New York.

[2]Supplied by Crystal Essence Corp., Plainfield, New Jersey.

applied in the areas where the microelectrode was attached to the mounting board.

After the micropipette was filled with 2M KCl, an Ag-AgCl electrode was introduced into its upper end in contact with the KCl. Electrical contact was checked through both anode and cathode. The upper end was then sealed with Dekophane. After coating the tip with Rhoplex AC35 and Formvar as described above, the electrodes were ready for calibration.

2. Electronic Circuitry:

The reference electrode was made of clean pure silver wire (1-2mm o.d.). AgCl was deposited on the wire by placing it in 0.1 N HCl and making the wire 1/5V positive with respect to a platinum cathode. In use, when the oxygen cathode was used as "open tip," the reference electrode was inserted into the periostium of the skull or introduced into the KCl filled shaft of the "internal reference" type electrode.

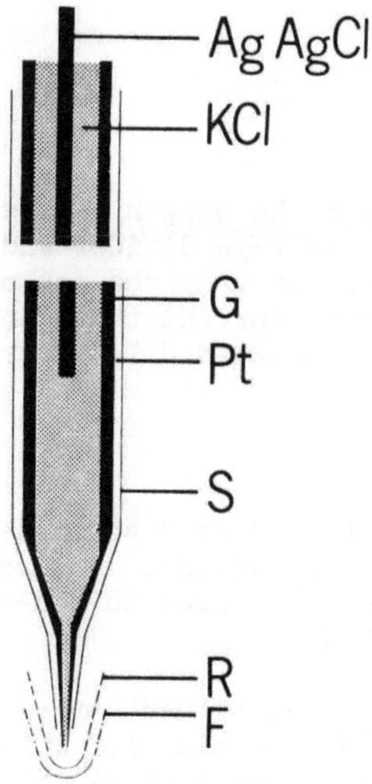

Figure 1a. AgAgCl: Silver silver-chloride reference electrode. KCl: Saturated potassium chloride solution. G: Glass pipette. Pt: Platinum layer covering glass. S: Seran F-310 resin coating. R: Rhoplex AC35 coating. F: Formvar coating.

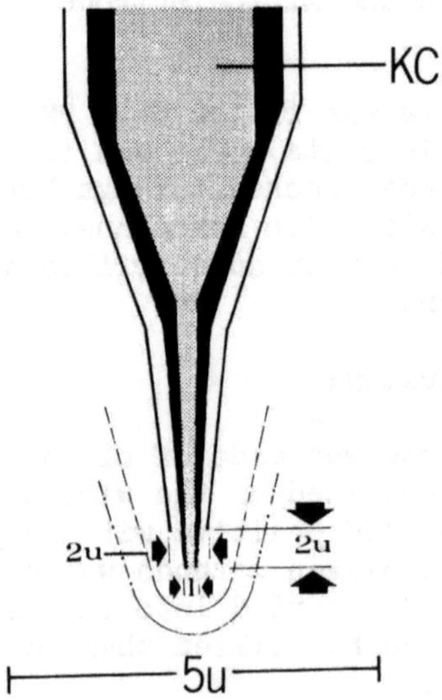

Figure 1b. Detail of microelectrode tip.

The e.m.f. to activate the oxygen cathode was applied from a cadmium battery through a voltage divider and indicated by a voltmeter. The current output from the cathode was measured with a Kiethley Model 416 picoammeter (Kiethley Instruments, Cleveland, Ohio) and recorded using a Beckman SII Dynograph (Beckman Instruments, Palo Alto, California).

3. Calibration:

Electrodes were calibrated as described by Silver (65) in saline solutions of known PO_2 values. Polarographic curves were also recorded placing the electrodes in non-respiring brain tissue, as described by Whalen (80).

The electrodes were "conditioned" by placing them in saline solution and applying 0.8V potential for 2 hours. After this treatment they were usually very stable. Stirring of the solution by a magnetic rotor did not change the oxygen current. The current reading at zero oxygen tension was very low (residual current) and the response of the microelectrode to changes of oxygen tension was very rapid (95 per cent response time of the order of 0.5 second).

The current voltage polarogram (fig. 2) showed a quasi plateau between 0.3 and 7V. The plateau, however, was not absolute and small differences could be detected at the extreme values. This type of plateau can be expected when using "open tip" electrodes (21). In these experiments a polarizing current of 0.6V was used. The relation between current output and oxygen tension was linear, the current per mm/Hg being of the order of magnitude of 0.6×10^{-11}A.

B. Recording of Blood PO_2:

Blood PO_2 was continuously recorded using a Beckman Mod 160 gas analyzer, through which the animal's blood was continuously circulated. This gas analyzer is equipped with an oxygen microelectrode (supplied by Beckman Co.) mounted in a special receptacle, in a constant temperature bath able to keep the environmental temperature at 37°C. The electrode has an equilibration time constant (τ) of 30 seconds. The output of the gas analyzer amplifier was fed into a D.C. channel of a Beckman SII dynograph recorder.

Constant blood circulation through the gas analyzer receptacle was established by cannulation of the femoral artery using oxygen impervious tygon tubing. A cannula from the artery conducted the blood to the gas analyzer inlet. After passing through a small chamber in which it came in close contact with the PO_2 electrode membrane, a second cannula returned the blood to the same femoral artery at a more distal point. The gas analyzer was placed as close as possible to the animal. The flow through this devise was excellent all the time that the blood pressure of the animal was above 80 mm Hg. The blood was prevented from clotting by the administration of Heparin (500 units /Kg i.v.).

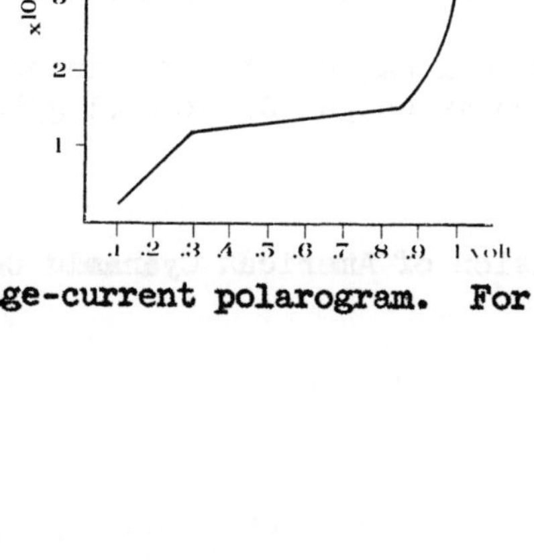

Figure 2. The voltage-current polarogram. For explanation, see text.

C. Blood Pressure and ECG:

Blood pressure was measured from a cannulated femoral artery using a Statham transducer. ECG (lead I) was recorded using subcutaneous stainless steel electrodes. All functions were simultaneously registered on a Beckman dynograph SII recorder.

D. Microcirculation Observations:

Microscopic observations of the bulbar conjunctiva microcirculation were made on 12 cats using a Zeiss stereoscopic microscope, at a magnification of X100.

E. Operation:

Observations were made on 40 cats. The animals were anesthetized with nembutal (60 mg/kg) given intra-peritoneally. Venous, arterial and tracheal cannulations were performed. The head of the animal was fixed in a David Kopf model 1730 stereotaxic instrument. The skin over the skull was incised and the periosteum retracted. A trephine hole was made in the skull over the frontal cortex and the dura removed. The exposed surface of the brain was covered with liquid paraffin. The electrode was fixed in a micromanipulator (David Kopf, Model 1760) and lowered onto the surface of the brain. Micro movements of the electrode within brain tissue were achieved using a David Kopf 1707B hydraulic microdrive. Measurements were made only in the upper 3 mm of brain cortex.

In order to measure the brain "reoxygenation time" (described later), the animal was curarized using Flaxedil[1] (1 mg/kg i.v.) and maintained under artificial respiration using a Harvard Apparatus (Dover, Massachusetts) Model 607 respiration pump. This procedure was necessary in order to exclude respiratory compensation when low PO_2 gases were administered instead of air. All gas mixtures were given to the animal through the inspiratory inlet of the respiration pump. All experiments were performed in an electrically shielded cage.

Blood samples for the sedimentation rate and hematocrit determinations were obtained from the cannulated artery.

Intravascular red cell aggregation was induced by the i.v. injection of high molecular weight Dextran (1 g/kg).

[1]Supplied by Division of American Cyanamid Co., Danbury, Connecticut.

De-aggregation of the red cells was achieved by the i.v. injection of substance "86," 200 mg/kg.

Controlled acute bleeding was carried out through the cannulated femoral artery. Roughly 20% of the calculated blood volume of the animal was slowly bled out. The slow bleeding allowed blood pressure levels to remain constant.

RESULTS

When the microelectrode was lowered into the brain substance, brain oxygen levels were found to vary sharply from point to point within very short distances. The electrode was usually moved in 10 μ steps, and this small position change sometimes resulted in 50 per cent alteration in the observed oxygen tension. The direction of the change was absolutely unpredictable. On the other hand, when the microelectrode was locked in position at a certain depth, and no additional maneuvers performed on the animal, the recorded PO_2 was remarkably constant over long periods of time, sometimes hours. The actual PO_2 levels recorded ranged from 95 down to but 1 or 2 mm Hg.

Definition of the Cortical "Reoxygenation Time"

When the respiratory gas administered to the animal was switched from air to 100 per cent nitrogen, a continuous decrease in the cortical oxygen level was recorded by the microelectrode (fig. 3, upper trace). After giving pure nitrogen for 60 seconds, the animal was again respired with air. A dual response was then

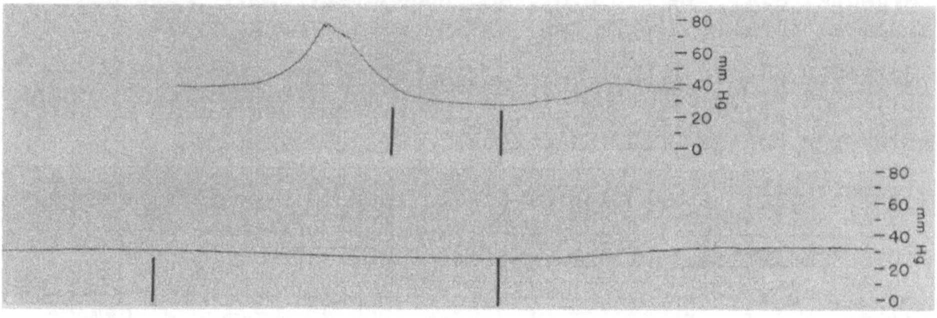

Figure 3. Recorded brain PO_2 during and after nitrogen administration. Upper trace, normal conditions; lower, after sludging. Bottom markers indicate "reoxygenation time". Paper speed 1 mm/sec (each medium size square equals 10 seconds.) Recording from right to left. For explanation, see text.

observed. First, there was a fast rise in PO_2 that restored the pre-nitrogen cortical levels in approximately 60 seconds, then a second, slow "overshoot phase" that continued for 1 to 2 minutes, during which the cortical PO_2 levels stayed above the original (pre-nitrogen) values. After the second phase, the PO_2 record returned to its original base line. These changes were very reproducible. We called the time (from the point of maximal depression under nitrogen) required to reach the original PO_2 level the "Reoxygenation Time."

This work was supported in part by United States Public Health Service Grants HE-04176, HE-05340, HE-07114, NB-08228, and FR-5420.

REFERENCES

1. Baker, S.B. de C.: The role of blood "sludging" in the renal ischaemic changes of experimental shock. J. Pathol. Bacteriol., 83:117, 1962.

2. Bergentz, S.E.: Studies on the genesis of posttraumatic fat embolism. Acta Chir. Scand., Suppl. 282, 1961.

3. Bergentz, S.E., Gelin, L.E., and Rudenstam, C.M.: Fats and thrombus formation. Thromb. Diath. Haemorrhag., 5:474, 1961.

4. Bergentz, S.E., Gelin, L.E., Rudenstam, C.M., and Zederfeldt, B.: Indications for the use of low viscous dextran in surgery. Acta Chir. Scand., 122:343, 1961.

5. Bicher, H.I.: Prevention of sludge induced myocardial damage by an anti-adhesive drug. Proc. V. Int. Conferance on Microcirculation, Gothenberg, (abstract), 1968.

6. Bicher, H.I.: Prevention of sludge induced myocardial damage by an anti-adhesive drug. Bibl. anat., in press, 1969.

7. Bicher, H.I.: 2-4 Butyl Methyl Ketolactone, a prototype anti-thrombotic drug preventing platelet and red cell aggregation. Circulation, in press, 1969.

8. Bicher, H. I., and Beemer, A. M.: Induction of ischemic myocardial damage by red cell aggregation (sludge) in the rabbit. J. Atherosclerosis Research, 7:409, 1967.

9. Bicher, H. I., and Licht, A.: K+ and the platelet membrane. Relationship to platelet aggregation and its prevention by anti-adhesive drugs. Proceedings of the International Union of Physiological Sciences, VII: 44, 1968.

10. Bigelow, W.G., Heimbeker, R..O., and Harrison, C.R.: Intravascular agglutination (sludged blood) vascular stasis and sedimentation rate of the blood in trauma. A.M.A. Arch. Surg., 59:667, 1949.

11. Bloch, E.H.: The IN VIVO microscopic intravascular and vascular reactions in acute poliomyelitis. Am. J. Med. Sci., 266:24, 1953.

12. Bloch, E.H.: Microscopic observations of the circulating blood in the bulbar conjunctiva in man in health and disesse. Ergeb. Anat. Entwicklungsgesch, 35:1, 1956.

13. Bloch, E.H.: Visual changes in the living microvascular system in man and experimental animals as they are related to thrombosis and embolism. Angiology, 10:6, 1959.

14. Bor, N.M., Stewart, P.A., Cho, Y.W., Kirkardesler, F., and Eren, S.: Capillary circulation. Characteristics during treatment of shock by means of artificial heart-lung procedures in the dog. J.A.M., 179:417, 1962.

15. Borgstrom, S., Gelin, L.E., and Zederfeldt, B.: The formation of vein thrombi following tissue injury. Acta Chir. Scand., Suppl., 247, 1959.

16. Branemark, P.I.: Experimental investigation of microcirculation in bone marrow. Angiology, 12:293, 1961.

17. Cater, D.: The measurement of PO_2 in tissue. In: Oxygen in Animal Tissue. New York, Macmillan Company, 1964, p. 239.

18. Cater, D., and Silver, I.: Electrodes and microelectrodes used in biology. In: Reference Electrode, J. Janz, ed. New York, Academic press 1961, chapter 11.

19. Clark, L.: Monitor and control of blood and tissue oxygen tension. Trans. Am. Soc. Artif. Internal Organs 2:41, 1956.

20. Cullen, C. F., and Swank, R.L.: Intravascular aggregation and adhesiveness of the blood elements associated with

alimentary lipemia and injections of large molecular substance. Effect on blood brain barrier. Circulation, 9:335, 1954.

21. Davies, P., and Brink, F.: Microelectrodes for measuring local oxygen tension in animal tissue. Rev. Sci. Instr., 13:524, 1942.

22. DeBono, A., and Gozetopoulos, N.: Experimental pulmonary embolism due to red cell aggregation. Thorax, 19:244, 1964.

23. DeNicola, P.: Aggregation of erythrocytes in small vessels due to occupational conditions. II. The role of the fatty meals as pre-disposing agents. Bibl. anat., 1:278, 1961.

24. Ditzel, J.: Angioscopic and hemodynamic changes in the smaller blood vessels in diabetes mellitus and their relationship to aging. Circulation, 14:386, 1956.

25. Ditzel, J., and Moinat, P.: Changes in serum proteins, lipoproteins, and protein - bound carbohydrates in relation to pathologic alterations in the microcirculation of diabetic subjects. J. Lab. Clin. Med., 54:843, 1959.

26. Fajers, C.M., and Gelin, L. E.: Kidney, liver and heart-damages from trauma and from induced intravascular aggregation of blood cells. Acta Pathol. Microbiol. Scand., 46:97, 1959.

27. Fatt, I.: An ultramicroelectrode. J. Applied Physiology, 19:326, 1964.

28. Finney, R.P., Jr.: Microscopic observation of locally burned blood and vessel walls. Am. J. Physiol., 163:711, 1950.

29. Gelin, L.E.: Studies in anemia of injury. Acta Chir. Scand. Suppl., 210, 1956.

30. Gelin, L.E.: Intravascular aggregation and capillary flow. Acta Chir. Scand., 113:463, 1957.

31. Gelin, L. E.: The significance of intravascular aggregation following injury. Bull. Soc. Intern. Chir., 18:4, 1959.

32. Girard-Mangin (Mme) and Henri, V.: Etude du phenomen d'agglutination. I. Agglutination des globules rouges par l'hydrate ferrique colloidal. Compt. Rend. Soc. Biol., 56:866, 1904.

33. Grehn, J.: Optische hilfmittel sur horizontalbeobachtung von settling - Erscheinungen. Bibl. anat., 1:260, 1961.

34. Harders, H.: Neue beobachtungen zum diatfehler. Verhandl. Deut. Gas. Inn. Med., 62:499, 1956.

35. Harders, H.: Intravasale erythrocyte aggregation. Thesis, Universitat Hamburg, Hamburg, Germany, 1956.

36. Harders, H.: Intravascular agglutination of erythrocytes. Schweiz. Med. Wschr., 87:11, 1957.

37. Heimbecker, R.O., and Bigelow, W.G.: Intravascular agglutination of erythrocytes (sludged blood) and traumatic shock. Surgery, 28:461, 1950.

38. Knisely, J. H.: Postburn circulatory pathologic physiology. In: Research in Burns, ed. by C. P. Artz. Am. Inst. Biol. Sci., Washington, D.C., 1962, p. 51.

39. Knisely, M.H.: Intravascular erythrocyte aggregation (blood sludge). In: Handbook of Physiology, Circulation III, ed. by W.F. Hamilton and P. Dow. Williams and Wilkins, Baltimore, 1965, chapter 63, p. 2249.

40. Knisely, M. H., Bloch, E.H., Brooks, F., and Warner, L.: Microscopic observations of the circulating blood of nine healthy normal horses, all of which had unagglutinated circulating blood cells and high IN VITRO erythrocyte sedimentation rates. Am. J. Med. Sci., 219:249, 1950.

41. Knisely, M. H., Bloch, E. H., Elliot, T. S., and Warner, L.: Sludged blood. Science, 106:431, 1947.

42. Knisely, M. H., Elliot, T. S., and Bloch, E.H.: Sludged blood in traumatic shock, Part 1 (Microscopic observations of the precipitation and agglutination of blood flowing through vessels in crushed tissue). Arch. Surg., 51:220, 1945.

43. Knisely, M.H., and Lockard, I.: By what mechanisms could known types of blood vascular and circulatory pathology directly cause or contribute to known types of signs and symptoms now related to the internal ear? Trans. Am. Otolo. Soc., 43:143, 1955.

44. Knisely, M. H., Warner, L., and Harding, F.: Antemortem settling. Microscopic observations and analyses of the settling of agglutinated blood-cell masses to the lower sides of vessels during life. Angiology, 11:535, 1960.

45. Knisely, W.H.: Microscopic observations of the architecture and behaviour of the smaller blood vessels in living lungs of mammals and related observations in other vessels of experimental animals and human patients. Ph.D. Thesis, Medical College of South Carolina, Charleston, S.C., 1954.

46. Lange, K., Weiner, D., and Boyd, L. J.: Frostbite, physiology, pathology and therapy. New Eng. J. Med., 237:383, 1947.

47. Lee, R.E.: Anatomical and physiological aspects of the capillary bed in the bulbar conjunctiva of man in health and disease. Angiology, 6:369, 1955.

48. Lister, J.: On the early stages of inflammation. Phil. Trans. Roy. Soc., London, 148:645, 1858.

49. Lofstrom, B.: Intravascular aggregation and oxygen consumption. Aggregation of red blood cells produced by high molecular weight dextran or by hypothermia. Acta Anaesthesiol. Scand., 3:41, 1959.

50. Lofstrom, B., and Zederfeldt, B.: Intravascular aggregation. Acta. Chir. Scand., 113, 272, 1957.

51. Long, D.M., Sanchez, L., Vargo, R. I., and Lillehei, C.W.; The use of low molecular weight dextran and serum albumin as plasma expanders in extracorporeal circulation. Surgery, 50:12, 1961.

52. Madow, B.P.: Use of antimalarial drugs as "de-sludging" agents in vascular disease processes. Preliminary report. J.A.M.A., 172:1630, 1960.

53. Meyer, J.S.: Localized changes in properties of the blood and effects of anti-coagulant drugs in experimental cerebral infarction. New Eng. J. Med., 258:151, 1958.

54. Miller, W.R., Willson, J.T., and Elliot, T.S.: Blood changes in the dog following trauma. Angiology, 10:375, 1959.

55. Mustard, J.F., Rowsell, H.C., Lotzf, and Hegardt, B.: The effect of adenine nuclotides on thrombus formation, platelet count and blood coagulation, Exp. Mol. Path., 5, 43, 1966.

56. O'Brien, J.R.: Platelet aggregation. Part II, some results from a new method of study. J. Clin. Path. 15, 452, 1962.

57. Odell, L.G. Aragon, and Pottinger, R.: Relationship between erythrocyte sedimentation rate, sludge blood and plasma proteins during pregnancy. Am. J. Obstet. Gynecol.,54;596,1947.

58. Opitz, E., and Schneider, M.: Ueber sauerstoffversorgung des gehirns und den mechanismus von mangelwirkungen. Ergeb. Physiol. Biol. Chem. Exptl. Pharmakol., 46:126, 1950.

59. Patterson, M.B., and Freyberg, R.H.: Blood sludge in patients with arthritis. In: Rheumatic Diseases (Proc. 7th Intern. Congr. Rheumatic Diseases). Saunders, Philadelphia, 1952, p. 345.

60. Piovella, C.: Sill'osservazione microscopica IN VIVO della circulazione renal nella rana durante le diverse fasi di un blocco renal circolazorio. Atti Congr. Soc. Ital. Cardiologia, 18e Trieste., 1956, p. 70.

61. Reneau, D.D., Bruley, D.F., and Knisely, M.H.: A mathematical simulation of oxygen release, diffusion, and consumption in the capillaries and tissue of the human brain. In: Chemical Engineering, in Medicine, and Biology, ed. by D. Hershey. Plenum press, New York, 1967, p. 135.

62. Reneau, D.D., Bruley, D.F., and Knisely, M.H.: Computer simulation showing multiple minute hypoxic areas in cerebral tissue as a result of sludged blood. Proc. International conference on microcirculation, Gothenburg, Sweden. Bibl. anat., in press.

63. Roizine, L., Abell, R.C., and Winn, J.: Preliminary studies of sludged blood in multiple sclerosis. Neurology, 3:250, 1953.

64. Saunders, E.A., and Knisely, M.H.: Living mesenteric terminal arterioles before and immediately after embolization. Measurement and comparison of internal diameter ; a contribution to the biophysics of disease. A.M.A. Arch. Pathol., 58:309, 1954.

65. Silver, I.: A simple microelectrode for measuring PO_2 in gas or fluid. Med. Electron. Biol. Engng., 1:547, 1963,

66. Silver, I.: Some observations on the cerebral cortex with an ultramicro, membrane covered oxygen electrode. Med. Electron. Biol. Engng., 3:377, 1965.

67. Silver, I.: The measurement of oxygen tension in tissue. In.: Oxygen Measurements in Blood and Tissue and Their Significance. Little, Brown and Co., Boston, 1968, p. 135.

68. Swank, R.L.: Suspension stability of the blood after injection of dextran. J. Appl. Physiol., 12:125, 1958.

69. Swank, R.L.: Effects of fat feeding on circulation. Bibl. anat., 1:287, 1961.

70. Swank, R.L.: Adhesiveness of platelets and leukocytes during acute exanguination. Amer. J. Physiol., 202:261, 1962.

71. Swank, R.L., and Cullen, C.F.: Circulatory changes in the hamsters cheek pouch associated with alimentary lipemia. Proc. Soc. Exptl. Biol. Med., 82:381, 1953.

72. Swank, R.L., and Nakamura, H.: Oxygen availability in brain tissues after lipid meals. Am. J. Physiol., 198:217, 1960.

73. Taub, W., and Cais, M.: The synthesis of ketolactones with potential pharmacodynamic properties. The Bull. Res. Counc. Israel., 11A, 18, 1962.

74. Thorsen, G., and Hint, H.: Aggregation, sedimentation, and intravascular sludging of erythrocytes. Interrelation between suspension stability and colloids in suspension fluid. Acta. Chir. Scand., Suppl., 154, 1950.

75. Thruansky, K., and Csetry, C.: A keringo ver intravascularis aggregatiojanak kozvetlen mikroskops vizsgalata macska retinan. Szemeszet, 92:81, 1955.

76. Walton, R.P., Richardson, J.A., and Brodie, O.J.: Cardiovascular actions of decaborane. J. Pharmacol. Exptl. Therap., 114:367, 1955.

77. Waltz, A.G., and Meyer, J.S.: Effects of changes in composition of plasma on pial blood flow. II. High molecular weight substance, blood constituents and toxicity. Neurology, 9:815, 1959.

78. Warner, L.: Some kinds of blood sludge plug kidney glomeruli. Anat. Record, 103:178, 1949.

79. Weille, F.L., Martinez, D.E., Gargano, S.R., and Irwin, J.W.: An experimental study of the small blood vessels of the spiral ligament and stria vascularis of living guinea pigs during anaphylaxis. Laryngoscope, 64:656, 1954.

80. Whalen, W., Riley, J., and Nair, P.: A microelectrode for measuring intracellular PO_2. J. Applied Physiology, 23:798, 1967.

81. Williams, A.V., Higginbotham, A.C., Melvin, M., and Knisely, M.H.: Increased blood cell agglutination following ingestion of fat. A factor contributing to cardiac ischemia, coronary insufficiency and anginal pain. Angiology, 8:29, 1957.

82. Wright, H.P.: The adhesiveness of blood platelets in normal subjects with varying concentrations of anticoagulants. J. Path. Bact., 53, 255, 1941.

83. Zerderfeldt, B.: Studies on wound healing and trauma with special reference to intravascular aggregation erythrocytes. Acta Chir. Scand. 7, Suppl. 224, p. 394, 1957.

84. Zilliacus, H.: Venous thrombosis and intravascular aggregation of erythrocytes. Acta Chir. Scandinav., 99:40, 1950.

85. Zilliacus, H., and Arajarvi, T.: Intravascular red cell aggregation in erythroblastosis foetalis. Acta Paediat., 41:267, 1952.

A DISTRIBUTED PARAMETER MATHEMATICAL ANALYSIS OF OXYGEN EXCHANGE FROM MATERNAL TO FETAL BLOOD IN THE HUMAN PLACENTA

Eric J. Guilbeau and Daniel D. Reneau
Department of Chemical Engineering
Louisiana Polytechnic Institute
Ruston, Louisiana

Melvin H. Knisely
Department of Anatomy
Medical University of South Carolina
Charleston, South Carolina

Editors Note: For these Proceedings, only a summary of the original paper is presented. However the original bibliography is presented in its entirety. For further information on this important topic, it is suggested that the authors be contacted directly.

SUMMARY

Based on the present state of knowledge concerning the anatomy and physiology of the placenta, a theoretical analysis of the transport of oxygen from maternal to fetal blood is presented in this paper. Distributed parameter mathematical models are derived which describe the convection, diffusion and reaction processes involved in the oxygen exchange. Steady state simulations of placental oxygen transport are obtained by solving the mathematical models for two cases. Case 1 is a simulation of oxygen transport as if it occurred from maternal blood flowing concurrently to fetal blood in a straight capillary. Case 2 simulated the process as if it occurred in a multivillous flow pattern with concurrent flow characteristic of each exchange unit.

Based on the guidelines and limitations set forth in this paper, the following conclusions can be drawn.

Case 1 Conclusions:

1. An increase above normal in fetal flow rate through the placenta increases oxygen transport from maternal to fetal blood but decreases fetal oxygen partial pressure leaving the placenta; a decrease below normal of fetal oxygen flow rate through the placenta increases fetal oxygen partial pressure leaving the exchange section of the placenta but decreases the quantity of oxygen transported from maternal to fetal blood.

2. An increase above normal of maternal flow rate through the placenta increases both the quantity of oxygen transported from maternal to fetal blood and the partial pressure of oxygen in fetal blood leaving the placenta; a decrease below normal of maternal flow rate through the placenta decreases both the quantity of oxygen transported from maternal to fetal blood and the partial pressure of oxygen in fetal blood leaving the placenta.

3. Fetal anemia can lead to large increases in the oxygen partial pressure of fetal blood leaving the placenta but results in a decrease in the total quantity of oxygen exchanged. Maternal anemia results in both decreased oxygen exchange and decreased oxygen partial pressure in fetal blood leaving the placenta.

4. Significant reductions in the diffusivities and/or solubilities of maternal and fetal blood can restrict fetal blood oxygenation.

5. Decreases in the arterial oxygen partial pressure of maternal blood from the normal value of 95.0 mm Hg to approximately 50 mm Hg only slightly affects fetal blood oxygenation. Further reductions in maternal oxygen partial pressure significantly decreases the degree of fetal oxygenation. It appears that significant increases in maternal oxygen partial pressure above 95.0 mm Hg only slightly increases fetal blood oxygenation. (Additional study is required to firm up this latter conclusion.)

Case 2 Conclusions:

1. The multivillous flow arrangement is more efficient than a straight concurrent flow arrangement.

2. The effect of the villi tissue (and its oxygen consumption), of dimensions described in this paper, on the transport of oxygen from maternal to fetal blood is negligible. This conclusion applies to normal conditions as outlined in this paper.

One overall conclusion needs to be stated. The results of these theoretical analyses appear to indicate that the placenta is operating at an optimum level under normal conditions.

BIBLIOGRAPHY

1. Aherne, W., Durrill, M.S., J. Path. Bact. 91:123-39 (1966).

2. Arts, N.F., Amer. J. Obstet. Gynec. 82: 147-66 (1961)

3. Barnes, A.C., "Intra-Uterine Development," Lea and Febiger, Philadelphia, 1968.

4. Bartels, H., Moll, W., Pflugers Archiv. 280: 165-77 (1964).

5. Bartels, H., Mill, W., Amer. J. Obstet. Gynec. 84: 1714-30 (1962).

6. Blum, J.J., Amer. J. Physiology 198: 991-998 (1960).

7. Borell, U., Ingemar, F., Ohlson, L., Wiguist, N., Amer. J. Obstet. Gynec. 93: 44-57 (1965).

8. Boyd, J.D., Hamilton, W.J., J. Obstet. Gynec. Brit. Comm. 74: 161-226 (1967).

9. Bruce, G.H., Peaceman, D.W., Rackford, H.H., Rice, J.D., J. Petr. Tech. 5: 70-92 (1953).

10. Bruley, D.F., Prados, J.W., Amer. Inst. Chem. Engr. J. 10: 612 (1964).

11. Bruns, P.D., Cooper, W.E., Droze, V.E., Amer. J. Obstet. Gynec. 82: 1079-89 (1961).

12. Corner, W.C., Clin. Obstet. Gynec. 6: 17-25 (1963).

13. Crank, J., Nicolson, P., Proc. Cambridge Phil Soc. 43: 50 (1947).

14. Crawford, J.M., Amer. J. Obstet. Gynec. 84: 1543-67 (1962).

15. Dawes, G.S., "Foetal and Neonatal Physiology," Year Book Medical Publisher, Chicago, (1968).

16. DeHaven, J.C., Deland, E.C, Assali, N.S., Manson, W., AD 459313. Available CFSTI 27 pages (1965).

17. Fuchs, F., Spackman, T., Assali, N.S., Amer. J. Obstet. Gynec. 86: 226-33 (1963).

18. Gonzales, L.O., Spencer, E.H., Chem. Engr. Sci. 18: 753-66 (1963).

19. Guilbeau, E.J., M.S. Thesis (1968).

20. Kirschbaum, T.H., DeHaven, J.C., Shapiro, N., Assali, N.S., Amer. J. Obstet. Gynec. 96 (1): 741-59 (1966).

21. Lapidus, L., "Digital Computation for Chemical Engineers," McGraw-Hill, New York, 1962.

22. Martin, C.B., Anesthesiology 26: 447-59 (1956).

23. Page, E.W., Clin. Obstet. Gynec. 3: 279-85 (1960).

24. Pappenheimer, J.R., Physiological Review 33: 387-423 (1953).

25. Ramsey, E.M., Amer. J. Obstet. Gynec. 84: 1649-63 (1962).

26. Ramsey, E.M., Corner, G.W., Donner, M.W., Amer. J. Obstet. Gynec. 86: 213-25 (1963).

27. Reneau, D.D., Bruley, D.F., Knisely, M.H., "In Chem. Engr. in Med. and Biol.," ed. by Hershey, D., Plenum, New York, 1967.

28. Stenger, V., Eitzman, D., Anderson, T., De Padua, C., Gessner, 1., Prystowsky, H., Amer. J. Obstet. Gynec. 88: 45-57 (1964).

29. Thews, G., Pflugers Archiv. 271: 197-226 (1960) (German).

30. Thews, G., Fischer, W.W., Vogel, H.R., Pflugers Archiv. 286: 257-74 (1965). (German).

31. Von Rosenberg, D.V., Durill, P.L., Spencer, E.H., unpublished paper, Esso Research Laboratories, Baton Rouge, Louisiana (1960).

32. Wells, R.E., Jr., New England J. Med. 270: 832-39 (1964).

33. Wulf, H., Amer J. Obstet. Gynec. 88 (1): 38-44 (1964).

This work was supported in part by United States Public Health Service Grants NB - 08228, HE - 04176, HE - 07114, HE - 05340, United States Navy Contract ONR: 441

THE EFFECTS OF INERT GASES, PRESERVATIVES AND STORAGE ON OXYGEN TRANSPORT IN BLOOD

Wolfgang H. Koch and Daniel Hershey

Department of Chemical and Nuclear Engineering

University of Cincinnati, Cincinnati, Ohio 45221

INTRODUCTION

Research in the area of membrane transport has concentrated on the determination of ionic fluxes across biological barriers (1, 2, 11). Erythrocyte membranes are frequent subjects of such studies since they are readily available (1, 11). The transport of oxygen through erythrocyte membranes in particular has received much attention recently. Some previous oxygenation studies have relied on a lumped model approach in which all resistances to oxygen transport are combined into one parameter (8, 12, 13, 14). Differences between predicted and experimental results for these models were explained in terms of possible membrane resistance. This study provides a method for the direct determination of membrane resistance from experimental data (4).

A study of the effects of inert gases and of preservatives and storage on the apparent erythrocyte membrane resistance to oxygen transport was also undertaken.

EXPERIMENTAL PROCEDURE

Figure 1 shows the experimental apparatus. A sample of sheep's blood is introduced into the reaction vessel and initially oxygenated by bubbling through it pure oxygen of known mixtures of oxygen with helium or nitrogen until the blood is saturated with respect to oxygen. When the partial pressure of oxygen in the plasma phase reaches a maximum steady state value, the sample is at equilibrium and the oxygen concentration inside the red cells

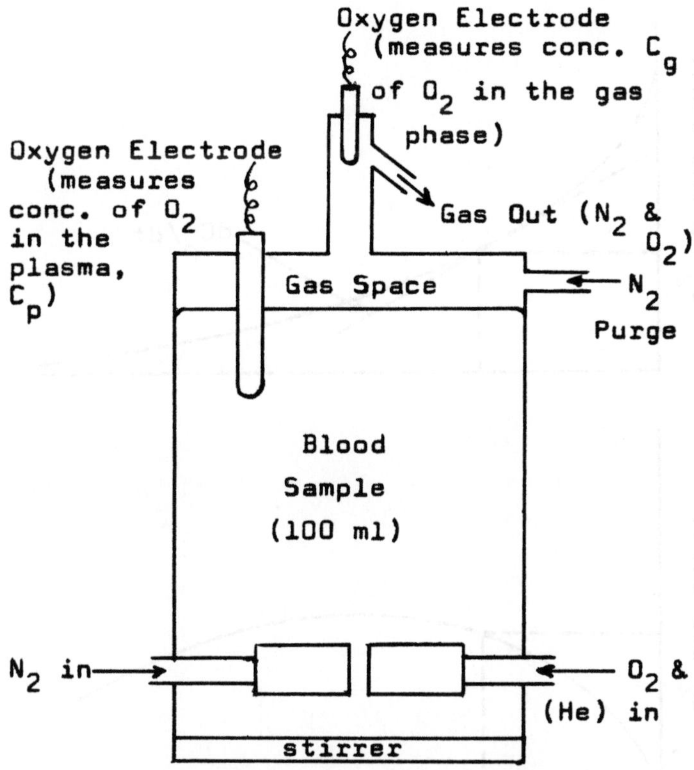

Figure 1. Deoxygenation Apparatus.

can be determined from solubility data. After the blood has been saturated, the gas space above the sample is quickly purged with nitrogen in order to sweep out residual oxygen. The blood sample is then deoxygenated by bubbling nitrogen through it. The oxygen partial pressures in the plasma and the gas phases are monitored continuously during deoxygenation. Typical concentration versus time curves are shown in figure 2.

Blood samples were studied with two preservatives and for different times of storage. This was done by adding either 15 ml Acid-Citrate-Dextross (ACD-A) or 14 ml Citrate-Phosphate-Dextrose (CPD-5) to 100 ml samples of blood (7, 6). The solutions were stored at 3°C and effective membrane resistances were determined over a four-week period.

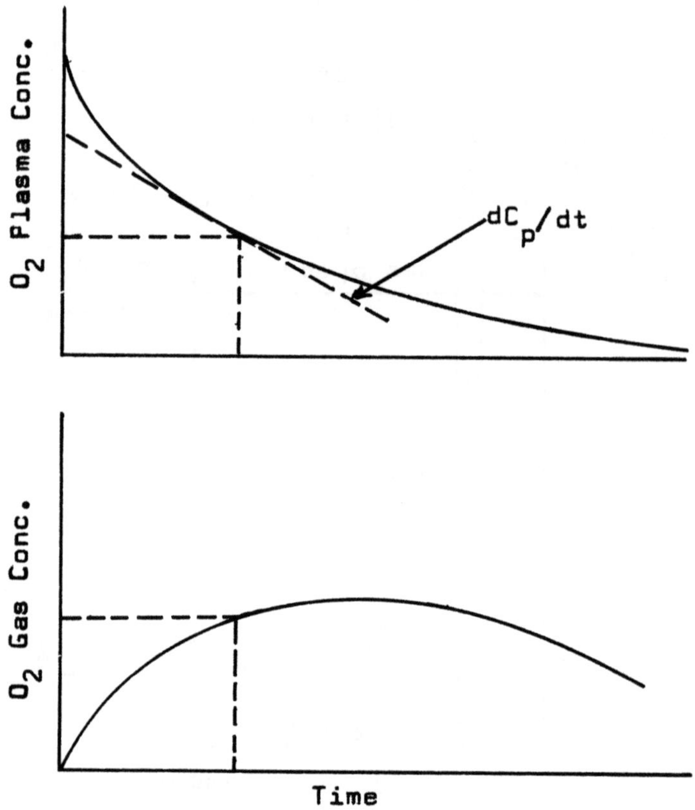

Figure 2. Sample Data Curves.

DETERMINATION OF THE EFFECTIVE MEMBRANE RESISTANCE BY MATERIAL BALANCES

Oxygen transport across the red cell membrane will be defined by:

$$N(t) = hA\ (C_i - C_p) \tag{1}$$

or

$$N(t) = A\ (C_i - C_p)/R \tag{2}$$

where

$N(t)$ = molar flow rate of oxygen out of red cells
$(C_i - C_p)$ = oxygen concentration driving force

C_i = oxygen concentration inside red blood cell membrane } Shown in Figure 3
C_p = oxygen concentration of plasma }
h = apparent mass transport coefficient
R = apparent red cell membrane resistance to oxygen transport
A = transport area (surface area of red blood cells in sample)

A total oxygen balance on the plasma will give:

$$\text{Input} = \text{Output} - \text{Rate of Depletion} \tag{3}$$

where

Input = $N(t)$ = molar flow rate of oxygen from red cells into the plasma
Output = vC_g = oxygen transported from plasma into gas phase above the plasma
v = flow rate of nitrogen for deoxygenation process
C_g = oxygen concentration of gas phase
Rate of Depletion = $V_p dC_p/dt$ = rate of change in plasma oxygen content
V_p = volume of plasma
C_p = oxygen concentration of plasma
t = time

Equation 3 may be rewritten as:

$$N(t) = vC_g - V_p dC_p/dt \tag{4}$$

Combining equations 2 and 4:

$$R = \frac{A\ (C_i - C_p)}{vC_g\ \ -V_p dC_p/dt} \tag{5}$$

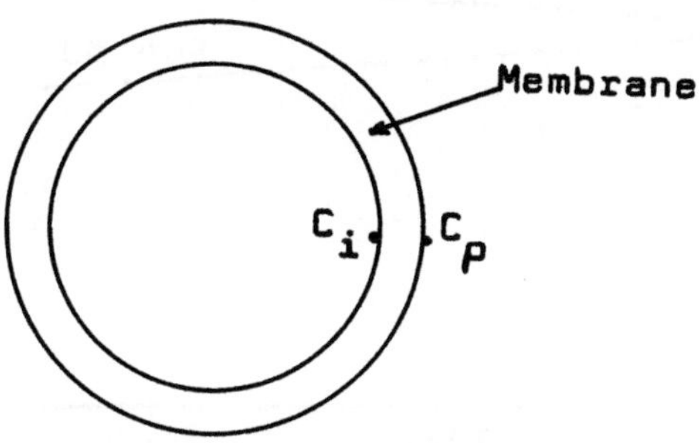

Figure 3. The Red Blood Cell.

All parameters of equation 5 are known as a function of time except the oxygen concentration inside the red cells, C_i. Equation 5 will be used to calculate the red cell membrane resistance, R, from experimental data at various times (using the initial value for C_i). A plot of R versus time, extrapolated to the initial time where C_i is known exactly will then give the desired value for the apparent membrane resistance. Typical curves are shown in Figure 4. It should be noted that curves for different samples converge to one value which is the apparent resistance under the given conditions.

RESULTS AND DISCUSSION

Effects of Inert Gas Atmospheres

Maio and Neville (9) have observed that tissue oxygen consumption drops in the presence of inert gases while the oxygen consumption of tissue homogenates is uneffected by inert gases. Their conclusion was that cellular membranes interact with inert gases to prevent efficient oxygen transport.

Assuming a simple diffusion process in which gas transport occurs through pores in the membrane by random molecular motion, it is possible that the membrane surface may be partially covered by inert gas molecules, thus reducing the number of diffusion sites available for oxygen and increasing the apparent resistance. Figure 5 shows our experimental curves of apparent resistances for helium-oxygen and nitrogen-oxygen. The curves indicate a trend

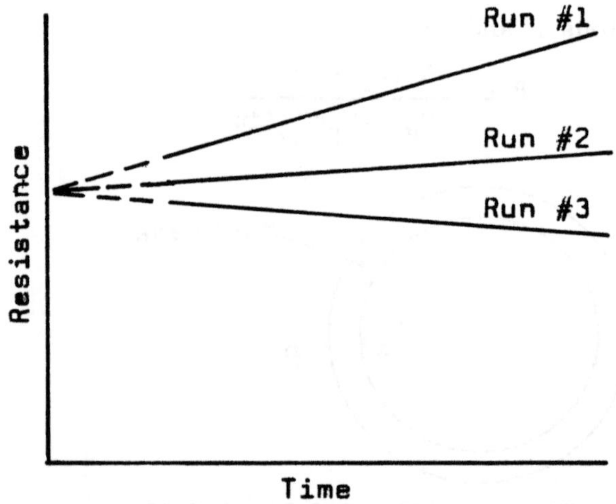

Figure 4. Resistance vs. Time Curves.

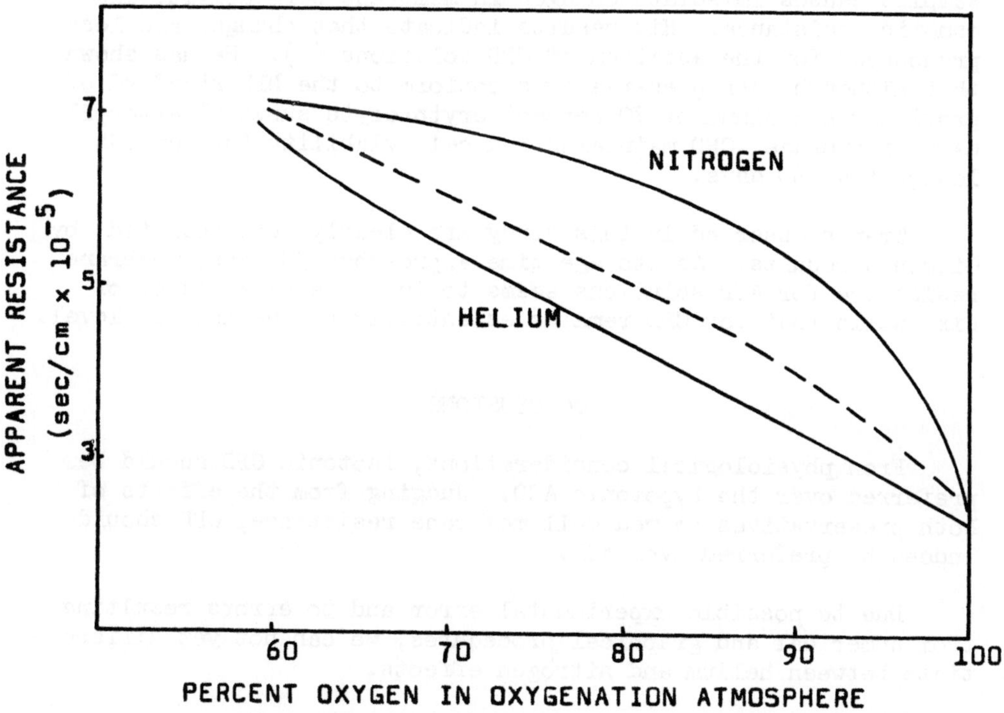

Figure 5. Apparent Membrane Resistances for Inert Gas Atmospheres.

of increased membrane resistance with increased amounts of inert gas. Results of our study indicate that the presence of 40 percent inert gases in the oxygenation atmosphere increases apparent membrane resistance threefold.

Effects of Preservatives and Storage

Figure 6 shows the effects of duration of storage on apparent membrane resistance for the addition of ACD and CPD solutions. Compared to the control sample without additives, resistance initially doubles for CPD and increases initially by about a factor of three for ACD. For the next 15 days, there appears to be no significant change in the two curves. After that time, resistance increases rapidly for ACD and seems to remain the same for CPD.

Solutions of CPD are isotonic while those of ACD are hypotonic. Both are acidic. Red blood cells subjected to these preservatives experience considerable shock due to the change in enviromental conditions. Gibson (6) has observed that addition of ACD to blood samples causes immediate changes in membrane permeability and osmotic resistance. His results indicate that changes are less pronounced for the addition of CPD solutions (5). He has shown that although both preservatives conform to the NIH standard of providing a minimum of 70 percent erythrocyte survival after 21 days of storage, CPD maintains red cell viability for periods longer than 30 days.

Trends observed in this study are clearly substantiated by Gibson's results. As storage time approaches 30 days, membrane resistance for ACD solutions seems to increase by a factor of six, while that for CPD remains essentially at the initial level.

CONCLUSIONS

From physiological considerations, isotonic CPD should be preferred over the hypotonic ACD. Judging from the effects of both preservatives on red cell membrane resistance, CPD should indeed be preferred over ACD.

Due to possible experimental error and to errors resulting from numerical and graphical procedures, we can not yet differentiate between helium and nitrogen effects.

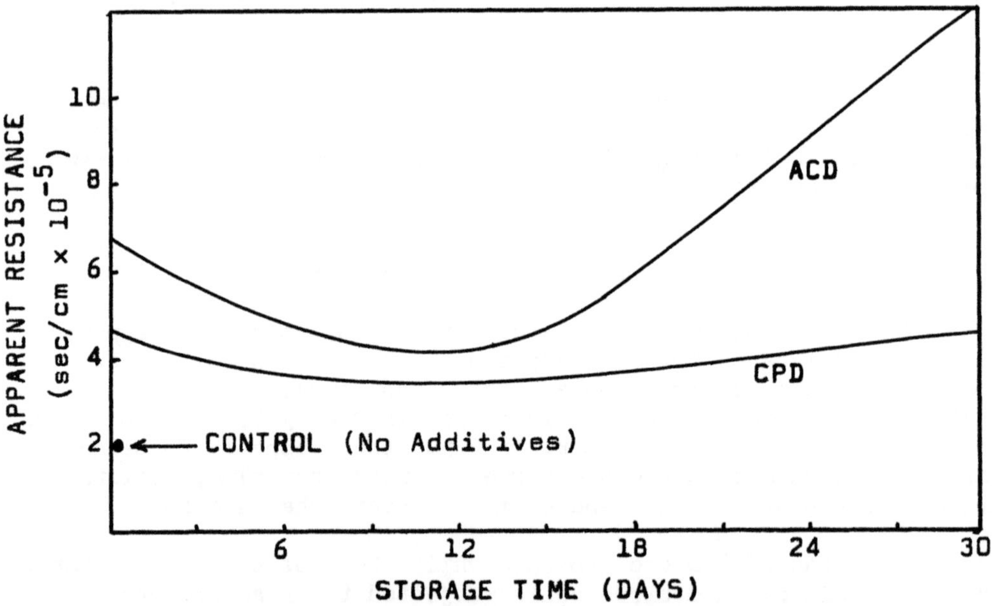

Figure 6. Effects of Storage Time and of Preservative on Apparent Membrane Resistance.

REFERENCES

1. Bishop, C., D.M. Surgenor, eds., The Red Blood Cell, Academic Press, New York, 1964.

2. Dowben, R.M., ed., Biological Membranes, Little, Brown and Co., Boston, 1969.

3. Fenn, W.O., "Inert Gases", Physiology in the Space Enviroment, Vol. II: Respiration, National Academy of Sciences, National Research Council, Washington, 1967.

4. Franco, R.S., "The Red Cell Membrane," Ph. D. Thesis, 1970.

5. Gibson, 2nd, J.G., S.B. Rees, T.J. McManus, W.A. Scheitlin, "A Citrate-Phosphate-Dextrose Solution for the Preservation of Human Blood", American Journal of Clinical Pathology, 28: 569-78, 1957.

6. ________, "Approaches to Red Cell Preservation in the Liquid State," in The Red Blood Cell, C. Bishop and D.M. Surgenor, eds., Academic Press, N.Y., 1964.

7. Handbook of Clinical Laboratory Data, The Chemical Rubber Co., Cleveland, 1968.

8. Kreuzer, F., W.Z. Yahr, "Influence of Red Cell Membrane on Diffusion of Oxygen", Journal of Applied Physiology, 15: 1123-29, 1960.

9. Maio, D.A., J.R. Neville, "The Effect of Chemically Inert Gases on Oxygen Consumption in Living Tissue" USAF School of Aero space Medicine, 1966.

10. Roth, E.M., "Inert Gas", in Compendium of Human Responses to the Aero space Enviroment, National Aeronautics and Space Administration, Washington, 1968.

11. Sirs, J.A., "The Egress of Oxygen from Sheep Erythrocyte" Biochimica et Biophysica Acta, 112: 538-49, 1966.

12. ________, "The Egress of Oxygen from Human HbO_2 in Solution and in the Erythrocyte", Journal of Physiology (London) 189: 461-473

13. Stein, W.D., The Movement of Molecules Across Cell Membranes, Academic Press, New York, 1967

14. Thews, G., "Ein Verfahren zur Bestimmung des O_2-Diffusionskoeffizienten, der O_2-Leitfaehigkeit und des O_2-Loeslichkeits koeffizienten im Gehirngewebe" Plüger Archiv der Gesamten Physiologie, 271: 227-44, 1960

THE CONTROL OF HEMOGLOBIN FUNCTION IN BLOOD STORED FOR TRANSFUSION PURPOSES

R. Ben Dawson

Blood Transfusion Division, US Army Medical Research

Laboratory, Fort Knox, Kentucky 40121

INTRODUCTION

Hemoglobin in the intact red cell has a lower oxygen affinity than hemoglobin in free solution. The main reason for this difference is 2, 3-diphosphoglycerate (2, 3-DPG), but this was not obvious for many years after 2, 3-DPG was found in unusually high concentrations in mammalian red cells (Greenwald, 1925) and in human red cells (Rapoport, 1941). In other cells 2, 3-DPG concentrations are very low. Because the association of hemoglobin with 2, 3-DPG has been established only recently, a brief review of important studies is given in the following paragraph.

Tiselius electrophoresis of red cell hemolysates revealed a slow-moving hemoglobin boundary (designated as component B), which decreased during blood storage in acid-citrate-dextrose (ACD). Further, experimental conditions known to increase red cell organic phosphate, caused relative increases in component B (Berry and Chanutin, 1957, 1958). After it was found that red cell 2, 3-DPG decreased rapidly during storage of blood in ACD (Bartlett and Barnet, 1960), component B was shown to be a complex of hemoglobin and organic phosphates, mainly 2, 3-DPG (Sugita and Chanutin, 1963; Chanutin and Curnish, 1964, 1965a). Chanutin then reported that 2, 3-DPG was capable of decreasing the affinity of hemoglobin for oxygen (1966). The low oxygen affinity of hemoglobin in the red blood cell and the high concentration of 2, 3-DPG could now be related. The effect of 2, 3-DPG on the oxyhemoglobin dissociation curve was confirmed (Chanutin and Curnish, 1967; Benesch and Benesch, 1967). Finally, the specific role of this compound as a cofactor for hemoglobin function has been indicated in studies which show strong binding of 2, 3-DPG to both oxyhemoglobin and deoxyhemoglobin, using two standard methods; equilibrium dialysis (Chanutin and Hermann, 1969) and ultracentrifugation (de Verdier and Garby, 1969), and two molecular filtration techniques (Luque, Diederich and Grisolia, 1969).

Before the role of 2, 3-DPG had been elucidated it was known that the oxygen affinity of blood increased during storage in ACD, as shown by a significant shift to the left of the oxyhemoglobin dissociation curve (Valtis and Kennedy, 1953, 1954; Gullbring and Strom, 1956). This impairment of hemoglobin function which develops during the first few days of storage was seen to persist throughout the storage period. It was suggested that the use of stored blood might impair the normal release of oxygen to the tissues until the metabolism of the recipient had corrected the defect. As an example, Valtis and Kennedy calculated that a transfusion of three units (pints) of seven day old ACD blood, which was sufficient to raise the hematocrit from 35 to 55%, actually decreased the oxygen delivered to the tissues for several hours after it was given. In most of their studies, the shift to the left in the recipients oxyhemoglobin dissociation curve, resulting from the transfusion, was still present 24 hours later. Valeri and Hirsch (1969) have shown that regeneration of 2, 3-DPG after transfusion of similarly stored blood is only 50% complete at 24 hours and that normal concentrations are not reached for several days. It is generally believed that this increased oxygen affinity of stored blood, seen as a shift to the left of the oxyhemoglobin dissociation curve, may be critical for a patient if survival depends on the immediate re-establishment of normal oxygen transport to the tissues.

The basic preservative solutions for storing blood at 4°C which are used in transfusion practices in the United States are ACD and citrate-phosphate-dextrose (CPD). Improvements in these have been directed toward maintaining the adequate ATP levels necessary for viability during storage. The human red cell depends entirely on anaerobic glycolysis for ATP (Fig. 1).

After extensive work on several compounds, adenine was shown to be most effective in concentrations that are small enough to be non-toxic to a recipient (Nakao, 1962; Simon, et al., 1962; de Verdier, 1964). No signs of toxicity were seen in humans who received the amount of adenine which would be contained in 15 transfusion units (Shields, 1967). Human studies have now shown that blood collected into ACD or CPD, containing adenine gives the

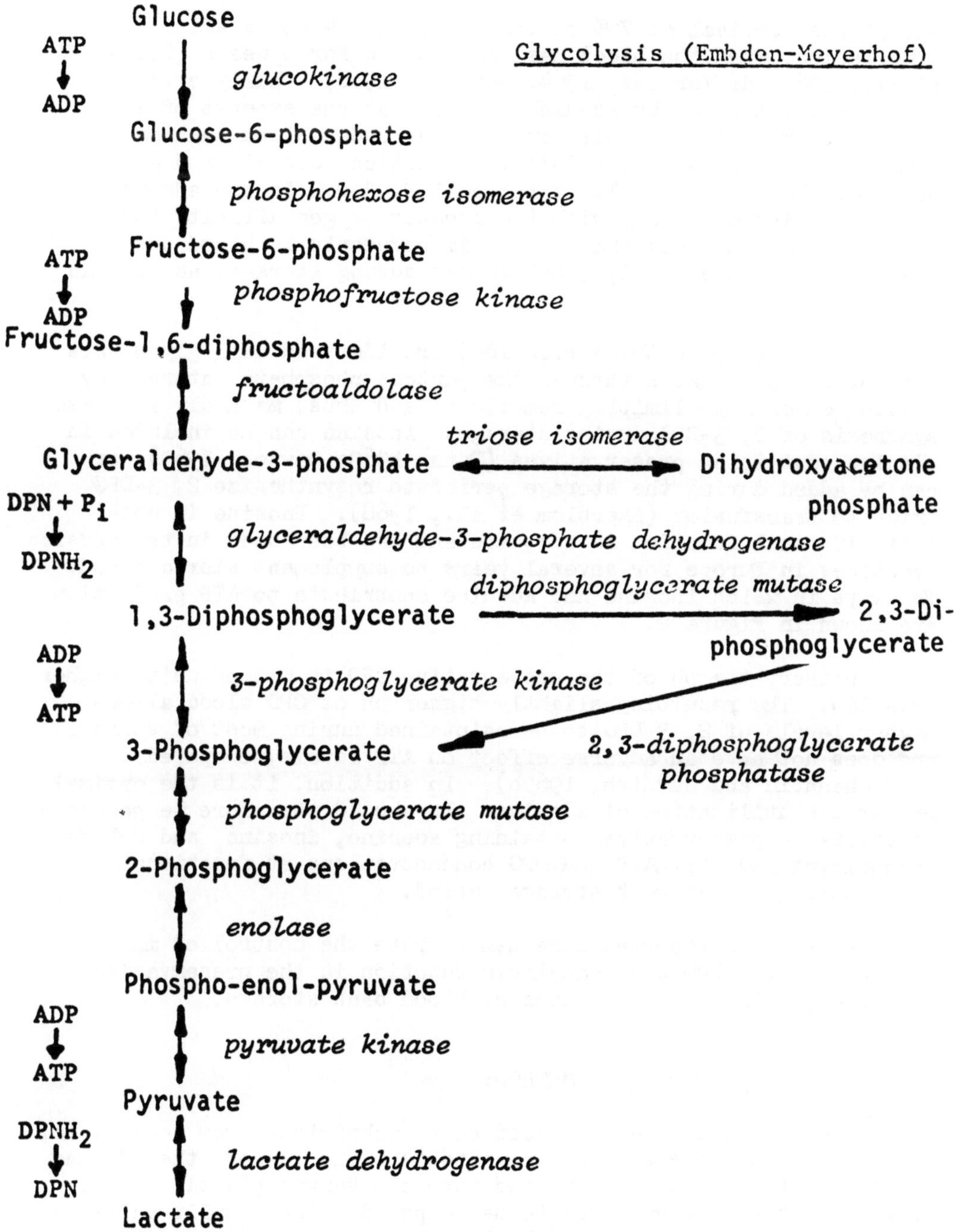

Fig. 1 The Metabolic Pathways in the Human Red Cell Which Provide ATP for Viability and 2, 3-DPG for Function. Two moles of lactate are produced from one mole of glucose. If the concentration of 2, 3-DPG remains constant, two moles of ATP are produced.

acceptable survival of 70% of the red cells 24 hours after transfusion, when the blood has been stored for 5 weeks (Simon et al., 1962; de Verdier, 1964; Shields, 1968). Early evidence that ATP maintenance by adenine occurred at the expense of 2, 3-DPG was confirmed by comparative studies under blood banking conditions of units, split during collection into ACD and ACD-adenine. The lower levels of red cell 2, 3-DPG in the adenine blood correlated closely with the greater oxygen affinity (Bunn, 1969). It seems clear that future improvements in blood preservation must maintain 2, 3-DPG levels during storage, as well as ATP.

Inosine (hypoxanthine riboside) provides red cell glycolysis with triose phosphates through the pentose phosphate pathway, bypassing major rate-limiting reactions; and thus, markedly increases synthesis of 2, 3-DPG during storage. Inosine can be included in adenine containing preservatives (Bunn, 1969; Dawson, 1969), or can be added during the storage period to resynthesize 2, 3-DPG prior to transfusion (Akerblom et al., 1968). Inosine is non-toxic if renal function is adequate and has been used in transfusion practices in Europe for several years to supplement stored blood. The ways in which inosine and adenine contribute to ATP production are shown in Figure 2.

Further, the pH of the preservative CPD is 0.5 pH units higher than ACD. The resulting slightly higher pH of CPD blood allows higher levels of 2, 3-DPG to be maintained during most of storage and does not have an adverse effect on ATP levels, compared to ACD (Chanutin and Curnish, 1965b). In addition, it is the optimal pH for the utilization of adenine. It should therefore be possible to utilize a preservative containing adenine, inosine, and CPD for maintaining red cell ATP and DPG concentrations at functional level during a five week storage period.

The studies reported here demonstrate the controlled maintenance of 2, 3-DPG and hemoglobin function in the preservatives discussed, using the conditions of blood bank storage.

METHODS

Blood was collected by blood bank technicians from young male, non-smoking volunteers as whole units (450 ml), except that during collection the blood was diverted into two Fenwal plastic bags*, containing the preservatives to be compared. These split-units were stored at 4°C in a blood bank refrigerator. Aliquots were

*Fenwal Laboratories, Morton Grove, Illinois.

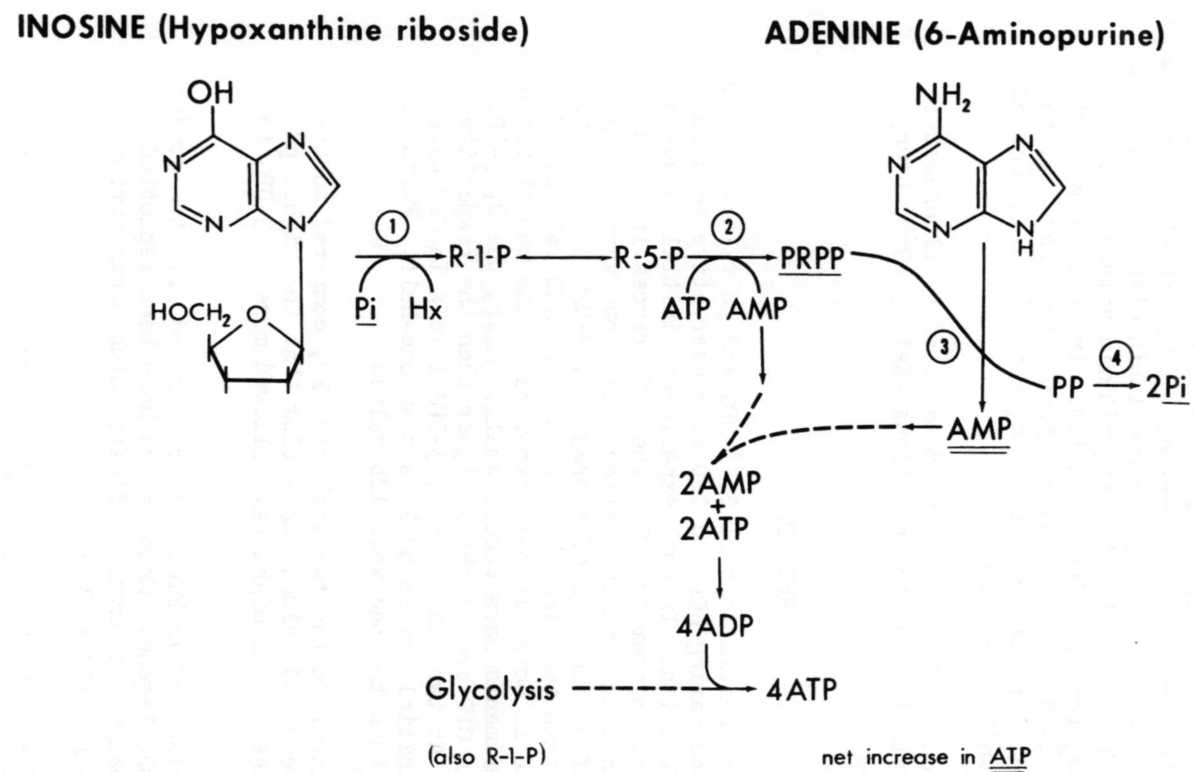

Fig. 2 Inosine and Adenine Incorporation into Red Cell Metabolism. Both compounds penetrate the membrane easily. The ribose phosphate resulting from reaction one can supply energy via glycolysis (see text), or it can be phosphorylated to phosphoribosyl pyrophosphate (PRPP), reaction 2. PRPP will combine with adenine, reaction 3, to synthesize a new molecule of AMP (adenosine monophosphate). The AMP is used as shown to effect a net increase in ATP.

removed at appropriate intervals for study. Anaerobic and asceptic techniques were used during collection, storage, and sampling.

Oxygen dissociation curves were done on whole blood using the Van Slyke manometric method as described by Bunn et al. (1968, 1969). In each case the sigmoid-shaped, oxyhemoglobin dissociation curve was redrawn as a linear plot using logarithmic values of per cent saturation and pO_2 from the Hill equation. In these studies 5 to 7 points were used to assure unambigious values for p_{50}, the pO_2 at which hemoglobin is 50% saturated, and "n" which denotes heme-heme on subunit interaction.

Concentrations of 2, 3-DPG were determined by fluorometric assays of enzymatically coupled reactions (Keitt, 1966; Lowry, 1964).

RESULTS

In table 1 concentrations of 2, 3-DPG and pH values are given for two studies comparing ACD and CPD as preservative solutions for hemoglobin function. In both experiments initial values of 2, 3-DPG and pH are higher in CPD. The differences in values between experiments one and two reflect the known variability among donors. After one week of storage, 2, 3-DPG concentrations in CPD are more than two times higher than the ACD values. This marked difference is noted in both experiments and the differences between the experiments have become small. Also, the 2, 3-DPG concentrations in CPD are slightly higher than the respective initial values. At two weeks the 2, 3-DPG levels in CPD have decreased from initial levels by less than one-half. Further, they are higher than the one week ACD values.

The pH is lower in the two week old CPD, compared to the respective one week ACD value, suggesting that the higher pH earlier in storage in CPD might have allowed more 2, 3-DPG to persist.

The data presented in Figure 3 are derived from two studies referred to in the legend. Mean values have been replotted to summarize the changes in oxygen affinity which occur during storage in these preservatives.

The effects of inosine on the hemoglobin function of CPD-adenine blood are given in Table 2, Values for p_{50} and "n" are shown for two units which were split to compare the effect of inosine when present at collection with its effect when added late in the storage period. Addition of inosine on day 25 results in higher p_{50} values for at least one week (through day 32), compared

TABLE 1

2, 3-DPG (uM/ml RBD) CONCENTRATIONS AND pH VALUES

(MEASURED AT 37°C) DURING STORAGE AT 4°C IN ACD AND CPD*

	Days of Storage at 4°C					
Experiment	0 - 1		6 - 8		13 - 15	
ACD (acid-citrate-dextrose)	DPG	pH	DPG	pH	DPG	pH
1	4.33	7.05	2.10	6.98	0.34	6.81
2	3.79	6.89	1.75	6.80	0.08	6.77
Mean Values	4.06	6.97	1.93	6.89	0.21	6.79
CPD (citrate-phosphate-dextrose)						
1	4.50	7.17	4.73	7.05	2.41	6.97
2	3.82	7.25	4.16	7.06	1.82	6.97
Mean Values	4.16	7.21	4.45	7.06	2.12	6.97

*Concentrations of 2, 3-DPG given for each experiment represent average values of duplicate analyses. Mean values for the two experiments are given for convenience. The pH determinations were made under the arterial blood condition of nearly complete oxygen saturation.

TABLE 2

OXYGEN AFFINITY OF HEMOGLOBIN IN BLOOD STORED AT 4° WITH INOSINE. VALUES OF "n" FOR EACH p_{50} ARE GIVEN IN PARENTHESIS

Experiment	Days of Storage at 4°C					
CPD-adenine-inosine	0	7	14	21	28	34
1	25.0(2.4)	24.0(2.6)	21.6(2.4)	17.1(2.5)	17.5(2.5)	16.6(2.5)
2	25.7(2.9)	24.0(2.8)	21.9(2.4)	17.2(2.7)	16.0(3.0)	15.4(2.7)
Average p_{50}	25.4	24.0	21.8	17.2	16.8	16.0
CPD-adenine; inosine added at day 25	25	26	32	39	40	
1	18.4(2.5)		19.3(2.7)		17.0(2.5)	
2		21.4(2.3)	17.7(2.3)	17.0(2.5)		
Average p_{50}			18.5			

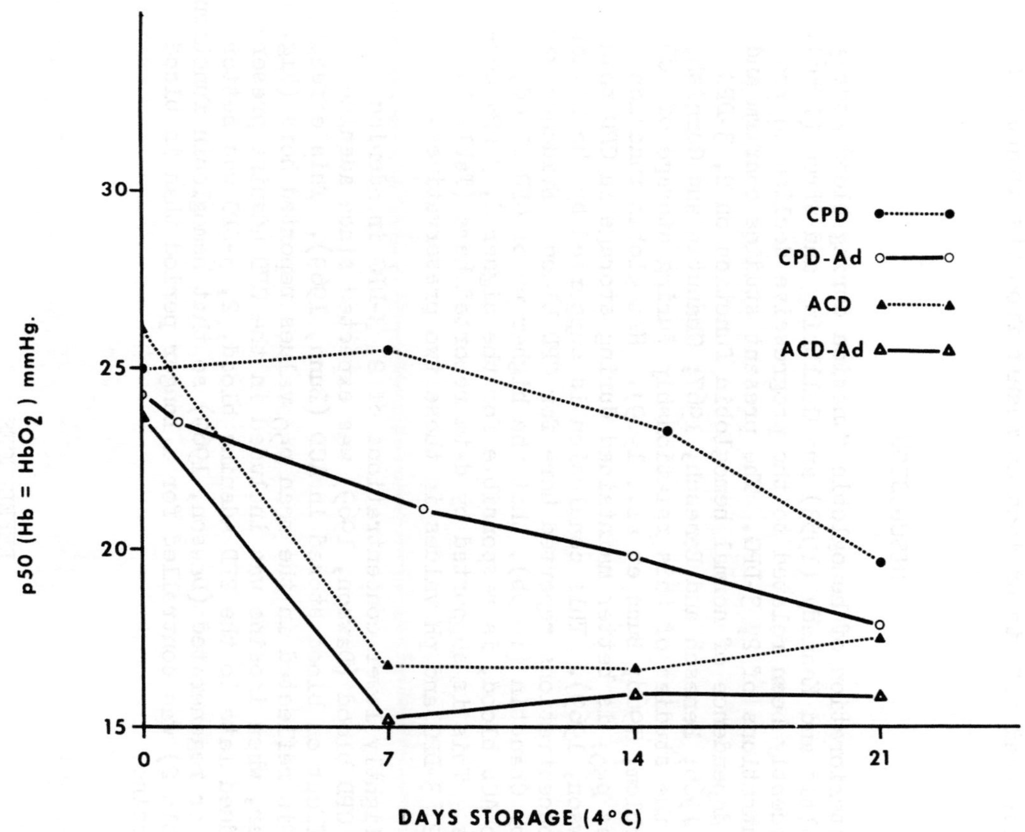

Fig. 3 Hemoglobin Function of Blood Stored in CPD and ACD, With and Without Adenine. The CPD and CPD-adenine values represent average p_{50} determinations taken from two comparative experiments (Dawson, 1969). The ACD and ACD-adenine values are mean p_{50} determinations calculated from the data of Bunn (1969). See text for discussion.

to the paired half units. Further, the p_{50} values at day 32 are higher than the respective 21 day values in CPD-adenine-inosine. The p_{50} one day after inosine addition (day 26) is similar to the paired 14 day value. The "n" values are given to show that normal heme-heme interaction is not impaired in these preservatives. The inosine was added in a small volume of saline, without incubation. Thus, the observed effects of inosine result from its metabolism at 4°C.

DISCUSSION

The deterioration of hemoglobin function during blood storage noted by Valtis and Kennedy (1953) and Gullbring and Strom (1956), has only recently been related to the progressive decline in red cell concentrations of 2, 3-DPG. The present studies confirm and extend the dependence of normal hemoglobin function on 2, 3-DPG (Chanutin, 1966; Benesch and Benesch, 1967; Chanutin and Curnish, 1967), and the studies of this relationship during storage of ACD blood (Akerblom, 1968; Bunn et al., 1969). Hemoglobin function expressed as p_{50}, is better maintained during storage in CPD than in ACD (Dawson, 1969). This conclusion is supported by the higher 2, 3-DPG concentrations reported here for CPD blood. Evidence was presented by Chanutin (1965b), that the higher pH of CPD blood, compared to ACD blood, is responsible for the higher 2, 3-DPG concentrations. This is supported by data reported here (Table 1) comparing 2, 3-DPG and pH values in these two preservatives.

The slightly lower concentrations of 2, 3-DPG in adenine containing CPD blood (Dawson, 1969) was expected since adenine had this effect on blood stored in ACD (Bunn, 1969). This effect of adenine is reflected in the mean p_{50} values reported here (Fig. 3). Further, when inosine was included in the CPD adenine preservative or added late to the CPD adenine blood, 2, 3-DPG was better maintained or regenerated (Dawson, 1969) so that hemoglobin function or p_{50} (Table 2) was controlled for a longer period than in blood lacking inosine.

REFERENCES

Akerblom, O. de Verdier, C.H., Garby, L. and Hogman, C. (1968), Scand. J. Clin. and Lab. Invest. 21, 245.

Bartlett, G.R. and Barnet, H.N. (1960), J. Clin. Invest. 39, 56.

Benesch, R. and Benesch, R.E. (1967), Biochem. Biophys. Res. Commun. 26, 162.

Berry, E.R. and Chanutin, A. (1957), J. Clin. Invest. 36, 225.

Berry,E.R. and Chanutin, A. (1958), J. Clin. Invest. 37, 974.

Bunn, H.F., May, M.H., Kocholaty, W.F. and Shields, C.E. (1969), J. Clin. Invest. 48, 311.

Bunn, H.F., May, M.H., Lenzi, D. and Taylor, J.F. (1968), USAMRL, Ft. Knox, Ky., Report #785.

Chanutin, A. and Curnish, R.R. (1964), Arch. Biochem. Biophys. 106, 433.

Chanutin, A. and Curnish, R.R., (1965a) Proc. Soc. Exp. Biol. (N.Y.) 120, 291.

Chanutin, A. and Curnish, R.R. (1965b), Transfusion 5, 254.

Chanutin, A. (1966), U.S. Army Report, R & D Command (Washington, D.C.), 30 June 1966.

Chanutin, A. and Curnish, R.R. (1967), Arch. Biochem. Biophys., 121, 96.

Chanutin, A. and Hermann, E. (1969), Arch. Biochem. Biophys., 131, 180.

Dawson, R.B. (1969), Intern Conf. Red Cell Metabolism and Function, Ann Arbor, Michigan, 1-3 October 1969, Proceedings: Plenum Press, N.Y.

de Verdier, C.H., Hogman, C., Garby, L. and Killander, J. (1964), Acta Physiol. Scand. 60, 141.

de Verdier, C.H. and Garby, L. (1969), Scand. J. Clin. Lab. Invest., 23, 149.

Greenwald, I. (1925), J. Biol. Chem. 63, 339.

Gullbring, B., and Strom, G. (1956), Acta. Med. Scand., 155, 413.

Keitt, A.S. (1966), Am. J. of Med., 41, 762.

Lowry, O., et al. (1964), J. Biol. Chem., 239, 18.

Lugue, J., Diederich and Grosolia, S. (1969), BBRC 36, 1019.

Nakao, K., Wada, T. and Kamiyama, T. (1962), Nature 194, 877.

Rapoport, S. and Guest, G.M. (1941), J. Biol. Chem. 138, 269.

Shields, C.E., Dennis, L.H., Eichelberger, J.W., and Conrad, M.E. (1967), Transfusion 7, 133.

Shields, C.E. (1968), Transfusion 8, 1.

Simon, E.R., Chapman, R.G., and Finch, C.A., (1962), J. Clin. Invest., 41, 351.

Sugita, Y. and Chanutin, A. (1963), Proc. Soc. Exp. Biol. and Med. (N.Y.), 112, 72.

Valeri, C.R. and Hirsch, H.M. (1969), J. Lab. Clin. Med. 73, 722.

Valtis, D.J., and Kennedy, A.C. (1953), Glasgow Med. J. 34, 521.

Valtis, D.J., and Kennedy, A.C. (1954), Lancet, 1, 119.

EFFECT OF PLASMA CONSTITUENTS ON OXYGEN DIFFUSIVITY

R.M. Navari, J.L. Gainer and K.R. Hall

Department of Chemical Engineering

University of Virginia, Charlottesville, Virginia

INTRODUCTION

Oxygen transport in the lungs from the alveoli to the hemoglobin is affected by a number of resistances: the pulmonary membrane, plasma, the red blood cell membrane, and the paracrystalline material inside the red blood cell (Figure 1). The pulmonary membrane which separates the alveolar gases from the blood stream consists of three parts: a thin alveolar membrane, interstitial fluid, and the capillary wall (44). The thickness of the total membrane is a few microns, and is, essentially, an aqueous medium. Plasma is an aqueous solution containing dissolved proteins, inorganic salts, and solutes of diverse organic compounds other than protein (54). The red blood cell consists of a biological semipermeable membrane surrounding a concentrated hemoglobin solution.

Many investigations have concentrated on resistances associated with the red blood cells in the oxygen multistage diffusion path (12, 17, 20, 21, 27, 31, 36, 48, 49). In addition, several theoretical and experimental studies have been performed on oxygen transport in hemoglobin solutions (7, 10, 11, 14, 18, 19, 24, 25, 30, 43, 45, 46, 52, 53, 56, 57, 60). At the present, the various authors are not in complete agreement on the magnitude of resistance provided by the red blood cell. Comparatively few studies have considered the resistances provided by the plasma and pulmonary membrane. Roughton (41) pointed out that these resistances may be as much as one-third to one-half the total resistance, and physiologists (42, 44) regard the respiratory gas transport through the plasma and pulmonary membrane as important steps in the diffusion path. These admonitions serve as driving forces in undertaking studies of oxygen transport

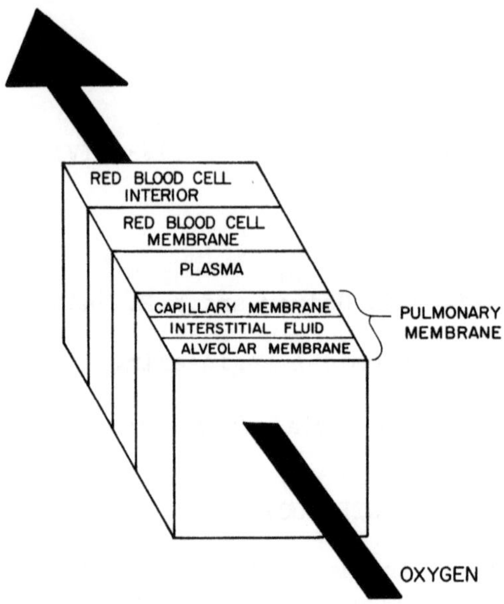

Figure 1. Diffusion pathway of oxygen in the lung from the alveolus to the red blood cell (44).

in plasma and protein solutions. In addition, recent articles by Metzner (35), Astarita (1), and Quinn and Blair (39) show that transport of gases through dilute aqueous-polymer solutions is greatly affected by the concentration of the polymer in the solution. Because plasma is a dilute protein solution, these investigations suggest that variations in the concentration of any plasma constituents over normal physiological ranges might significantly affect oxygen transport. Table 1 summarizes previous studies of oxygen transport in plasma and in aqueous solutions of albumin, the major plasma protein. The fact that these few published values show little agreement may be explained in part by differences in experimental and theoretical approaches and in part by variations in the concentration levels of the constituents of the plasma employed. To compare data of different investigations, Spaeth and Friedlander (46) applied the Stokes-Einstein correction to the data of Yoshida and Oshima (59), Goldstick (14) and Keller and Friedlander (25) for the diffusion of oxygen in water and their results at 25.0°C and 37.5°C are the last entries in the table. Goldstick's (14) and

Table 1

Previous Investigations of Oxygen Diffusion in Plasma

Investigator	System	Temperature(°C)	Diffusivity (10^5 cm^2/sec)
Hershey and Karhan (20)	Sheep Serum	25.0	1.95-2.0
Yoshida and Oshima (59)	Ox Serum	37.0	1.89
Gertz and Loeschcke (13)	Serum	37.0	2.54
Kreuzer (26)	Albumin-Water (8.5g/100 ml.)	25.0	1.61
Goldstick (14)	Albumin-Water (3.7g/100 ml.)	25.0	1.88
Spaeth and Friedlander (46)	Oxygen-Water Data of Yoshida and Oshima (59),	25.0	1.17*-1.24*
	Goldstick (14), Keller and Friedlander (25)	37.5	1.57*-1.66*

*Stokes-Einstein correction for temperature and viscosity.

Kreuzer's (26) studies of oxygen diffusion in albumin-water solutions are the only data which show the effect of protein concentration on diffusivity. However, their systems only contained one of the major proteins of plasma and therefore their values should not be applicable to plasma.

In order to establish the concentration dependency of the diffusivity of oxygen in plasma, a series of experiments were performed in this study (37) with an aqueous solution containing all the constituents of human plasma. Oxygen diffusion coefficients were measured as a function of the concentration of one of the plasma constituents; the concentration levels of the remaining constituents were the same as the average levels in normal human plasma. The experimental technique employed to measure the diffusivity of oxygen in the fabricated plasma at various physiological concentrations of the constituents was a steady-state diaphragm cell method developed by Tham, Bhatia and Gubbins (47). Results obtained using this method have been demonstrated to be in excellent agreement with literature values for a number of systems. Himmelblau (22) has compared experimental methods available for studying the diffusion of dissolved gases in liquids. In this article and in another by Unver and Himmelblau (50), the laminar liquid jet technique is favored over other techniques. The data of Tham, Bhatia, and Gubbins obtained with the steady-state diaphragm cell, however, agree very well with data on similar systems obtained with the laminar liquid jet (47). An entirely different unsteady state method has been recently developed by Krieger, Mulholland and Dickey (28), but this technique appears to involve a number of experimental difficulties.

The steady-state diaphragm cell method was chosen over the liquid-laminar jet in this investigation for a number of reasons. The laminar liquid jet has been criticized for some systems because of the requirement for a detailed knowledge of the fluid mechanical behavior of the jet (2, 39). In addition the steady-state diaphragm cell method requires significantly less time for measurement, and allows the use of a highly sensitive gas chromatograph analysis for oxygen.

EXPERIMENTAL

A. Fabricated Plasma

The fabricated plasma was prepared by dissolving the various plasma constituents in completely degased distilled water at the system temperature. Table 2 lists the various plasma constituents (3, 4, 6, 54, 58). The non-protein compounds were stored at a constant room temperature and the proteins at the refrigeration

Table 2

Constituents of Normal Human Plasma

Constituent	Normal Range	Average Value
PROTEINS:	(g./100 ml.)	(g./100 ml.)
Albumin	2.80-4.50	3.50
α_1-Globulins:		
α_1-Lipoprotein (1.093 g./ml.)	0.05-0.13	0.09
α_1-Lipoprotein (1.149 g./ml.)	0.30-0.40	0.35
α_2-Globulins	0.40-0.90	0.65
β-Globulins:		
β_1-Lipoprotein (0.99 g./ml.)	0.13-0.20	0.17
β_1-Lipoprotein (1.03 g./ml.)	0.20-0.25	0.23
β_1-Metal Binding Globulin	0.40	0.40
γ-Globulins	0.70-1.50	1.10
Fibrinogen	0.30	0.30
NONPROTEIN NITROGEN:	(mg./100 ml.)	(mg./100 ml.)
Urea	20-30	25
Bilirubin	0.2-1.4	0.8
Creatine	0.2-0.9	0.6
Creatinine	1.0-2.0	1.5
Taurine	0.4-0.8	0.6
Uric Acid	2.0-6.0	4.0
Amino Acids:		
Alanine	3.0-3.7	3.3
α-Aminobutyric Acid	0.2-0.4	0.3
Arginine	1.2-1.9	1.6
Asparagine	0.5-0.7	0.6
Aspartic Acid	0.01-0.07	0.04
Cysteine	1.10-1.30	1.20
Glutamic Acid	0.4-1.2	0.8
Glutamine	5.0-12.0	8.0
Glycine	1.3-1.7	1.5
Histidine	0.8-1.5	1.1
Isoleucine	0.7-1.3	1.0
Leucine	1.4-2.3	1.8
Lysine	2.5-3.0	2.75
Methicnine	0.3-0.4	0.35
Ornithine	0.6-0.8	0.7
Phenylalanine	0.7-1.0	0.85
proline	1.8-3.3	2.5

Table 2 (Continued)

Serine	1.1-1.2	1.15
Threcnine	1.2-1.7	1.45
Tryptophan	1.0-1.2	1.1
Tyrosine	0.8-1.5	1.1
Valine	2.4-3.7	3.0
CARBOHYDRATES:	(mg./100 ml.)	(mg./100 ml.)
Glucose	65-90	77
Fructose	6.0-8.0	7.0
Glycogen	5.0-6.0	5.5
Hexose	70-105	88
Gluchrenic Acid	0.4-1.4	0.9
Pentose	2.0-4.0	3.0
ORGANIC ACIDS:	(mg./100 ml.)	(mg./100 ml.)
Citric Acid	1.4-3.0	2.2
α-Ketoglutaric Acid	0.2-1.0	0.6
Malic Acid	0.1-0.9	0.5
Succinic Acid	0.1-0.6	0.4
Acetoacetic Acid	0.8-2.8	1.8
Lactic Acid	8.0-17.0	13.0
Pyruvic Acid	0.4-2.0	1.2
LIPIDS:		
Cholesterol	130-260	200
Lecithin	100-200	150
Cephalin	0-30	15
Sphingomyelin	10-50	30
Plasmacogens	7.0-8.0	7.5
IONS:	(meq./1000 ml.)	(meq./1000 ml.)
Anions:		
Bicarbonate	24-30	27
Chloride	100-110	105
Phosphate	1.6-2.7	2.1
Sulfate	0.7-1.5	1.1
CATIONS:		
Calcium	4.5-5.6	5.0
Magnesium	1.6-2.2	1.9
Potassium	3.8-5.4	4.6
Sodium	132-150	141

temperature suggested by the supplier, Nutritional Biochemicals Corporation. The constant temperature storage and dissolution were necessary in order to prevent denaturing of the compounds. Following dissolution, the solution was buffered to the desired pH of 7.41 (44, 54, 58) with a dry buffer salt supplied by Fisher Scientific Co. Measurements of pH were made throughout all the diffusion measurements with a Fisher pH meter having calomel and glass electrodes. No changes in pH were noted over the duration of the diffusivity experiments. An additional check on the stability of the solution was made by measuring the density of the solution with a Gay-Lussac pycnometer which was calibrated with distilled water. Again, no changes were noted in the value of the density before and after each diffusion measurement. These pH and density measurements, as well as the observation that the solution retained its original slightly yellowish transparent color, indicate that no significant denaturation or degradation of the solutions occurred.

B. Steady State Diaphragm Cell Method

The diaphragm cell, shown in a sketch of the apparatus in Figure 2, was immersed horizontally in a thermostated water bath which was controlled to 0.01°C. The cell contained three fritted glass diaphragms which had a mean pore size of 5 microns. The diffusion diaphragm, 90 mm. in diameter, divided the cell into two compartments of approximately 425 ml. in volume. The two smaller diaphragms, each 20 mm. in diameter, were used at the gas inlets to each cell compartment to obtain efficient dispersion of the gases. Glass stirrers with bar magnets at one end were used inside the cell compartments and were rotated by small horseshoe magnets outside the cell.

The cell was filled with the plasma at a constant temperature and under vacuum according to the procedure recommended by Gordon (15). Oxygen was passed through one of the compartments saturating this side of the cell. The carrier gas used for the Beckman GC-2A gas chromatograph was helium and this was passed continuously through the other cell compartment, removing any diffused oxygen from solution and keeping the oxygen concentration in this compartment very close to zero. When the helium-oxygen gas stream passed through the compartment, it was vented either through a soap bubble meter for measuring the flow rate or passed into the gas sampling valve of the gas chromatograph for analysis of oxygen content.

After steady state had been achieved, evidenced by a constant amount of oxygen in the helium-oxygen gas stream to the gas chromatograph, the flow rate of the helium-oxygen stream was measured by the soap bubble meter. At the end of the experiment, samples of the liquid from each of the cell compartments were analyzed for

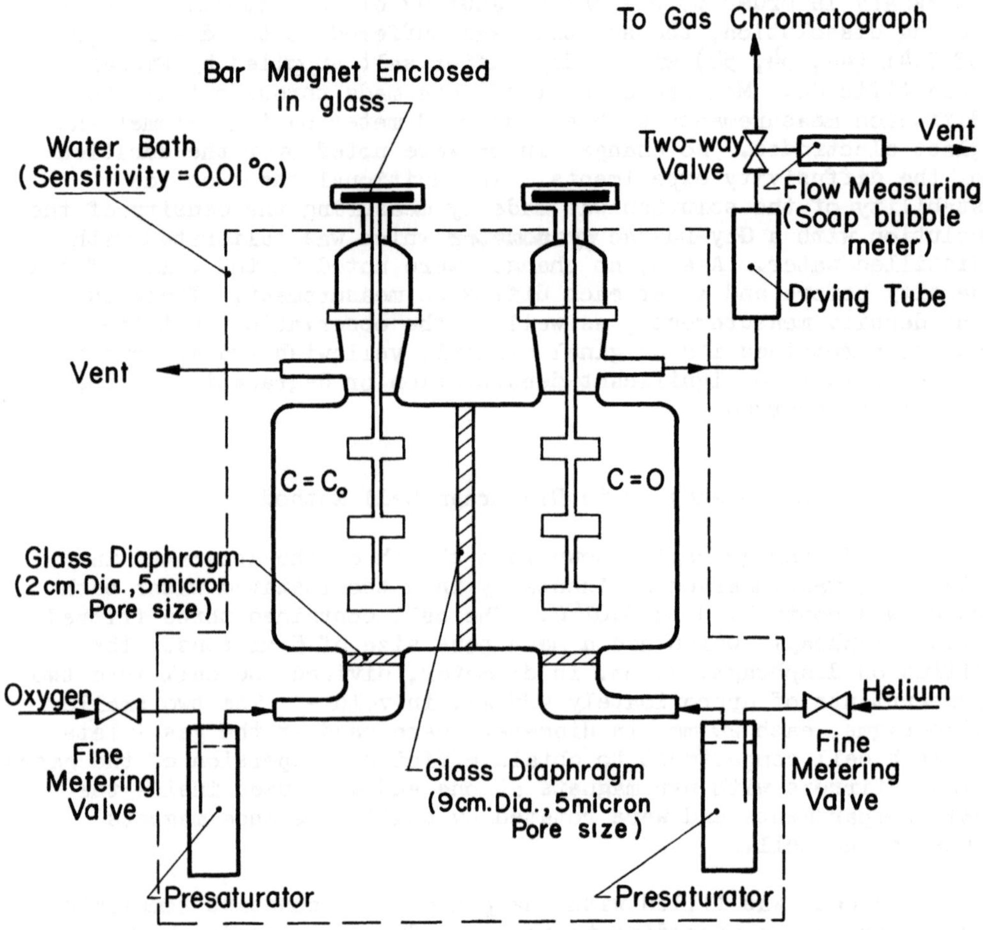

Figure 2. Schematic of Apparatus.

dissolved gas by gas chromatography and no measurable oxygen was found in the samples taken from the helium side of the cell.

C. Analysis

The diffusional resistance of the plasma in the glass diaphragm is equivalent to a cylinder of stagnant liquid of length L and cross-sectional area A. The volume of the theoretical cylinder is equal to the volume of the plasma contained in the frit. At a steady state condition, the flow of the oxygen across the cylinder of stagnant plasma will be (5)

$$N = \left(\frac{A}{L}\right)(C_o)(D)$$

where

N = Flow of oxygen (cm.^{3} of oxygen/sec.)
A = Cross-sectional area (cm.^{2} of plasma)
L = Length (cm. of plasma)
C_o = Oxygen solubility in plasma $\left(\frac{\text{cm.}^3\ \text{oxygen}}{1\ \text{cm.}^3\ \text{plasma}}\right)$
D = Diffusivity (cm.^{2} plasma/sec.)
$\frac{A}{L}$ = Cell constant

The cell was calibrated by using the value of oxygen diffusivity in water at 25.0°C. and was checked with the value of oxygen in water at 37.0°C. Recent investigations (2, 9, 14, 20, 23, 38, 51) determining the diffusivity of oxygen in water by a number of experimental methods have eliminated the differences in recorded values which once existed. The cell constant was checked a number of times and no changes in the value was noted.

Results

Oxygen diffusion runs were performed on plasma solutions which contained the average concentration of each constituent found in human plasma (Table 2). The constituent whose concentration was to be varied over its normal physiological range was initially withheld from the solution and then added in equal increments after each diffusivity determination was made. In this manner, values of oxygen diffusivity as a function of the concentration of one of the plasma constituents were obtained. In the present study, albumin and gamma-globulins were chosen as the constituents whose concentrations would be varied. Albumin and gamma-globulins are present in large concentrations in the plasma relative to the other constituents. Variations in their concentrations have become increasingly important in treating certain blood diseases. Figures 3 and 4 show the changes in diffusivity at 25.0 and 37.0°C. as the concentration of albumin and gamma globulins is increased. Note that for the two temperatures the diffusion coefficient of oxygen decreases approximately 55-70% from its mean value as the albumin concentration is varied over its normal physiological range (2.0-4.5 g/100 ml.); the diffusivity decreases 45-55% from its mean value as the gamma-globulins concentration is varied over its physiological range (0.7-1.5 g/100 ml.). It is difficult to compare these data with previously published values since only Kreuzer (26) and Goldstick (14) measured a protein concentration as well as oxygen diffusion. However, the values found in the present study (37) for different protein concentrations are in the range of values obtained from previous investigations.

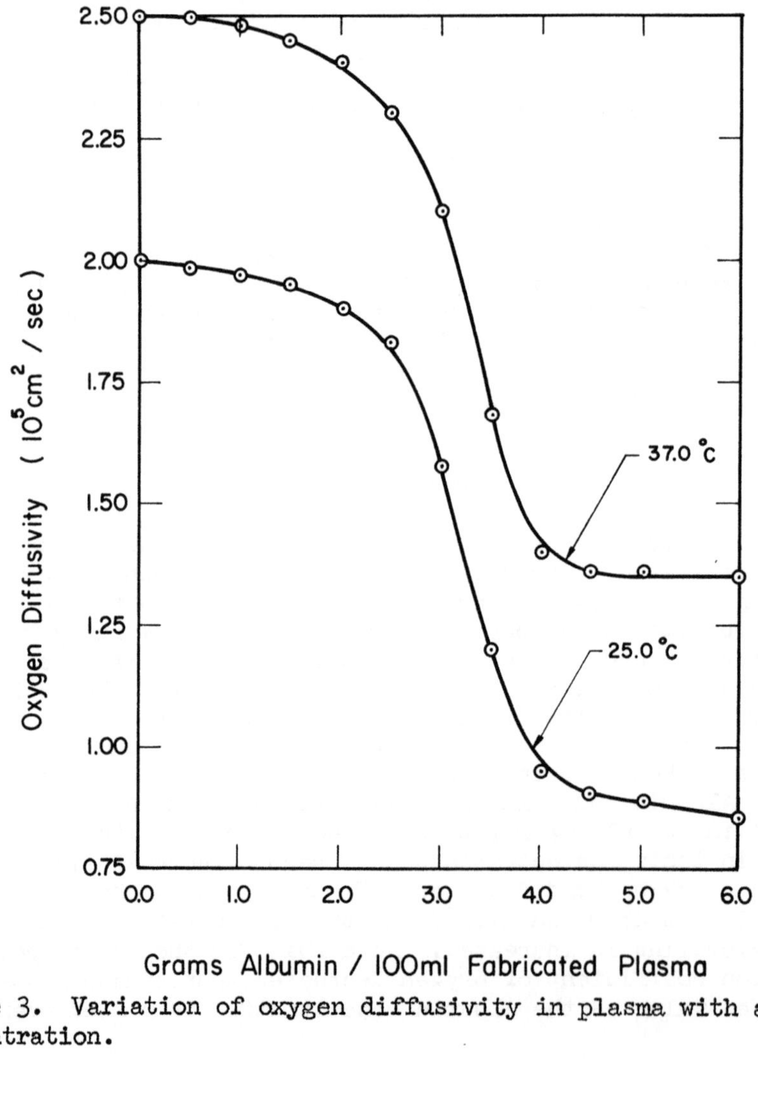

Figure 3. Variation of oxygen diffusivity in plasma with albumin concentration.

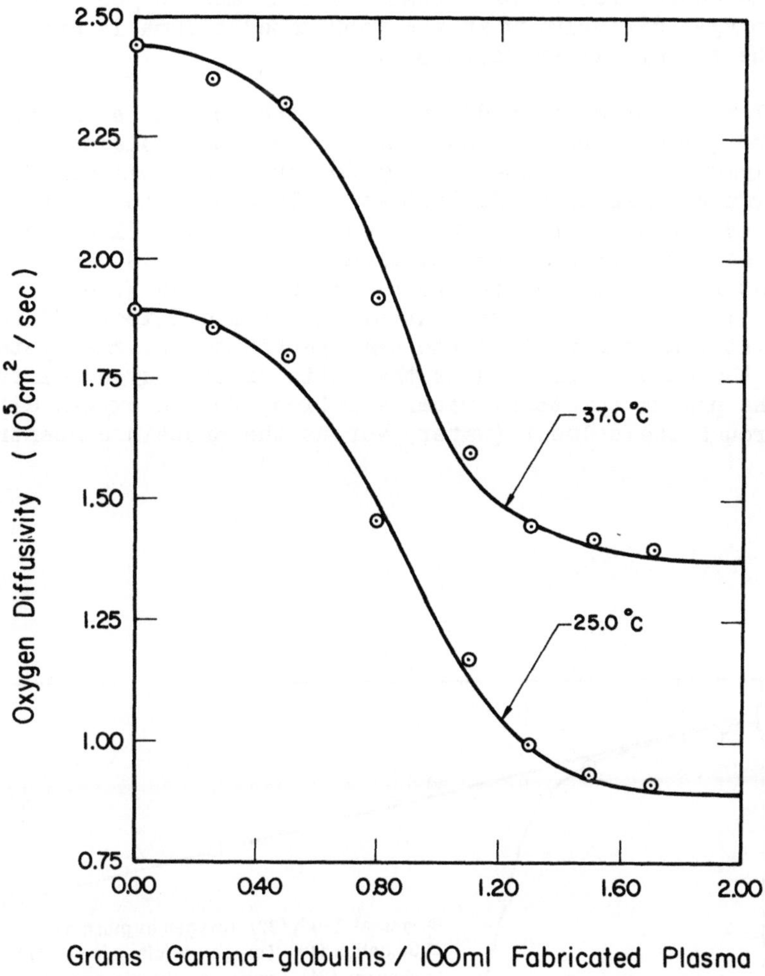

Figure 4. Variation of oxygen diffusivity in plasma with gamma-gobulin concentration.

The data of Goldstick (14) and Kreuzer (26) for oxygen diffusion in albumin solutions at 25.0°C. are plotted in Figure 5 with the oxygen diffusivity-albumin data of this study. A comparison of the data shows the large effect of the constituents of plasma other than albumin on the oxygen diffusivity. This comparison shows that as the concentration of albumin is increased beyond 2.0 g/100 ml., the oxygen diffusivity in the fabricated plasma is much smaller than in the albumin water solutions.

The large decreases in diffusivity with increases in the albumin and gamma-globulins concentration may be explained by the recent polymer solutions theory proposed by Li and Gainer (29) who showed large changes in diffusivities of liquids into dilute polymer solutions as the polymer concentration was varied. Plasma closely approximates a dilute polymer solution because of its composition and the nature of protein structure. Figures 6 and 7 show the data taken in this study with that of Connor and Gainer (8) for dextrose-water diffusing into the same fabricated plasma systems. Note that the data is plotted as the ratio of the oxygen diffusivity through the plasma (proteins-water solution) to the oxygen diffusivity through the solvent (water) versus the protein concentration.

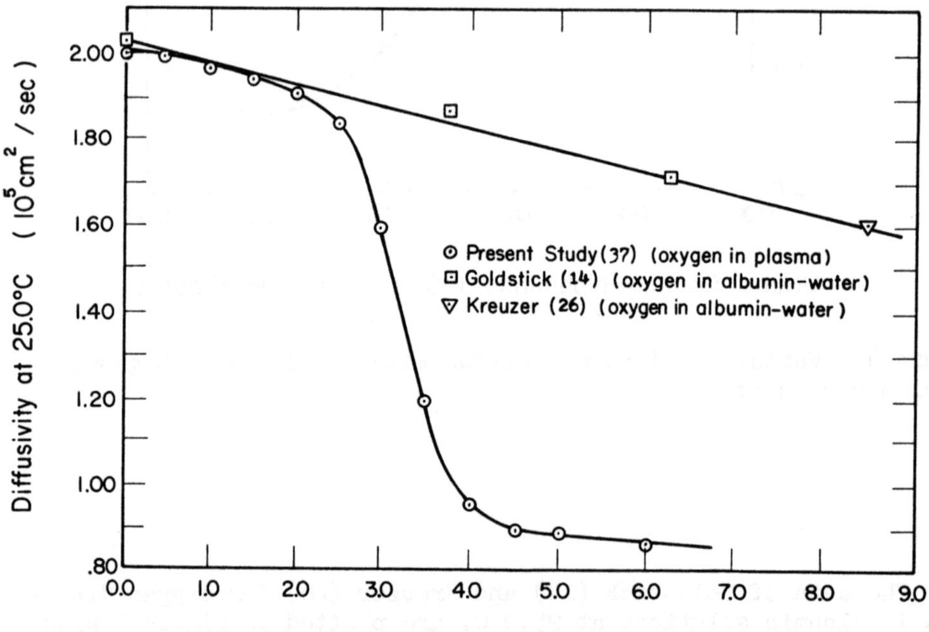

Figure 5. Variations of oxygen diffusivity in albumin solutions with albumin concentrations.

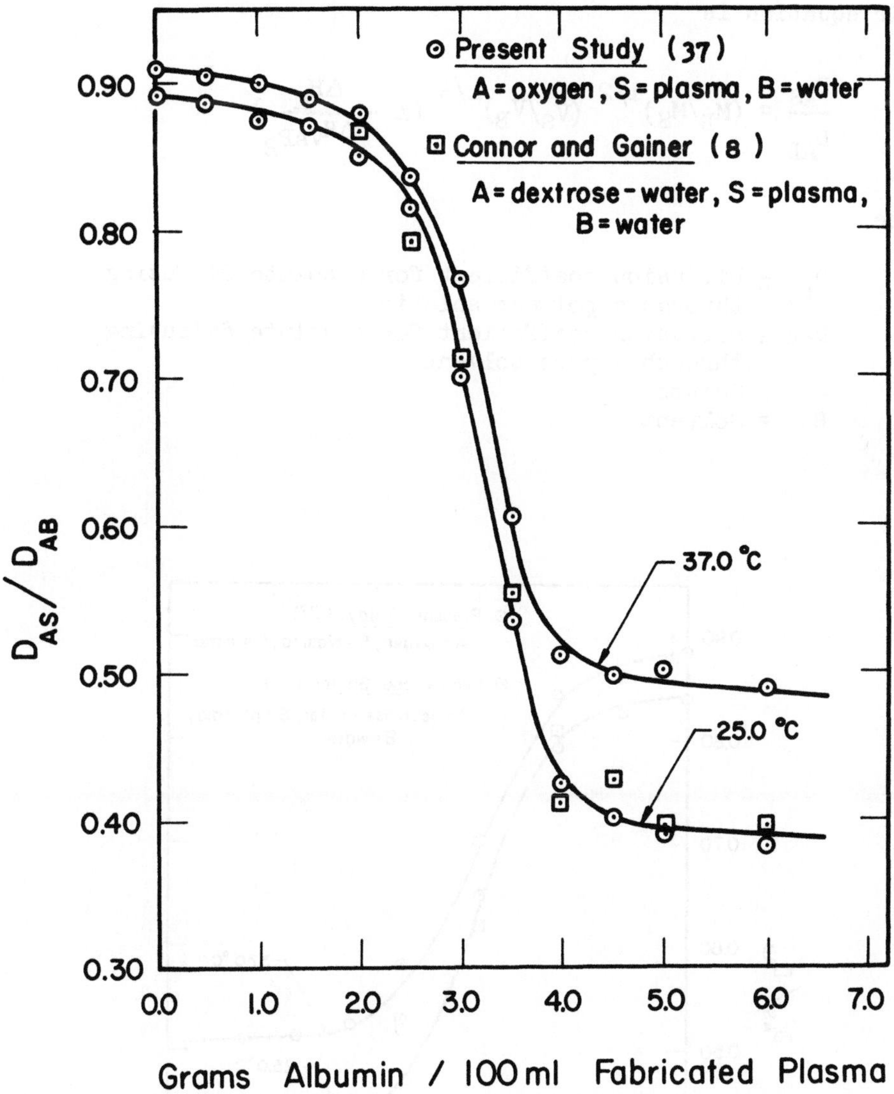

Figure 6. Variation of ratios of diffusivities in plasma with albumin concentration.

The diffusion ratio is used because Li and Gainer (29) have proposed an equation which predicts the diffusion coefficient for a gas or liquid diffusing through a polymer solution.

Their equation is

$$\frac{D_{AS}}{D_{AB}} = (M_B/M_S)^{1/2} (V_S/V_B)^{1/3} \left(1 - \frac{\Delta H_M}{\Delta H_{VAP_B}}\right)$$

where

D_{AS} = Diffusion coefficient for a solute diffusing through a polymer solution
D_{AB} = Diffusion coefficient for a solute diffusing through a pure solvent
A = Solute
B = Solvent

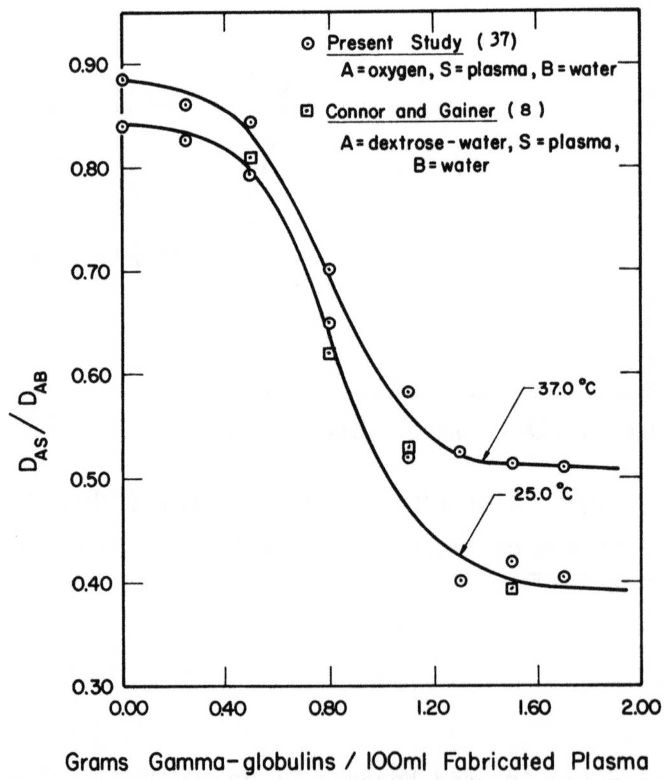

Figure 7. Variation of ratios of diffusivity in plasma with gamma-gobulins concentration.

S = Polymer solution
M = Molecular weight
V = Molar volume
ΔH_M = Heat of mixing of the polymer with the solvent
ΔH_{VAP_B} = Heat of vaporization of the pure solvent.

The equation predicts that the ratio of diffusivities should be independent of the diffusing solute. The fact that Figures 6 and 7 show that the oxygen diffusion coefficient ratios have the same values as the dextrose-water ratios within experimental error supports this proposition.

In order to check further on the similarity between the fabricated plasma and normal human plasma, viscosity and density measurements were made on the fabricated plasma which contained all the constituents at an average concentration. Table 3 shows that both the viscosity and density measurements compare very favorably with the literature values. A capillary viscometer was used for the viscosity measurements.

Table 3

Viscosity and Density Measurements of the Fabricated Plasma

	Temperature (°C)	Fabricated Plasma	Human Plasma
Viscosity (cp.)	25	1.70	1.67 (34)
	37	1.47	1.50 (40)
Density (grams/ml.)	25	1.031	1.031*(44)
			1.038*(55)

*Temperature correction of literature value.

Conclusions

Variations in the concentration of albumin and gamma-globulins over normal physiological ranges caused a large decrease in the diffusivity of oxygen through the plasma solutions. The magnitude of the decrease was 45-70% from the mean value, and these large reductions in oxygen diffusivity suggest that plasma may provide a major resistance in the path of the oxygen to the hemoglobin. This becomes more significant when one considers that previous investigations have suggested that plasma and the pulmonary membrane may be from one-third to one-half the total resistance of the oxygen path to the hemoglobin inside the red blood cell.

These measurements have pointed out the importance of knowing protein concentration levels when oxygen diffusion determinations into plasma are made. This was not done by previous investigators and explains why there are large discrepancies in previously published data. In addition, comparison of the data in this study with oxygen diffusion values in albumin-water solutions show the large effect of constituents other than albumin on the diffusivity and that albumin-water solutions are not good approximations of plasma.

It is important to note that the diffusivity values undergo the largest reduction in the physiological concentration ranges of both albumin and gamma-globulins. This is further evidence for small protein concentration changes causing large changes in the oxygen diffusivity.

Plotting the data as a ratio of diffusivities in a solution to that in the pure solvent, and comparing it to data for a similar system using a completely different diffusing solute showed that this diffusivity ratio is not a function of the diffusing solute. This lends credence to the polymer solution theory proposed by Li and Gainer (29) for interpreting the data.

Future work includes varying other constituents in plasma to observe the effect on oxygen diffusivity, and measuring diffusivities in normal human plasma, determining the protein concentration by electrophoresis. In addition, the polymer solution theory of Li and Gainer (29) will be extended to plasma solutions.

Cited Literature

1. Astarita, G., I.E.C. Fund., 4, 236 (1965).

2. Baird, M.H.I., J.F.Davidson, Chem. Eng. Science, 17, 473 (1962).

3. Bell, G.J., J.N. Davidson, and H. Scarborough (eds.), "Textbook of physiology and Biochemistry," ch.5,22-24, William and Wilkins, Co., Baltimore (1961).

4. Bennhold, H., "Transport Function of the Serum Proteins", in "Transport Function of Plasma Proteins", P. Desgrez, P.M. DeTraverse (eds.), pp. 1-5, Elsevier Publishing Co., N.Y.(1966).

5. Carslaw, H.S., J.C. Jaeger, "Conduction of Heat in Solids", p. 99, Oxford University Press, London (1959).

6. Cohn, E.J., J.L. Oncley, L.E. Strong, W.L. Hughes, S.H. Armstrong, J. Clin. Invest., 23, 417 (1944).

7. Collins, R.E., Science, 133, 1593 (1961).

8. Connor, E.D., J.L. Gainer, "Transport in Biol. Systems", C.E.P. Symposium Series, in press (1969).

9. Davidson, J.F., E.J. Cullen, Trans. Inst. Chem. Eng. (London), 35, 51 (1957)

10. Fatt, I., R.C. LaForce, Nature, 195, 505 (1962).

11. Fatt, I., R.C. Laforce, Science, 133, 1919 (1961).

12. Forester, R.E., F.J.W. Roughton, F. Kreuzer, W.A. Briscoe, J. Appl. Physiol., 11, 260 (1957).

13. Gertz, K.H., H.H. Loeschcke, Z. Naturfersch, 9, 1 (1954).

14. Goldstick, T.K., Ph.D. Dissertation, University of California, Berkeley (1966).

15. Gordon, A.R., Ann. N.Y. Acad. Sci., 46, 285 (1945).

16. Groth, C., Acta. Chir. Scand., 131, 290 (1966).

17. Hartridge, H.H., F.J.W. Roughton, J. Physiol. (London), 62, 232 (1927).

18. Hemmingsen, E., Science, 135, 733 (1962).

19. Hemmingsen, E.P. Scholander, Science, 132, 1370 (1960).

20. Hershey, D., T. Karhan, A.I.Ch.E.J., 14, #6, 969 (1968).

21. Hershey, D., C.J. Miller, R. Menke, J.F. Hesselberth, "Chemical Engineering in Medicine and Biology", D. Hershey, (ed.), p. 117, Plenum Press, N.Y. (1967).

22. Himmelblau, D.M., Chem. Rev., 64, 527 (1964).

23. Jordon, J., J. Am. Chem. Soc., 78, 2979 (1956).

24. Katchalsky, A., P.F.Curran, "Nonequilibrium Thermodynamics in Biophysics", p. 203, Harvard University Press, Cambridge, Mass. (1965).

25. Keller, K.H., S.K. Friedlander, J. Gen. Physiol., 49, 663 (1966).

26. Kreuzer, F., Helv. Physiol. Pharmacol. Acta., 8, 505 (1950).

27. Kreuzer, F., W.Z. Yahr, J. Appl. Physiol., 15, 1117 (1960).

28. Krieger, I.M., G.W. Mulholland, C.S. Dickey, J. Phys. Chem., 71, 1123 (1967).

29. Li, S.U., J.L. Gainer, I.E.C. Fund., 7, 433 (1968).

30. Markus, G., Science, 132, 95 (1960).

31. Marx, T.I., et al., J Appl. Physiol., 15, 1123 (1960).

32. Mayer, G.A., Can Med. Ass. J., 91, 951 (1964).

33. Mayer, G.A., J. Fridrich, J. Newell, J. Sznek, Biorheol., 3, 177 (1966).

34. Merrill, E.W., G.C. Cokelet, A. Britten, R.E. Wells, Circulation Res., 13, 48 (1963).

35. Metzner, A.B., Nature, 208, 267 (1965).

36. Mochizuski, M., J. Fukoka, J. Physiol. Soc. Japan, 8, 206 (1958).

37. Navari, R.M., Ph.D. Dissertation, University of Virginia (1970).

38. Pircher, L., Helv. Physiol. Pharmacol. Acta., 10, 110 (1952).

39. Quinn, J.A., L.M. Blair, Nature, 214, 907 (1967).

40. Rand, P.W., E.A. LaCombe, H.E. Hunt, W.H. Austin, J. Appl. Physiol., 19(1), 117 (1964).

41. Roughton, F.J.W., Prog. in Biophysics, 9, 56 (1959).

42. Ruch, T.C., H.D. Patton, "Physiology and Biophysics", pp 761-787, W.B. Saunders Co., Philadelphia (1966).

43. Scholander, P., Science, 131, 585 (1960).

44. Selkurt, E.E., "Physiology", 2nd. ed., pp. 213-228, 427-470, Little, Brown, and Co., Boston (1966).

45. Snell, F.M., J. Theor. Biol., 8, 469 (1965).

46. Spaeth, E.E., S.K. Freidlander, Biophys. J., 7, 827 (1967).

47. Tham, M.J., K.K. Bhatia, K.E. Gubbins, Chem. Eng. Science, 22, 309 (1967).

48. Thews, G., Pfluger Arch. Ges. Physiol., 268, 308 (1959).

49. Thews, G., W. Niesel, Pfluger Arch. Ges. Physiol., 268, 318 (1959).

50. Unver, A.A., D.M. Simmelblau, J. Chem. Eng. Data, 9, 428 (1964).

51. Vivian, J.E., A.I.Ch.E.J., 10, #2, 220 (1964).

52. Wang, J.H., Science, 133, 1770 (1961).

53. Weissman, M.H., L.F. Mockros, J. Eng. Mech., 93, 225 (1967).

54. White, A., P. Handler, E.L. Smith, "Principles of Biochemistry", 3rd ed., pp.91-161, 627-672, McGraw Hill Book Co., N.Y. (1964).

55. Whitmore, R.L., "Rheology of the Circulation", pp. 57-61, Pergamon Press (1968).

56. Wise, D.L., G. Houghton, Biophys. J., 9, 36 (1969).

57. Wittenberg, J.B., J. Biol. Chem., 241, 104 (1966).

58. Wuhrman, F., C. Wunderly, "The Human Blood Proteins", pp. 1-72, Grune and Stratton, N.Y. (1960).

59. Yoshida, F., N. Oshima, J. Appl. Physiol., 21(3), 915 (1966).

60. Zilversmit, D.B., Science, 149, 874 (1965).

INTERNAL VISCOSITY OF THE RED BLOOD CELL AND THE CAPILLARY INVERSION PHENOMENA AS FACTORS IN OXYGEN TRANSPORT

Leopold Dintenfass

Hemorheology Unit, Sydney Hospital, Department of Medicine, University of Sydney, Sydney, Australia

In human circulation blood flows through a series of vessels of diminishing (if measured at increasing distance from the heart) diameters but the total cross section of these vessels progressively increases. Finally, in an enornous capillary network, blood flows through the tissues bringing nourishment (i.e., oxygen) and taking away metabolites. Water, electrolytes, gases, proteins, carbohydrates and fats are being continuously exchanged between the blood and the tissues. The vascular system serves not only for irrigation or collection of refuse, but it is also a highway for transit of cells and chemical agents required for control or defence of the integrity of the circulation and, thus, the life itself.

Continuation of an adequate flow of blood becomes a crucial factor. Oxygenation of blood and tissues is controlled by a number of mechanisms or agencies, including the rheology of blood itself; what is more, there exists a reciprocal effect of oxygenation of the tissues or blood on the rheology of blood and rheology of the red cell itself.

There is a great confusion of thought regarding mechanisms of diseases based on circulatory deficiency or failure. Both physiologists and clinicians have paid great attention to the driving pressure and to the bore of arteries but have neglected both the nature and the role of the fluid which this pressure drives through the vessels; and neglected, per force, the interactions between the rheology of blood and the rheology of the small vessels.

Is the blood pressure or the capillary perfusion more relevant to the health of tissues? and is not such a capillary perfusion more dependent on the viscosity of blood and on the fluidity of the red cell than on the blood pressure?

The newly postulated concept of the critical capillary radius at which the INVERSION phenomenon does take place, and at which a sudden and dramatic elevation of the resistance to flow occurs, brings new insight into the problems of tissue perfusion.[1]

The importance of the rheological properties of the whole blood and of the blood cells in the control of circulation has been largely overlooked by physiologists and clinicians alike, and most attention has been given to capillary "collapse", vasodilatation and vasoconstriction. This attitude was quite understandable as long as one could presume that capillaries (or small vessels) can flutter, collapse, or open, depending on the pressure drop existing in the particular area of the microcirculation. This attitude is not so secure now, especially due to the works of Fung and Zweifach [2,3], and more recent theoretical work of Oka [4].

Blood circulation and tissue perfusion are also affected by the formation of microthrombi, microemboli and platelet aggregates. Consequently, no discussion of tissue perfusion would be complete without consideration of the role of blood flow velocity and velocity gradient on the clotting times, viscosities and morphologies of thrombi [5, 6, 7], and especially their microproducts.

EDITOR'S NOTE

Extensive blood flow data and analysis has been deleted from the original paper. However the bibliography at the end of this paper has been left intact.

PLATELET AGGREGATES, MICROEMBOLI AND BLOOD COAGULATION

The presence of red cell aggregates or platelet aggregates can be due to a number of causes, one of them being blood coagulation. In principle, the intermediate products of fibrinogen polymerization increase aggregation of the red cells. However, formation of platelet aggregates can be induced, or enhanced, not only by products of static coagulation but also by the velocity gradient.

The key role of the flow velocity on the clotting time of blood was recognized in 1962/63 [21, 30, 39], although a vague and diffuse knowledge that shaking affects clotting time has been prevalent for a long time. Using thromboviscometer (VFTV [6, 15, 16, 58]) it was possible to define clotting times, viscosities and morphologies of coagula formed at predetermined shear rates.

The "casting" shear rate determines the final properties of clots and thrombi. Coagula formed at very low rates of shear exhibit gel-like and highly viscous and thixotropic properties. Red and white artificial thrombi, formed at shear rates between 20 and 80 sec^{-1}, exhibit both viscous and thixotropic properties, but their viscosity is rather low, in the order of one poise. Suspension coagula (this new term introduced to describe the new type of coagulum) formed at shear rates above 400 sec^{-1} show viscosities of the order of the viscosity of uncoagulated blood.

Simultaneously, increasing shear rates used resulted in decreasing coagulation times. While clotting time at near zero shear rates was of about 10 minutes, the clotting time of suspension coagula was in the order of one to ten seconds.

Microscopic examinations revealed that an increase in the velocity gradient influenced the morphological structure of blood coagula [5, 6, 59, 60]. When coagulation of blood takes place at near zero shear rate, a red clot is formed in which the red cells are embedded in a random mesh of fibrin. When coagulation proceeds at higher shear rate, an artificial thrombus is formed, this containing large masses of platelet aggregates. Finally, suspension coagula contain suspended particles composed of platelet aggregates and fibrin.

Platelet aggregation is greatly enhanced at increasing flow velocities. There is a quantitative relationship between the velocity gradient and the degree of platelet aggregation. This relationship depends however on the intrinsic characteristic of the blood of each donor.

Usually, in suspensions of reversibly aggregated cells or colloidal perticles, an increase in the rate of shear leads to a disaggregation of such aggregates. That is not the case with platelets. A progressive process of platelet aggregation, at simultaneously increasing shear rates, suggests an irreversible phenomenon; it would appear that the surface properties of platelets undergo a profound change and that we are dealing with a new entity in each subsequent period of time.

Dintenfass and Stewart [7] observed that an artificial thrombus can "dissolve" or disintegrate spontaneously within about 30 seconds, releasing a shower of platelet aggregates. This phenomenon was found to be peculiar to patients suffering from renal failure, hypertension, and peripheral arterial disease.

While it is well realized that formation of an intramural or mural thrombus must affect flow of blood in larger vessels, realization that platelet aggregates and microemboli may have a profound effect on the microcirculation is rather more recent.

OXYGEN TRANSPORT AND ISCHEMIA AS AFFECTED BY RHEOLOGIC FACTORS. INFARCTION AND THROMBOSIS

Such rheological factors as the internal viscosity of the blood cells, aggregation of the red cells, viscosity of plasma, or velocity gradient-induced aggregation of platelets will play a profound role in the stability of the circulation and efficiency of the oxygen transport.

An increase in the rigidity of the red cell, due either to the genetic aspects, tonicity of plasma, hydrogen ion concentration, antigen-antibody phenomena, drug action, age of the cell, or any other acquired or hereditary factors, will be expected to impair the proper and sufficient tissue perfusion. A possible natural defense might exist in a filtration and sequestration of the more rigid cells by spleen, lungs or liver.

The degree of aggregation of the red cells can be enhanced by infection, fever, excess of fibrinogen, presence of macroglobulins or toxins, while the aggregation of platelets might be increased by catecholamines [61] or viral infection [62, 63]. Excessive aggregation of the red cells leads to an enhanced displacement of platelets from the main stream towards the high shear rate zone adjacent to the vessel wall, thus enhancing their adhesive properties [44].

An increase of blood viscosity, due to any of the above mentioned causes, will decrease the flow rate of blood. It has been recognized already (see Guyton and Richardson [64], or Burton [65]) that efficiency of oxygen transport depends on hematocrit, and that for maximum efficiency in oxygen transport the hematocrit value should be such as to allow the maximum flow of hemoglobin. Burton [65] showed that the point of greatest efficiency depends on the size and shape of the red cell, and will differ in different species.

An increase in hematocrit, or increase in the internal viscosity or aggregation of the red cells, will slow down the flow of blood and may lead to localized stasis; this, in turn, leads to an increased internal viscosity and aggregation of the red cells through localized hypoxia and acidosis. Stasis in capillaries may lead to an increased permeability of the capillary wall, leakage of plasma, and further hemoconcentration. An increased aggregation of the red cell, even only as a direct result of the slow down of blood flow, will lead to enhanced plasma skimming and, thus, hemoconcentration in some segments of the microcirculation.

The inter-relationship between the critical capillary radius of the inversion phenomenon and the rheology of blood and cells, discussed earlier, indicates that even small changes in the rheology of blood might result in a multifold decrease of the rate of flow and, thus, of the oxygen transport.

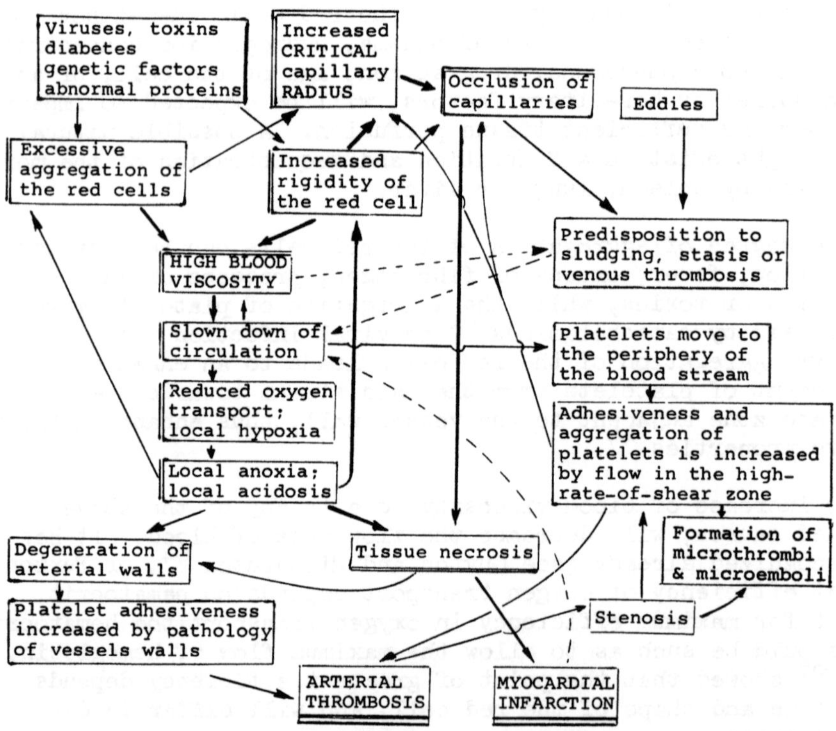

Fig. 1 A diagram representing the role of blood viscosity in pathogenesis of arterial thrombosis and myocardial infarction. The sequence of events is indicated by arrows. Double frame indicates more important steps.

Some of the vicious circles are illustrated in Figure 1. Another one would be due to the Bohr effect [66]. More recently, Naera, Petersen, Boye and Severinghaus [67] showed that increase in CO_2 tension, apart of induced thereby pH changes, decreases the oxygen affinity of human blood in vitro. The increased rigidity of the red cell and the decreased affinity for oxygen are therefore parallel phenomena, and one may wonder if or how they interconnect on molecular levels.

A hypoxic and damaged red cell can release clotting factors and initiate intravascular coagulation. As already stated, hypoxia leads to increased aggregation of platelets. Both phenomena may lead to formation of microthrombi or microemboli which can impair the microcirculation. The importance of these particles in the inversion phenomenon was discussed.

A capillary stasis, a capillary occlusion or a capillary thrombosis will lead to localized areas of infarction and tissue necrosis. If this phenomenon does take place in the heart, it leads to heart failure. If it happens in the vasa vasorum of large arteries, it will lead to formation of a mural thrombus. A deposition of the platelets on the injured endothelium, and consequent decrease of the lumen, may lead to increased local blood flow velocities, greatly increased local shear rates, and thus enhanced deposition and or aggregation of platelets.

One result of the blood viscosity elevation may be a decrease in the cardiac output. This can be directly due to the induced pathology of the heart itself, or to the increased peripheral resistance; and the latter can be due to the increased blood viscosity, or to the presence of rigid blood cells or aggregates. A slow down of the peripheral circulation can cause, directly, ischemia and failure of the peripheral tissues or organs. The rheological factors might act in a synergistic manner, and superimposition of one or more of them on the already weakened circulation can lead to a heart failure [57].

The one-way bridge of the high blood viscosity syndrome [15, 41, 45, 68, 69] represents the role of blood rheology in the etiology of ischemic diseases. A number of causes, such as diabetes, macroglobulinemia, cancer, infection, trauma, genetic factors, etc., will affect one or more of the rheological parameters and will increase the viscosity of blood and its subphases. Whether the viscosity elevation is due to one or the other cause, the fact remains that it is the blood viscosity elevation, general or localized, which is directly responsible for the slow down of the circulation, and directly responsible for the development of ischemia, necrosis and infarction. Because, were the coagulation factors (in the orthodox sense, and not in the dynamic coagulation

(VFTV) sense) the primary cause, thrombosis and coronary heart diseases or thrombosis of other organs would not have been observed in patients with hemophilia and other coagulation defects.

The one-way bridge of the high blood viscosity syndrome (Fig. 2) illustrates this and indicates that vicious circles are important not only in the progressive increase of the viscosity

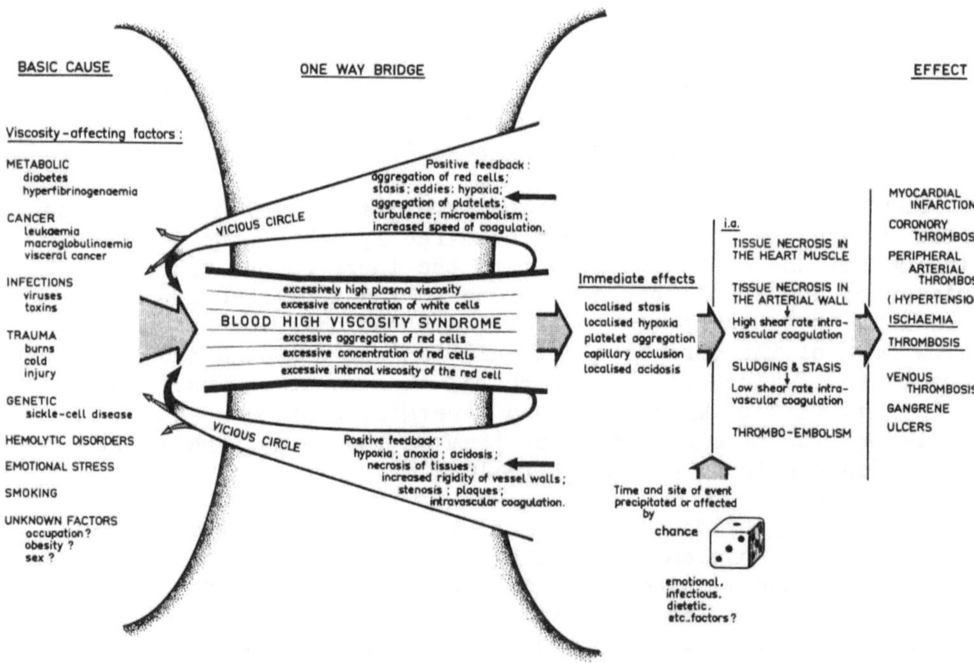

Fig. 2 One-way bridge of the high blood viscosity syndrome. This diagram illustrates the role of blood viscosity in the development of ischemia, infarction and thrombosis. For details see text.

of blood (that is, increase in viscosity of any subphase of blood, and at any shear rate), but can impact also on the original cause of the viscosity elevation and, thus, reinforce factors responsible for this blood viscosity elevation.

The particular locations or sites of infarction or ischemia might be due to chance; not a blind chance, but a chance affected by genetic abnormalities, hydrodynamics of flow, or sensitization of a particular tissue by drugs or diet. Precipitation or onset of sudden ischemic episode might be due to a sudden additional elevation of one or more of the rheological parameters of blood viscosity (this including the aspects due to dynamic coagulation and dynamic thrombus formation in the VFTV sense) as in cases of sudden infection or emotional trauma. As already discussed, even a small change of rigidity of blood cells or blood cell aggregates will have a pronounced effect on the critical vessel radius of the inversion phenomenon.

CONCLUSIONS

Finally, we can suggest that blood oxygenation and ischemic diseases are influenced by a series of rheological factors and a series of vicious circles based on these rheological factors, the latter influencing also the biochemical processes.

The following points would be of importance:

1. A sufficient oxygenation of the tissue can proceed only under satisfactory condition of blood flow; any increase in blood viscosity (from any causes) resulting in a decrease of blood flow will lead to a decrease in the oxygen transport and to tissue ischemia;
2. The paramount effect in the tissue perfusion is due to the internal viscosity of the blood cells and blood cell aggregates; even a slight increase of internal viscosity (from any causes) will lead to a greatly amplified decrease in tissue perfusion as a result of the inversion phenomenon;
3. An increase in the internal viscosity of the red cell, as caused i.a. by a decrease of blood pH, might, through the causal or parallel processes, also affect the hemoglobin-oxygen affinity;
4. A decrease in the oxygen transport, resulting in hypoxia or anoxia, leads to an increase in the internal viscosity of the red cells, an increase in the aggregation of red cells, and an increase in platelet aggregation, all which cause a further slow-down of the blood flow, intensifying the tissue ischemia;
5. A slow-down of blood flow might lead to the trapping of blood cells and stasis in certain segments of circulation, these cells not forming any further an integral part of the functional circulation; a severe metabolic acidosis results not only in

viscosity increase of the blood and the blood cells in the stasis regions, but due to the consequent damage to the cell membrane it might initiate a second pathway of intravascular coagulation;
6. The effects of an interaction of oxygen and electrolytes on the blood rheology, and vice versa, are not yet well understood; but unless studied, it would be difficult to imagine that respiratory processes will be satisfactorily comprehended and controlled; respiration involves not only biochemical but also rheological processes;
7. A treatment of heart failure and failure of other organs as a result of ischemia, should be based on considerations including the rheological behaviour of blood which would, at the least, compound and, at the most, dwarf the effects of the vascular restrictions or constriction due to vessel anatomy, pathology and rheology.

ACKNOWLEDGMENTS

This study was supported by grants from the National Heart Foundation of Australia, and grants-in-aid from the Sydney Hospital Medical Research Fund.

REFERENCES

1. Dintenfass, L., "Inversion of the Fahraeus-Lindqvist phenomenon in blood flow through capillaries of diminishing radius", Nature, 215, 1099 (1967)

2. Fung, Y.C., "Theoretical considerations of the elasticity of red cells, and small blood vessels", Fed. Proc., 25, 1761 (1966)

3. Fung, Y.C., Zweifach, B.W., and Intaglietta, M., "Elastic environment of the capillary bed.", Circulation Res., 19, 441 (1966)

4. Oka, S., "Theoretical approach to the tension in blood vessel walls", presented at the Symposium on Biorheology, Institute of Physical and Chemical Research, Tokyo, 1968, and at the Second Internal Hemorheology Conference, Heidelberg, 1969

5. Dintenfass, L., and Rozenberg, M., "The influence of the velocity gradient on in vitro blood coagulation and artificial

thrombosis", J. Atheroscler. Res., 5, 276 (1965)

6. Rozenberg, M., and Dintenfass, L., "Platelet aggregation in the variable-frequency thromboviscometer", Nature, 211, 525 (1966)

7. Dintenfass, L., and Stewart, J.H., "Formation, consistency and degradation of artificial thrombi in severe renal failure. Effect of A B O blood groups", Thromb. diath. Haemorrh., 20, 267 (1968)

8. Dintenfass, L., "An application of a cone-in cone viscometer to the study of viscosity, thixotropy and clotting of blood", Biorheology, 1, 91 (1963)

9. Dintenfass, L., "A trolley viscometer for estimating viscosity and clotting of blood in hospital wards", Lancet, 2, 567 (1964)

10. Dintenfass, L., "A ring-in-ring adapter for the cone-in-cone rotational viscometer", Biorheology, 2, 221 (1965)

11. Dintenfass, L., "A coaxial rhombo-spheroid viscometer: a further development of the cone-in-cone viscometer", Biorheology, 6, 33 (1969)

12. Dintenfass, L., "Guard rings, surface films and artefacts in the viscometry of human blood", Nature, 213, 179 (1967)

13. Dintenfass, L., "Some observations on the role of guard-rings in the cone-in-cone viscometer. Observations on the surface-film in water, oil and human blood", Rheol. Acta, 7, 197 (1968)

14. Dintenfass, L., "Viscosity of blood at high haematocrits measured in microcapillary (parallel-plate) viscometers of r = 3 - 30 microns", in Hemorheology, Proceedings of the First Conference, Reykjavik 1966, ed. A.L. Copley, p. 197, Pergamon Press, Oxford and New York, 1968

15. Dintenfass, L., "Blood viscosity, internal fluidity of the red cell, dynamic coagulation and the critical capillary radius as factors in the physiology and pathology of circulation and microcirculation. Med. J. Australia, 1, 688 (1968)

16. Dintenfass, L., "Dynamics of blood coagulation. Introducing a new coagulation factor: "velocity gradient".", Haematologia, 1, 387 (1967)

17. Dintenfass, L., "A study of thixotropy of concentrated pigment suspensions", Kolloid Ztschr., 48, 163 (1959)

18. Roscoe, R., "The viscosity of suspensions of rigid particles", Brit. J. Appl. Physics, 3, 267 (1952)

19. Haynes, R.H., "Physical basis of the dependence of blood viscosity on tube radius", Amer. J. Physiol., 198, 1193 (1960)

20. Langsjoen, P.H., "Blood viscosity changes in acute myocardial infarction", Postgraduate Med. 39, No. 5, A42 (1966)

21. Dintenfass, L., "Viscosity and clotting of blood in venous thrombosis and coronary occlusions", Circulation Res., 14, 1 (1964)

22. Dintenfass, L., "Viscosity of the packed red and white blood cells", Experimental and Molecular Pathology, 4, 597 (1965)

23. Dintenfass, L., and Rozenberg, M., "Some observations on the viscosity of blood in various diseases", Angiologica, 4, 116 (1967)

24. Dintenfass,L., "Considerations of the internal viscosity of red cell and its effects on the viscosity of whole blood", Angiology, 13, 333 (1962)

25. Taylor, G.I., "The viscosity of a fluid containing small drops of another fluid", Proc. Royal Soc., London, A138, 41 (1932)

26. Oldroyd, J.G., "The elastic and viscous properties of emulsions and suspensions", Proc. Royal Soc., London, A218, 122 (1953)

27. Oldroyd, J.G., "The effect of interfacial stabilizing films on the elastic and viscous properties of emulsions", Proc. Royal Soc., London, A232, 567 (1955)

28. Karnis, A., Goldsmith, H. L., and Mason, S.G., "Axial migration of particles in Poiseuille flow", Nature, 200, 159 (1963)

29. Rumscheidt, F.D., and Mason, S.G., "Particle motions in sheared suspensions. XI. Internal circulation in fluid drops", J. Colloid Sci., 16, 210 (1961)

30. Dintenfass, L., "A study in flow, viscosity and clotting of human blood", Med. J. Australia, 1, 575 (1963)

31. Dintenfass, L., "Molecular and rheological considerations of the red cell membrane in view of the internal fluidity of the red cell", Acta Haemat., 32, 299 (1964)

32. Dintenfass, L., "Internal viscosity of the red cell and a blood viscosity equation", Nature, 219, 956 (1968)

33. Dintenfass, L., "Rheology of the packed red blood cells containing haemoglobins A-A, S-A, and S-S.", J. Lab. & Clin. Med., 64, 594 (1964)

34. Dintenfass, L., and Burnard, E.D., "Effect of hydrogen ion concentration on in vitro viscosity of packed red cells and blood at high haematocrits.", Med. J. Australia, 1, 1072 (1966)

35. Burnard, E.D., and Dintenfass, L., "Effect of hydrogen ion concentration on viscosity of red cells.", Bibl. Haemat., No. 29, Part 4, 1155 (1968) (Proc. XI Congr. Blood Transf.)

36. Chien, S., Usami, S., Dellenback, R.J., and Gregersen, M.I., "Blood viscosity: influence of erythrocyte deformation.", Science, 157, 827 (1967)

37. Goldsmith, H.L., "The microrheology of red blood cell suspensions.", J. Gen. Physiol., 52, 5 (1968)

38. Wells, R., and Schmid-Schonbein, H., "Red cell deformation and fluidity of concentrated cell suspensions.", J. Appl. Physiol., 27, 213 (1969)

39. Dintenfass, L., "Thixotropy of blood and proneness to thrombus formation.", Circulation Res., 11, 233 (1962)

40. Dintenfass, L., Julian, D.G., and Miller, G., "Viscosity of blood in normal subjects and in patients suffering from coronary occlusion and arterial thrombosis.", Amer. Heart J., 71, 587 (1966)

41. Dintenfass, L., "A preliminary outline of the blood high viscosity syndrome.", Arch. Internal Med., 118, 427 (1966)

42. Dintenfass, L., "Blood rheology in cardiovascular diseases.", Nature, 199, 813 (1963)

43. Dintenfass, L., "Rheologic approach to thrombosis and atherosclerosis", Angiology, 15, 333 (1964)

44. Dintenfass, L., "Some rheological factors in pathogenesis of thrombosis", Lancet, 2, 370 (1965)

45. Dintenfass, L., "Fluidity (internal viscosity) of the erythrocyte and its role in physiology and pathology of circulation. Haematologia, 2, 19 (1968)

46. Dintenfass, L., "The internal viscosity of the red cell and the structure of the red cell membrane. Considerations of the liquid crystalline structure of the red cell interior and membrane from rheological data.", Molecular Crystals, Liquid Crystals, 8, 101 (1969)

47. Goldsmith, H. L., and Mason, S.G., "The microrheology of dispersions", in Rheology, ed. F.R. Eirich, vol. 4, p. 86, Academic Press, New York and London, 1967

48. Chien, S., Dellenback, R.J., Usami, S., and Gregersen, M.I., "Plasma trapping in hematocrit determination. Difference among animal species.", Proc. Soc. exp. Biol. Med., 119, 1155 (1965)

49. Economou-Mavrou, C., and Tsenghi, C., "Plasma trapping in the centrifugated red cells of children with severe thalassaemia.", J. Clin. Path., 18, 203 (1965)

50. Chien, S., Dellenback, R.J., Usami, S., Seaman, G.V.F., and Gregersen, M. I., "Centrifugal packing of suspensions of erythrocytes hardened with acetylaldehyde.", Proc. Soc. exp. Biol. Med., 127, 982 (1968)

51. Whitmore, R.L., "The flow behaviour of blood in circulation", Nature, 215, 123 (1967)

52. Whitmore, R.L., "A theory of blood flow in small vessels.", J. Appl. Physiol., 22, 767 (1967)

53. Kondo, S., and Oka, S., "Flow of blood in capillary.", J. Society of Materials Science, Japan, 16, No. 166, 523 (1967)

54. Dintenfass, L., "Micro-rheological classification and analysis of complex multiphase suspensions.", Proc. Intern. Symp. on Second Order Effects in Elasticity, Plasticity and Fluid Dynamics, Haifa, p. 764, Pergamon Press, 1964

55. Dintenfass, L., "Thixotropy and dilatancy in complex multiphase emulsions and suspensions.", Proc. IV Intern. Congr. Rheology, p. 489, John Wiley & Sons, New York, 1964

56. Dintenfass, L., "Viscometry of human blood for shear rates of 0 - 100,000 sec^{-1}.", Nature, 211, 525 (1966)

57. Dintenfass, L., and Read, J., "Pathogenesis of heart failure in acute-on-chronic respiratory failure.", Lancet, 1, 570 (1968)

58. Dintenfass, L., "Effect of velocity gradient on the clotting time of blood and on the consistency of clots formed in vitro.", Circulation Res., 18, 349 (1966)

59. Rozenberg, M., and Dintenfass, L., "Thrombus formation in vitro: a rheological and morphological study.", Australian J. exp. Biol. Med. Sci., 42, 109 (1964)

60. Dintenfass, L., and Rozenberg, M., "Effect of temperature, velocity gradient and i.v. heparin on in vitro blood coagulation and platelet aggregation.", Thromb. diath. Haemorrh., 17, 112 (1967)

61. Ardlie, N.G., Glow, G., and Schwartz, C.J., "Influence of catecholamines on nucleotide-induced platelet aggregation,", Nature, 212, 415 (1966)

62. Mustard, J.F., Glynn, M.F., Nishizawa, E. E., and Packham, M.A., "Platelet-surface interactions: relationship to thrombosis and hemostasis.", Fed. Proc., 26, 106 (1967)

63. McKay, D.G., and Margaretten, W., "Disseminated intravascular coagulation in virus diseases.", Arch. Int. Med., 120, 119 (1967)

64. Guyton, A.C., and Richardson, T.Q., "Effect of hematocrit on venous return.", Circulation Res., 9, 157 (1961)

65. Burton, A.C., "The mechanics of the red cell in relation to its carrier function.", in Ciba Foundation Symposium on Circulatory and Respiratory Mass Transport, p. 67, ed. G.E.W. Wolstenholme and Julie Knight, J. & A. Churchill Ltd., 1969

66. Bohr, C., Hasselbach, K., and Krog, A., "Uber einer in biologischer Beziehung wichtigen Einfluss, den die Kohlen-Saurespannung des Blutes auf dessen Sauerstoffbindung ubt.", Skand. Arch. Physiol., 16, 402 (1904)

67. Naeraa, N., Petersen, E.S., Boye, E., and Severinghaus, J.W., "pH and molecular CO_2 components of the Bohr effect in human blood.", Scand. J. Clin. & Lab. Investig., 18, 96 (1966)

68. Dintenfass, L., "Blood rheology in pathogenesis of the coronary heart diseases.", Amer. Heart J., 77, 139 (1969)

69. Dintenfass, L., "Blood rheology as a factor in the pathogenesis of coronary heart disease.", Israel J. Med. Sci., 5, 652 (1969)

THE OXYGENATION OF BLOOD IN ARTIFICIAL MEMBRANE DEVICES

Edmund E. Spaeth

Department of Chemical Engineering, Washington

University, St. Louis, Missouri

INTRODUCTION

The use of an extracorporeal "heart-lung machine" system to replace heart and lung function in a surgical patient is an increasingly common event in medicine today. The most widely-used gas exchange devices (artificial lungs) in the heart-lung systems are still of the type that contact the blood directly with the gas phase, even though it is well known that prolonged exposure of blood to this high-energy interface causes toxic degradation (8, 30, 37) resulting in an unusually high mortality rate for procedures longer than a few hours duration. Presumably the reason for the continued use of gas contact devices is that rapid surgical procedures have been developed and taking into account factors such as reliability, cost, and convenience, surgeons find the gas contact devices to be adequate. For the long term support of a patient, however, it is agreed that direct gas contact devices cannot be used (15, 25, 29, 37).

In order to significantly reduce the toxic degradation of blood in an artificial lung, it is necessary to avoid direct blood-gas contact. The most popular approach to this problem employs the method used in living organisms-the separation of the blood and gas phases by a thin, semi-permeable membrane. Associated with the use of a membrane, however, is the additional resistance to mass transfer of both the membrane and the slowly moving blood adjacent to it. The development of a practical artificial membrane lung has been difficult because these additional transfer resistances increase the surface area required for adequate gas transfer, and this in turn results in increased size and priming volume.

The development of highly permeable membranes has resulted in most membrane lungs becoming limited by the mass transfer resistance of the blood itself. In order to understand the performance of these devices, then, it becomes necessary to understand the fundamental nature of convective diffusion in blood. In this paper the current status of the various theories describing gas exchange with flowing blood is presented, and an analysis is made of the behavior of artificial membrane lungs operating under various conditions. This analysis is based upon studies performed with well-defined experimental systems which avoid artifacts in the blood flow due to irregularities in geometry and structural properties.

CONVECTIVE DIFFUSION THEORY APPLIED TO GAS EXCHANGE IN ARTIFICIAL MEMBRANE LUNGS

Since an artificial membrane lung generally has characteristic dimensions much greater than the size of the red blood cell, it is reasonable to consider blood flowing through such a device to act as a homogeneous fluid. The conventional equations of fluid flow can then be used as the basis for a general mass transfer model. This approach must be used with caution, however, since both diffusion and chemical reaction can occur in the fluid, along with flow, and blood is clearly non-homogeneous on the microscopic scale of these processes. What is generally done in a situation like this is to describe the diffusion and chemical reaction using the standard equations for a homogeneous media, but to substitute "effective", or integral values for the phenomenological coefficients. The basic differences between most models for convective diffusion in blood result from the different approaches taken to describe this "effective" behavior. At present there are three such approaches.

Macroscopic Model

The most widely used approach is to treat blood as a homogeneous, incompressible fluid with uniformly distributed chemical reactions. This point of view does not take into account the fact that red cells are discrete bodies which can contain important chemicals not present in plasma. The appropriate general equation for the convective diffusion of a species i in such a homogeneous, incompressible fluid is (7):

$$\frac{D}{Dt}(n_i) = -\nabla \cdot \vec{J} + R_i \quad (1)$$

Rate of change in a flowing element	Net diffusion from the element	Net generation by reaction in the element

where $\frac{D}{Dt}(\ )$ is the substantial derivative $\frac{\partial}{\partial t}(\ ) + \vec{v}\cdot\nabla(\ )$. If this equation is written for each of a special set of species k=1, 2, ... , r, where n_{i1} refers to the concentration of physically dissolved i , and $n_{i2}, \ldots, n_{ir}$ refer to the concentration of i present as the reversibly reacted forms k, and if these r equations are summed, the resulting equation is a total balance on all pertinent forms of species i . Since everything is written in terms of the amount of species i, and the total amount of i in the set k=1 ...r is unchanged by the chemical reactions, the sum of the reaction terms must vanish, giving:

$$\frac{D}{Dt}(n_T)_i = \frac{D}{Dt}\left(n_{i1} + \sum_{k=2}^{r} n_{ik}\right) = -\nabla\cdot(\vec{J}_T)_i \tag{2}$$

where

$$(\vec{J}_T)_i = \vec{J}_{i1} + \sum_{k=2}^{r} \vec{J}_{ik} \tag{3}$$

Substituting the solubility-partial pressure product for n_{i1}, and defining

$$N_i = \sum_{k=2}^{r} n_{ik} \tag{4}$$

$$N_i^* = N_i \quad \text{at chemical equilibrium} \tag{5}$$

equation (2) can be rearranged to give:

$$\frac{D}{Dt}(\alpha_i P_i + N_i^*) = -\nabla\cdot(\vec{J}_T)_i + \frac{D}{Dt}(N_i^* - N_i) \tag{6}$$

Rate of change of total species i assuming chemical equilibrium in a flowing element.	Net diffusion from the element	Rate of change of the displacement from chemical equilibrium in the element.

In order to solve this general "macroscopic" equation for gas transfer to blood flowing in an artificial lung, the following parameters must be specified:

(1) the macroscopic velocity field, $\vec{v}$.
(2) the effective solubility in blood, α_i.
(3) the chemical equilibrium relation N_i^*.
(4) the total "effective" diffusional flux in blood, $(\vec{J}_T)_i$.
(5) the local displacement from chemical equilibrium, $(N_i^* - N_i)$.

Coupled Macroscopic-Microscopic Model

A more detailed and correspondingly more difficult approach is to consider blood to flow as an incompressible, macroscopically homogeneous fluid, as above, but with the red cells acting as uniformly distributed chemical source/sinks. The behavior of an individual sink is in turn described using the same homogeneous fluid equations, applied on a microscopic scale to the red cell interior. The two equations (macroscopic and microscopic) are then coupled using the boundary conditions at the red cell-plasma interface. This general method of treating heterogeneous, reacting systems by coupling macroscopic equations (containing "effective" phenomenological coefficients) with microscopic equations describing the particulate phase is an established technique in the analysis of catalytic reactors (44, 42).

For convective diffusion in blood, this approach differs from the "macroscopic" theory in that the convective diffusion equation is not summed over the entire set k=1, ..., r of physically dissolved and reversibly reacted forms of i, but rather is summed only over those forms present in the plasma (i.e. the set k, plasma). This results in a net chemical reaction term in the macroscopic equation describing the blood:

$$\frac{D}{Dt}\left(\alpha_i P_i + \sum_{\substack{k,\text{plasma}\\ k\neq 1}} n_{ik}\right) = -\nabla\cdot\left(\vec{J}_{i1} + \sum_{\substack{k,\text{plasma}\\ k\neq 1}} \vec{J}_{ik}\right) + \sum_{k,\text{plasma}} R_i \quad (7)$$

For the case of oxygen diffusion in blood only the physically dissolved form exists in the plasma, so equation (7) reduces to:

$$\frac{D}{Dt}\left(\alpha_{O_2} P_{O_2}\right) = -\nabla\cdot\vec{J}_{O_2 1} + R_{O_2} \quad (8)$$

For carbon dioxide, both physically dissolved and reversibly reacted forms exist in the plasma.

The net chemical reaction term in equation (7) is then evaluated by solving the equation for diffusion and chemical reaction applied on a "microscopic" scale to the interior of the red cell source/sinks:

$$\frac{D}{Dt}\left(\alpha'_i P_i + \sum_{\substack{k,\text{cell}\\ k\neq 1}} n'_{ik}\right) = -\nabla\cdot\left(J'_{i1} + \sum_{\substack{k,\text{cell}\\ k\neq 1}} J'_{ik}\right) + \sum_{k,\text{cell}} R'_i \quad (9)$$

In this case, k, cell refers to the set of physically dissolved and reversibly reacted forms of species i present within the red cell.

These equations are generally non-linear since the reaction terms contain mixed products of the n'_{ik} 's. In addition to Equation (9) a significant mass transfer resistance in the red cell membrane would require inclusion of the equation:

$$\frac{D}{Dt}\left(\alpha_i'' P_i + \sum_{\substack{k,\text{membrane}\\ k\neq 1}} n''_{ik}\right) = -\nabla\cdot\left(\vec{J}_{i1} + \sum_{\substack{k,\text{membrane}\\ k\neq 1}} \vec{J}''_{ik}\right) \tag{10}$$

at the surface of the red cell source/sinks. Equation (9) and (10) are coupled at the inside surface of the membrane by continuity of flux and chemical potential. This set of microscopic equations is in turn coupled to the macroscopic equation by the continuity of flux and chemical potential at the outer surface of the cell membrane:

$$\sum_{k,\text{membrane}}\left[n''_{ik} = \beta_k\, n_{ik}\right] \tag{11}$$

$$\sum_{k,\text{plasma}} R_i = C_{rc} \sum_{k,\text{membrane}} \iint_S \vec{J}''_{ik}\cdot d\vec{S} \tag{12}$$

where β_k is the partition coefficient for species k between the plasma and the membrane, and C_{rc} is the number of red cells per unit volume.

In order to solve the set of equations comprising the coupled macroscopic-microscopic model, the following parameters must be specified:

(1) the macroscopic velocity field, $\vec{v}$.
(2) the effective solubility in blood, α_i.
(3) the total effective diffusional flux in blood, $\sum_{k,\text{plasma}} \vec{J}_{ik}$.
(4) the solubility in the red cell interior, α_i'.
(5) the total diffusional flux in the red cell interior, $\sum_{k,\text{cell}} \vec{J}'_{ik}$.
(6) the chemical reaction rates in the red cell, $\sum_{k,\text{cell}} R_i'$.
(7) the solubility in the cell membrane, α_i''.
(8) the total diffusional flux in the membrane, $\sum_{k,\text{membrane}} \vec{J}''_{ik}$.

(9) the partition coefficient, β_k.

Statistical Models

A somewhat different approach to the modeling of convective diffusion in blood is to consider the behavior of blood at the microscopic level, using a statistical analysis to take into account the net behavior of individual red cells which are exchanging material with the surrounding plasma. Since individual red cells have been observed to follow an erratic path in macroscopically smooth blood flow, red cells flowing in an artificial lung may be exposed to significantly different plasma concentration histories. Any lack of equilibration between the cells and the plasma will generate a distribution function which characterizes the probability of finding a cell with a specific solute concentration at a given location. Such a probability distribution function can be incorporated into a macroscopic conservation of species balance, using the "microscopic" equations presented above to characterize the rate of solute exchange between the plasma and individual red cells. The inherent complexity of this approach restricts its use to practical situations in which major simplifying assumptions can be made, and in which adequate experimental data can be obtained for use in developing the form of the probability distribution function.

Discussion

In order to use any of the general models described above, it is first necessary to define the exact form of the various terms in the equations. This involves specifying the nature of the blood flow, the gas content of blood, the nature of diffusion in blood, and the behavior of the chemical reactions in blood.

The macroscopic velocity field v is most easily obtained by solving the equations of flow for a Newtonian fluid, and neglecting the fact that blood exhibits non-Newtonian behavior both at low shear rates (33, 34), presumably due to the breakup of red cell clusters, and at high shear rates (22), presumably due to red cell deformation. This latter effect is generally small, so in strongly sheared systems (shear rate $>$ 100 sec^{-1}) it is reasonable to assume the viscosity of blood to approach a constant "Newtonian" value (35). This approximation is a particularly reasonable one in artificial membrane lungs since the mass transfer is generally controlled by diffusion near the membrane surface, where the shear is greatest.

Another effect which may be of more importance to convective diffusion is the microscopic motion of the red cells. It has been observed that in a "macroscopically" uniform blood flow, such as low Reynolds number flow through tubes, individual red cells undergo both rotational and translational movements. At the high hematocrits of whole blood, the rotational movements are significantly retarded, but Goldsmith (17) has observed net radial translational movements on the order of 5 microns (over a one second interval) during flow through a 100 micron tube of a 0.42 hematocrit suspension of red cells and reconstituted ghosts in plasma. These results were obtained for wall shear rates up to only 20 sec^{-1}.

Changes in mass transfer rates due to microscopic red motion or non-Newtonian rheological behavior have not been observed experimentally in studies of inert gas and organic solute transfer to blood flowing through membrane devices (46, 5). Although neither of these experiments operated at extremely high shear rates, shear rates on the order of 10^2 -10^3 sec^{-1} were used.

The solubilities and chemical equilibrium relationships for most of the gases of physiological interest are available in the literature (19). The relationship between oxygen tension (P_{O_2}) and total oxygen content (i.e. the sum of the physically dissolved and reversibly reacted forms) in normal whole blood at 37° is shown in Figure 1. The curve is shifted to the right by a decrease in pH (the Bohr effect) or an increase in temperature. Above an oxygen tension of 150mm Hg further increase in oxygen content is almost entirely due to the increase in physically dissolved oxygen, and the oxygen content curve for blood is essentially parallel to that for water. The relationship between carbon dioxide tension (P_{CO_2}) and total carbon dioxide content for normal whole blood at 37°C is shown in Figure 2. This curve is shifted to right by increases in oxyhemoglobin concentration.

The nature of diffusion in blood has been the object of considerable study in recent years, and much of this information has just recently become available (5, 48, 21). Since blood is a highly heterogeneous fluid on a microscopic scale, the simple Fick's law relationship employing molecular diffusivities cannot be used directly. The usual procedure is to introduce an effective diffusivity which, for the case of dissolved gases diffusing through blood, is defined by:

$$\vec{J}_i = - D_{i,b} \alpha_i \nabla P_i \qquad (13)$$

The effective diffusivity $D_{i,b}$ is a function of hematocrit and the relative permeabilities of the suspended and continuous phases, as well as a function of temperature and viscosity.

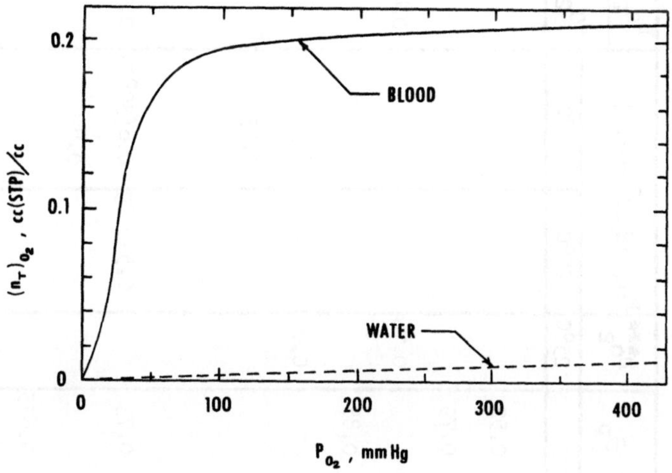

Figure 1. The total oxygen content of water and whole blood at pH 7.4 and 37°C.

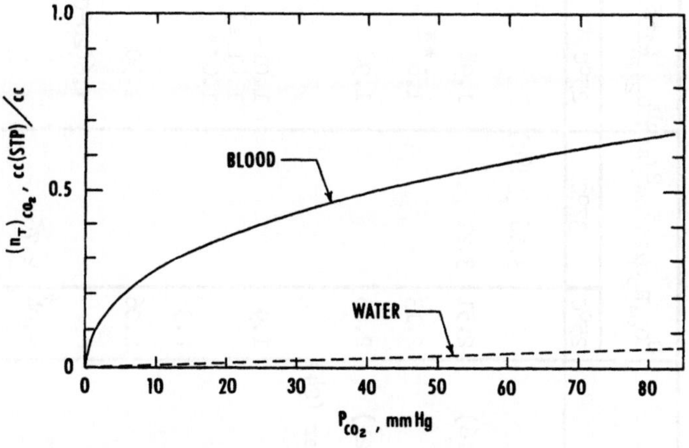

Figure 2. The total carbon dioxide content of water and whole blood at pH 7.4 and 37°C. (f = 0.6)

Table I. The Diffusivity of Oxygen in Water, Plasma, Concentrated Hemoglobin Solution, and Packed Red Cell.

Reference	D_{O_2,H_2O} ($\times 10^5 cm^2/sec$)		$D_{O_2,p}$ ($\times 10^5 cm^2/sec$)		θ_p	$D_{O_2}^{***}$	$D_{O_2,rc}$ ($\times 10^5 cm/sec$)		$D_{O_2,rc}/D_{O_2,p}$
	25°C	37°C	25°C	37°C		25°C	25°C	37°C	
Buckless et al (1)		3.56 (38°C)		1.86 (38°C)	0.52				
Gertz & Loeschke (16)	2.51	3.21	1.98	2.54(s)	0.79				
Goldstick (18)	2.13		1.8**			0.76			0.43
Hershey & Karhan (21)	2.25		1.98		0.88				
Keller & Friedlander (24)						0.86			
Kreuzer (26)	1.9		1.01-1.41(s) 1.6**					0.43	
Pircher (38)	2.3								
Stein (48)	2.06		1.50		0.73		0.5-0.79	0.6(30°C)	0.33-0.53
Thews (49)			1.6(20°C)					0.8	
Yoshida et al (53)	2.04*	2.76		1.86(s)	0.675				

* Stokes-Einstein correction of value measured at 37°C.
** 7.0 gm/100ml protein solution (eg. Bovine Serum Albumin).
*** (x 10^5 cm^2/sec for 33% Hb)

Effective diffusivities in heterogeneous media can be estimated using equations based on Maxwell's treatment of conduction in a suspension of non-interacting spheres (2, 6). The equations presented by Fricke (13) for a suspension of oblate spheroids have been applied to the diffusion of gases in blood with some success (11, 19). These equations are generally written in terms of ψ, the relative permeability of the heterogeneous media (blood) to the suspending medium (plasma):

$$\frac{\alpha_{i,b} D_{i,b}}{\alpha_{i,p} D_{i,p}} \equiv \psi = f\left(\text{shape}, H, \frac{\alpha_{i,rc} D_{i,rc}}{\alpha_{i,p} D_{i,p}}\right) \tag{14}$$

Colton (5,4) has in turn applied this type of equation to diffusion in plasma, treating the plasma as a dilute suspension (in saline) of impermeable prolate ellipsoids (proteins) which are capable of reversible solute binding and diffusion. For the case of a non-interacting solute, such as oxygen, this analysis gives:

$$\frac{D_{i,p}}{D_{i,H_2O}} \equiv \theta_p \cong 0.95 \tag{15}$$

The diffusivities of oxygen in blood and blood constituents have been measured by a number of workers, and reasonably consistent values are emerging, as shown in Table I. The measured values of θ_p are somewhat lower than the predicted value in equation (14). The data for $D_{O_2,rc}$ is still limited and lacking in agreement. Values for $D_{O_2,rc}/D_{O_2,p}$ calculated using both the oxygen diffusivity in the red cell and the oxygen diffusivity in a concentrated hemoglobin solution are in reasonable agreement, which is consistent with the finding of Kreuzer and Yahr (27) that the resistance of the red cell membrane to oxygen diffusion is small.

Direct experimental evidence of the applicability of heterogeneous media theory to diffusion through blood has recently been reported, as shown in Figure 3. The solid data points are measurements by Stein (48) of the diffusivity of oxygen through immobilized suspensions of red cells in agar. In one experiment (circles) the agreement of the data with the curve predicted by heterogeneous media theory is excellent. In a second experiment (triangles), under essentially the same conditions, the data are scattered about the theoretical curve, and a significantly different curve is predicted due to the much lower measured red cell permeability(H=1.0). The open data points in Figure 3 are the results of an investigation by Colton (5) of the diffusion

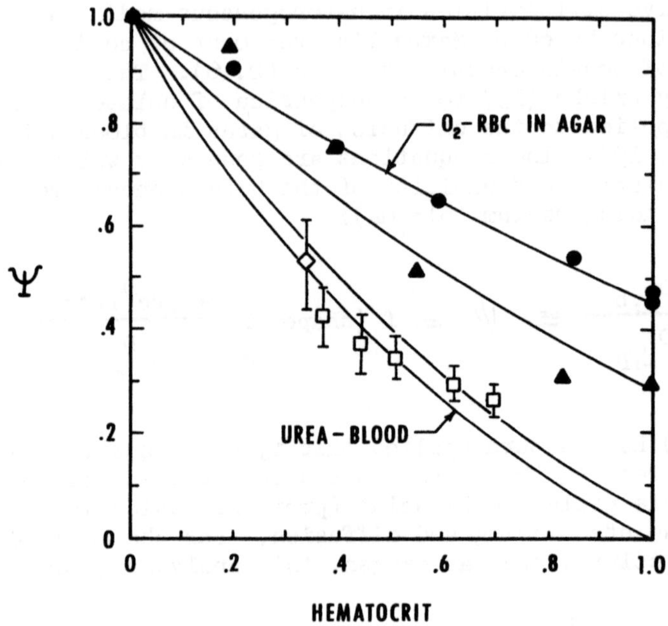

Figure 3. A comparison of theoretically predicted heterogeneous media diffusivities (curves) with experimentally measured values. The solid data points are taken from experiments by Stein (48) using red cells suspended in agar, and the open data points are taken from experiments by Colton (5) using defibrinated and ACD blood.

of urea in stagnant defibrinated blood (squares) and flowing fresh whole blood containing ACD (diamond). These data all lie within experimental error of the curves predicted by heterogeneous media theory for the two limiting values of red cell permeability to urea, although a somewhat different curve shape would fit the data better. Although these studies do not indicate complete agreement between experiment and the heterogeneous media theory, from the point of view of artificial organ design the use of the heterogeneous media theory will not result in much error.

The diffusion of oxygen in unsaturated hemoglobin solutions has been found to be "facilitated" due to the simultaneous diffusion of oxyhemoglobin along with the dissolved oxygen, under conditions of near chemical equilibrium (24, 45, 52). The degree of facilitation is related to the slope of the blood oxygenation curve (Figure 1) which is clearly a function of oxygen partial pressure. Because of this, the effective diffusivity of oxygen through blood could be expected to depend on the oxygen partial pressure. Recent measurements of the diffusivity of hemoglobin molecules in hemoglobin solutions at concentrations comparable to the red cell interior, however, show a dramatic fall in diffusivity with increasing concentration (36). Calculations of the effective diffusivity of oxygen in blood, using the heterogeneous media equations and these lower values of the hemoglobin diffusivity, indicate a maximum increase of $D_{O_2,b}$ by only 10-15% as a result of the "facilitation" effect (46). Even this increase is questionable on the basis of the theoretical analysis of Friedlander and Keller (14) which shows that the near-equilibrium behavior necessary for facilitation to occur might not be established in hemoglobin layers as thin as the red blood cell. Recent measurements of the diffusion of oxygen through packed red cells and red cells suspended in agar (48) have demonstrated that there is indeed a negligible dependence of $D_{O_2,b}$ on the oxygen partial pressure, so it appears that facilitation of oxygen diffusion by the simultaneous diffusion of oxyhemoglobin is not an important mechanism in blood.

According to Table I a reasonable value of $D_{O_2,rc}/D_{O_2,p}$ would be 0.4. Using this value in the heterogeneous media theory, and assuming the oxygen solubility to be the same in plasma and the red cell interior, results in the curve for the effective diffusivity of oxygen in blood as a function of hematocrit, presented in Figure 4. Superimposed on this curve is the data of Hershey and Karhan (21) for oxygen diffusion in stagnant, defibrinated sheep's blood. The fact that their data lies consistently above and parallel to the theoretical curve could be due to a different value for $D_{O_2,rc}$ in sheep's blood, or the result of convective effects caused by red cell sedimentation during their measurements.

The diffusion of carbon dioxide in blood is also of great importance to the performance of an artificial lung, but far less is known about the behavior of this system. The combination of carbon dioxide with blood not only involves several chemical species, but the rapid reaction of the most prevalent form, bicarbonate ion, requires an enzyme catalyst (carbonic anhydrase) found only within the red cells. Because of this, the dissolved carbon dioxide must diffuse into the red cell before significant reaction takes place, complicating the nature of the diffusion process. Several workers have shown that carbon dioxide diffusion is "facilitated" in aqueous solutions of carbonic anhydrase (10, 50, 43) in a manner analogous to the facilitation of oxygen diffusion in hemoglobin solutions. The carrier species in this case is the bicarbonate ion. The facilitated diffusion of carbon dioxide in blood can be important, unlike oxygen, because the diffusivity of the carrier species is relatively large, and the phenomenon is not restricted to the interior of individual red cells.

Experimental evidence of the existence of facilitated carbon dioxide diffusion in flowing whole blood has been presented by Spaeth and Friedlander (46), in experiments with a silicone rubber membrane rotating disk device. Measured carbon dioxide transfer rates were about 25% greater than predictions made assuming local chemical equilibrium. The facilitation of carbon dioxide diffusion in a stagnant red cell suspension (H = 0.02) has also been observed by Enns (11), who found that the degree of facilitation increased five-fold in the same solution with the cells lysed, indicating rate limiting effects of diffusion into and out of the red cells. Spinak (47) on the other hand has analytically studied the equilibration of squat cylinder and sphere models of red cells, and concludes that the diffusion of carbon dioxide only affects the equilibration rate in a small way. Enns also found that carbonic anhydrase itself can facilitate carbon dioxide diffusion without significant bicarbonate ion present (10), but this effect was confined to partial pressures below about 23 mmHg. Washed ghost cells were also found to facilitate carbon dioxide diffusion presumably due to carbonic anhydrase which either was a part of the membrane structure or attached to the membrane surface.

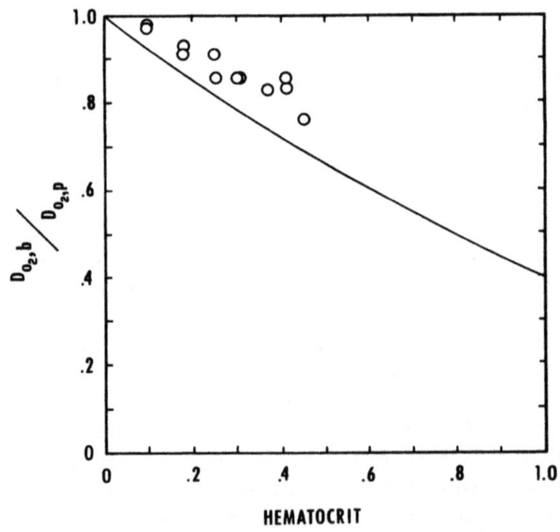

Figure 4. The effective diffusivity of oxygen in blood calculated from heterogeneous media theory, using $D_{O_2,rc}/D_{O_2,p} = 0.4$. The data are taken from Hershey and Karhan (21) for oxygen diffusion in stagnant defibrinated blood.

Along with the nature of blood flow and the nature of diffusion in blood, the behavior of the chemical reactions in blood must also be specified in order to use the various convective diffusion models described above. The simplest approximation is to assume that the red cells are locally near chemical equilibrium with the surrounding plasma, and use the appropriate chemical equilibrium relationship from the literature (such as Figure 1 or Figure 2). The error resulting from this approximation can be estimated using measurements of the equilibration rates of red cells. The results of numerous red cell equilibration rate studies are available for oxygen transfer to and from red cells, but few studies of this type have been performed for carbon dioxide transfer (12, 41, 40). The measured equilibration rates of red cells with both oxygen and carbon dioxide are an order of magnitude smaller than the pure chemical reaction rate in free solution, indicating important rate limiting effects of diffusion in these experiments.

The equilibration of red cells with oxygen has been found to approximate an exponential process. This fact can be used to estimate the displacement from chemical equilibrium ($N_{O_2}^* - N_{O_2}$) for the macroscopic model. For a flowing system it is important to determine this displacement locally in the flow field, since the velocity of the blood and thus the effective residence time of the cells as they pass through the concentration gradient, can vary widely with position. A relatively simple way to estimate the local displacement from chemical equilibrium in flowing blood is to determine the distance the cell would travel during the time required for re-equilibration, following a step perturbation in plasma concentration. The characteristic time required for the cell to return to a specified fraction of equilibrium following the step perturbation is easily calculated for an exponential process, and the product of this characteristic time and the local velocity gives the "re-equilibration length." This re-equilibration length should be small with respect to the characteristic diffusional distance of the system (eg. boundary layer thickness) if a local equilibrium assumption ($N_{O_2} = N_{O_2}^*$) is used.

A similar analysis cannot be made as easily for carbon dioxide transfer, since this equilibration involves several simultaneous processes. The uptake of carbon dioxide by red cells has been found to occur in two stages - a fast initial rate followed by a slower rate. The exchange of bicarbonate and chloride ions between the red cells and the plasma seems to be a relatively slow process, and may be responsible for the secondary, slow equilibration rate. Because the overall equilibration rate of red cells with carbon dioxide is so poorly understood, it is difficult to estimate the displacement from chemical equilibrium in a quantitative manner. Using carbonic anhydrase catalyzed Michaelis-Menten kinetics at 37°C, in membraneless red cell models (cylinder, sphere), Spinak(47) has estimated the time required for a red cell to respond to a step change in plasma concentration to be on the order of 0.1 - 1.0 msec. These times may be short due to the omission of chloride diffusion and cell membrane resistance in the models. Measurements of red cell equilibration in the Hartridge-Roughton rapid reaction mixing apparatus (41) indicate equilibration times on the order of 0.1 second for both oxygen and carbon dioxide transfer. These times are probably too long due to diffusional boundary layer resistances in the plasma adjacent to the cell membranes.

The literature contains a number of analyses of gas exchange in blood, each of which applies a specific set of assumptions about the nature of flow, diffusion, and chemical reaction in blood to one of the basic convective diffusion models described above. The macroscopic model is by far the most extensively studied at this time. Marx, et al (32) investigated the unsteady oxygen uptake rate of a stagnant horizontal blood film separated from a flowing oxygen gas stream by a silicone rubber membrane. They simplified equation (6) by assuming local chemical equilibrium (i.e. $N_{O_2} = N_{O_2}^*$) and a constant value for $D_{O_2,b}$:

$$\frac{\partial}{\partial t}\left(\alpha_{O_2} P_{O_2} + N_{O_2}^*\right) = D_{O_2,b}\frac{\partial^2}{\partial x^2}\left(P_{O_2}\right) \qquad (16)$$

$D_{O_2,b}$ was evaluated from the experimental measurements, and found to be 1.3×10^{-5} cm^2/sec at 27°C. Since the diffusion occurred mostly through saturated blood, with the oxyhemoglobin reaction confined to a thin "advancing front," it was difficult to determine whether the chemical reaction had any effect on $D_{O_2,b}$. A possible source of error in the interpretation of the data was the neglect of red cell sedimentation. This could easily have been significant during the 1-20 minute period required for the measurements.

Buckles et al.(1) studied oxygen transfer to blood flowing through a straight silicone rubber membrane tube, using the macroscopic model with local chemical equilibrium and constant $D_{O_2,b}$, and using the velocity profile predicted from the Casson equation to account for non-Newtonian effects. The appropriate form of equation (6), neglecting axial diffusion, is:

$$v_z(r)\frac{\partial}{\partial z}\left(\alpha_{O_2} P_{O_2} + N_{O_2}^*\right) = D_{O_2,b}\,\alpha_{O_2}\left(\frac{1}{r}\frac{\partial}{\partial r}\left(r\frac{\partial P_{O_2}}{\partial r}\right)\right) \qquad (17)$$

This model was tested experimentally for the case of oxygen uptake by reduced blood at 38°C ($P_{O_2} = 0$, $P_{CO_2} = 36$mm Hg) from a gas mixture (P_{O_2} = 686mm Hg, P_{CO_2} = 36mm Hg). The data agreed moderately well with transfer rates predicted from equation (17) using a value of $D_{O_2,b} = 1.38 \times 10^{-5}$ cm^2/sec obtained from heterogeneous media theory (H = 0.28), although the shape of the theoretical curve did not fit the data well. The importance of using the non-Newtonian velocity profile was not clear, since the corresponding transfer rates for Poiseuille flow were not presented.

Weissman and Mockros (51) also studied oxygen uptake by blood flowing through silicone rubber membrane tubes, but used a simpler version of the macroscopic model. They neglected the transfer resistance of the tube wall (membrane) and assumed Newtonian flow behavior of the blood, along with assuming constant $D_{O_2,b}$ and local chemical equilibrium. Using this model they were able to fit data for oxygen transfer from a gas mixture (P_{O_2} = 667mm Hg, P_{CO_2} = 43mm Hg) to freshly drawn cattle blood at 38°C (initially at 75% oxyhemoglobin saturation), using a value of 0.89×10^{-5} cm^2/sec for $D_{O_2,b}$. The theoretical curve fit the data from experiments performed with tubes of varying wall thickness-diameter ratios, indicating that the transfer resistance of the membrane was small, as assumed, and the gas exchange process was fluid limited.

The macroscopic model for convective diffusion in blood was tested experimentally over a wide range of blood gas conditions by Spaeth and Friedlander (46), using a membrane rotating disk device. Measurements were made of the uptake of inert gas (Kr-85), the removal of oxygen, and the removal of carbon dioxide from freshly-drawn citrated whole human blood (H = 0.41) at 37.5°C. The exchange rate of the inert gas was found to be quantitatively predictable using the macroscopic model with Newtonian fluid mechanics and a constant effective diffusivity calculated from heterogeneous media theory. The exchange rate of oxygen was found to be up to four and a half times that of an inert gas (i.e. $N_{O_2} \equiv 0$), but could also be described well, as shown in Figure 5, by the macroscopic model assuming local chemical equilibrium, Newtonian fluid mechanics, and constant $D_{O_2,b}$ calculated from heterogeneous media theory. The value of $D_{O_2,b}$ used in these calculations was 1.33×10^{-5} cm^2/sec. For steady state oxygen transfer in the rotating disk system, equation (6) becomes:

$$v_z\left[\alpha_{O_2} + 1.34\, C_{Hb}\left(\frac{df}{dP_{O_2}}\right)\right]\left(\frac{dP_{O_2}}{dz}\right) = D_{O_2,b}\,\alpha_{O_2}\,\frac{d^2P_{O_2}}{dz^2} \qquad (18)$$

where

$$\left[\alpha_{O_2} + 1.34\, C_{Hb}\left(\frac{df}{dP_{O_2}}\right)\right] = \frac{d}{dP_{O_2}}\,(n_T)_{O_2} \qquad (19)$$

Use of a "facilitated" oxygen diffusivity in the red cell interior:

$$D_{O_2^*,rc} = D_{O_2,rc} + D_{Hb,rc}\left(\frac{1.34\ C_{Hb}}{\alpha_{O_2}}\right)\frac{df}{dP_{O_2}} \qquad (20)$$

to calculate $D^*_{O_2,b}$ from the heterogeneous media theory, results in the "facilitated" local equilibrium macroscopic equation:

$$v_z\left[\alpha_{O_2} + 1.34C_{Hb}\left(\frac{df}{dP_{O_2}}\right)\right]\frac{dP_{O_2}}{dz} = \alpha_{O_2}\frac{d}{dz}\left(D^*_{O_2,b}\frac{dP_{O_2}}{dz}\right) \qquad (21)$$

For oxygen transfer, the predicted effect of facilitated diffusion in blood was found to be small, and as mentioned earlier, more recent measurements of oxygen diffusion in packed red cells do not demonstrate any facilitation effect at all. The rate of carbon dioxide desorption from blood near venous conditions was found to be more than twice the transfer rate of inert gas, but in this case could not be accounted for by a local chemical equilibrium model using a value of $D_{CO_2,b} = 1.3 \times 10^5$ cm^2/sec, as shown in Figure 6. This data suggests an important facilitation

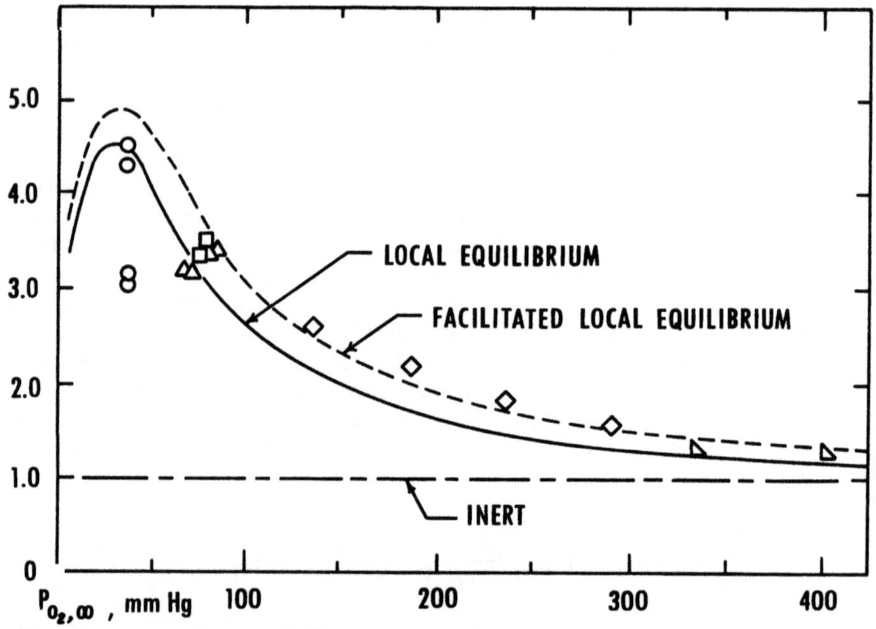

Figure 5. A comparison of the measured oxygen removal rates from freshly drawn citrated blood (pH = 7.4, $\sqrt{\ } = 0.03$ cm^2/sec, C_{Hb} = 0.136 g/ml, H = 0.41, T = 37°C) with the transfer rates predicted by the inert and local equilibrium macroscopic models ($D_{O_2,b}$ = 1.33×10^{-5} cm^2/sec, $\alpha_{O_2,b} = 3.0 \times 10^{-5}$ cc(STP)/cm^{3}mmHg).

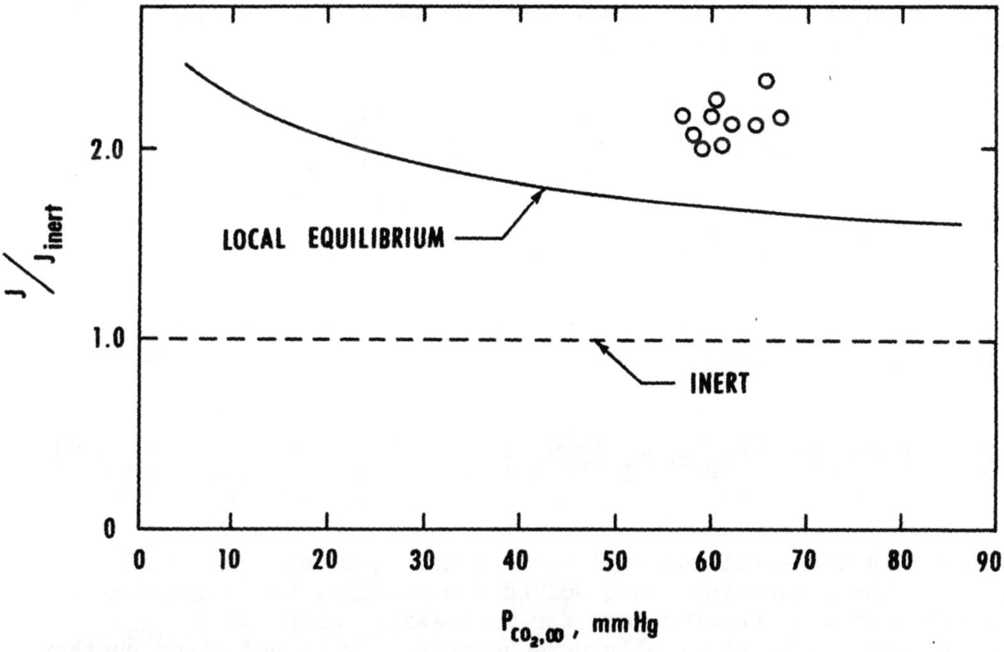

Figure 6. A comparison of the measured carbon dioxide removal rates from freshly drawn citrated blood (P_{O_2} = 40 mmHg, pH = 7.22, ν = 0.026 cm^2/sec, T = 37°C) with transfer rates predicted by the inert and local equilibrium macroscopic models ($D_{CO_2,b}$ = 1.3 x 10^{-5} cm^2/sec, $\alpha_{CO_2,b}$ = 6.3 x 10^{-4} cc(STP)/cm^3mmHg).

of carbon dioxide diffusion in blood, presumably due to simultaneous diffusion of bicarbonate ion.

Other examples of the application of the macroscopic model to gas transfer in blood are available in the literature, but they use essentially the same local equilibrium assumption.

An example of the coupled macroscopic-microscopic model is the analysis by Hershey and Hesselberth (20) of oxygen transport in a falling film of blood. They considered the cells to be homogeneously distributed membraneless spherical sinks. For steady state oxygen absorption in a falling blood film, the macroscopic equation (7) becomes:

$$v_z \alpha_{O_2} \frac{P_{O_2}}{z} = D_{O_2,b} \alpha_{O_2} \frac{\partial^2 P_{O_2}}{\partial y^2} + R_{O_2} \qquad (22)$$

flow being in the z-direction. For the spherical red cells, they used the microscopic equation:

$$v_z \alpha'_{O_2} \frac{\partial P_{O_2}}{\partial z} = \left(\frac{D_{O_2,rc} \alpha'_{O_2}}{r^2} \right) \frac{\partial}{\partial r} \left(r^2 \frac{\partial P_{O_2}}{\partial r} \right)$$

$$+ k_f \alpha'_{O_2} P_{O_2} C'_{Hb} - k_r C'_{HbO_2} \qquad (23)$$

and coupled the two equations by assuming uniform plasma P_{O_2} over the entire cell surface, and:

$$R_{O_2} = -S D_{O_2,rc} \alpha'_{O_2} \left(\frac{\partial P_{O_2}}{\partial r} \right) \qquad (24)$$

where S is the total red cell surface area per unit volume of blood. These equations were solved numerically, but a maximum occurred in the transfer rate for increasing values of $D_{O_2,rc}$, which seems to be physically unreasonable. It is not clear whether this was the result of the model or the numerical computation.

An example of a statistical model of oxygen transfer in blood is the analysis by Martisovits and Veis (31). They assumed the oxygenation of the red cell to be described by the exponential equation

$$f - f_g = (f_o - f_g)\, e^{-\gamma t} \qquad (25)$$

where f is the fractional saturation of the red cell, f_g is the value of f at the P_{O_2} of the gas stream ($f_g = 1$ for $P_{O_2,g} \geq 100$mm Hg), f_o is the initial saturation, and γ is the red cell oxygenation rate constant (dependent on both the kinetics of hemoglobin oxygenation and oxygen diffusion). By considering a four-dimensional space where the coordinates x,y,z specify the position of the red cell, and the coordinate f specifies the oxygen saturation of the cell, and by defining a probability density function ξ (x,y,z,f,t) which characterizes the probability of the occurrence of a red cell with a specific value of f, they set up a mass balance for the red cells and arrive at the condition:

$$\frac{\partial \xi}{\partial t} + \bar{v} \cdot \nabla \xi + \gamma (f_g - f) \frac{\partial \xi}{\partial f} = \gamma \xi \qquad (26)$$

which can be solved for the distribution function $\varsigma(x,y,z,f,t)$ knowing the blood velocity $\vec{v}(x,y,z,t)$ and the oxygen distribution, defined using $\gamma(x,y,z,t)$ and $f_g(x,y,z,t)$. Unfortunately these quantities are difficult to specify in an actual oxygenator. In order to use this method to analyze a "general" oxygenator operating at steady state, they found it necessary to divide the oxygenator into one-dimensional stages connected in series, in each of which γ and f_g were assumed to be constant. The resulting expressions were used to describe the operating characteristics of a disk oxygenator, but experimental verification of these results was not presented.

APPLICATIONS TO THE DESIGN OF ARTIFICIAL MEMBRANE LUNGS

One conclusion that can be drawn from the work presented in the preceeding section is that the Newtonian, local-equilibrium macroscopic model describes convective oxygen transfer in blood with an accuracy sufficient for design purposes. It is now of interest to use this model to investigate the design of artificial membrane lungs.

One important consideration in the design of an artificial lung is the relative transfer rate of oxygen to carbon dioxide. An artificial lung must not only add oxygen to blood and remove carbon dioxide, but it must do this at the proper relative rates in order to maintain physiological balance. The factors that determine the overall relative transfer rates of oxygen and carbon dioxide in an artificial membrane lung are the relative transfer rates in both the blood and the membrane phases, and the fraction of the total mass transfer resistance of the lung attributable to each of these phases. From the experiments described earlier it is clear that the major resistance to gas transfer can be the blood film adjacent to the membrane and not the membrane itself.

The relative permeabilities of oxygen and carbon dioxide in blood can be determined from the experimental data obtained by Spaeth and Friedlander (46) using the rotating disk system. Since the relative permeability measured under local equilibrium conditions is related only to the properties of the fluid, relative permeabilities evaluated from data taken in the rotating disk system should apply to other systems as well. The experimental measurements in the rotating disk system were for oxygen removal rather than uptake, so it is first necessary to numerically calculate the oxygen uptake rates in the rotating disk system using the macroscopic local equilibrium model. The results of such a calculation are presented in Figure 7, along with the experimentally verified curve describing the deoxygenation rates.

It is of interest to note the difference between the rate of oxygen uptake and the rate of oxygen removal for similar values of the net driving force, $|P_{O_2,g} - P_{O_2,\infty}|$, the rate of oxygen uptake being considerably greater than the corresponding rate of oxygen removal. This can be explained as follows. Since the enhancement of the oxygen transfer rate over that of an inert gas is due to the convective term $\vec{v}\cdot\nabla(N^*_{O_2})$ in equation (6), the enhancement effect will be greater in regions of higher fluid velocity. Since $\nabla N^*_{O_2}$ is large when $P_{O_2} < 100$ mm Hg, the greatest enhancement will occur when the low oxygen concentrations occur in a region of high fluid velocity. Since $\vec{v}$ is small near the membrane surface, and this is the region where $P_{O_2} < 100$ mm Hg during deoxygenation, the deoxygenation enhancement effect is suppressed, and drops off very quickly for net driving forces greater than 100 mm Hg. Conversely, during oxygenation, the region where $P_{O_2} < 100$ mm Hg is far away from the membrane, and thus is generally in a region where $\vec{v}$ is large, resulting in a strong enhancement of the oxygenation rate. The net effect, then, is that the oxygenation transfer rate is faster than the deoxygenation transfer rate.

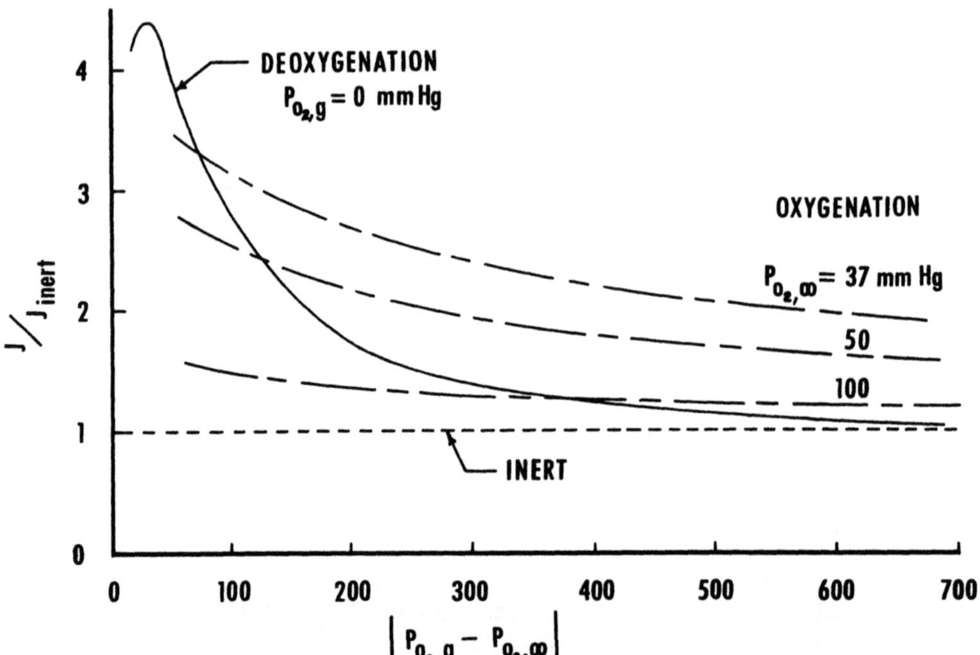

Figure 7. The rates of oxygen uptake and removal by normal whole blood in a rotating disk system. $P_{O_2,g}$ is the oxygen partial pressure of the gas, and $P_{O_2,\infty}$ is the oxygen tension of the blood outside the diffusion boundary layer. The gas and blood are separated by a 1.0 mil silicone rubber membrane.

Using the numerically calculated data for oxygen uptake in the rotating disk system, and the experimental data for carbon dioxide removal, the relative transfer rates of oxygen and carbon dioxide in blood can be calculated as a function of the concentration boundary layer thickness adjacent to the rotating disk membrane. The results of such calculations for the case of pure oxygen gas exchanging with venous blood are presented in Figure 8. This analysis shows that when the blood film resistance is controlling, too much carbon dioxide is removed, and when the silicone rubber membrane resistance is controlling, too little carbon dioxide is removed. Thus, it can be predicted that a membrane lung which has a large blood film resistance must be operated in such a way as to reduce the carbon dioxide transfer rate. This indeed has been found to be necessary in a number of membrane lungs (15, 39, 23) in which carbon dioxide is added to the gas stream to reduce the P_{CO_2} gradient. A more efficient approach to correcting the imbalance in gas exchange rates would be to operate the lung hyperbarically, thus selectively increasing the oxygen transfer rate, since only the oxygen concentration in the pure gas stream would be affected. The effect of this on the relative gas exchange rates is shown in Figure 8, by the curve for $P_{O_2,g}$ = 1470 mm Hg. If proven feasible, this method of correcting unbalanced transfer rates may be well worth applying to existing oxygenators which are blood limited, since a doubling

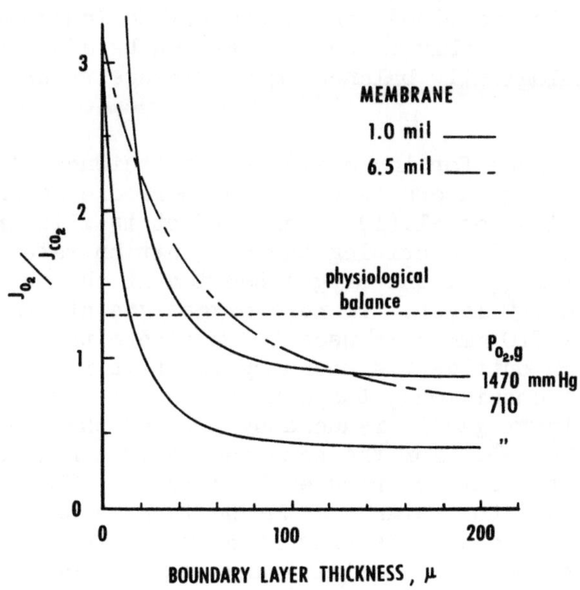

Figure 8. The relative rates of oxygen and carbon dioxide transfer as a function of the concentration boundary layer thickness adjacent to the silicone rubber membrane in a rotating disk system. Wet oxygen gas is assumed to be exchanging with venous blood at 37°C.

of the operating pressure should approximately double the oxygen transfer rate. The other, and perhaps more feasible approach to balancing the oxygen and carbon dioxide transfer rates is to reduce the effective blood film thickness adjacent to the membrane. This has been done with some success by inducing secondary flow patterns in the blood flowing past the membrane. Weissman and Mockros (51) found that by coiling silicone rubber tubes, secondary flows are produced which mix the blood sufficiently to change the process from blood limited to membrane limited. Drinker et al (9) found that secondary flows can also be induced in a spirally coiled tube if the coil is made to undergo torsional oscillations about its axis of curvature. Using a 0.25-inch diameter silicone rubber tubes of 5-mil wall thickness they find this system becomes membrane limited, with an oxygen transfer rate approximately twice the carbon dioxide transfer rate. A third apparently successful attempt at reducing the blood film resistance is presented by Illickal et al (23) who used a membrane-disk rotating in a housing (4 mm spacing) at up to 1000 rpm. They found the oxygen transfer rate to increase linearly with rpm, and apparently had not yet reached a membrane limited situation. An important point regarding these secondary flow systems is that the introduction of mixing to the extent of making the devices membrane limited is undesirable, since the device will then have too small a carbon dioxide transfer rate, as shown in Figure 8. Devices such as the rotating disk and oscillating coil are promising in this respect, since the degree of mixing in the blood can be controlled independently of the flow rate so presumably these devices can be adjusted to operate at physiologically balanced gas exchange rates while in use.

Another technique for increasing the efficiency of blood-limited membrane oxygenators is to employ periodic remixing, as suggested by Buckles, et al.(1). This effect is illustrated by the curves in Figure 7 describing the oxygenation rates for various values of $P_{O_2,\infty}$, the oxygen tension at the outer edge of the blood boundary layer. If pure oxygen gas at atmospheric pressure ($P_{O_2,g}$ = 710 mmHg) is used to oxygenate the blood, and the oxygen tension outside the boundary layer is raised just 13 mmHg above venous conditions, the enhancement of oxygen transfer over that of an inert gas is reduced by 30% and the overall transfer rate is reduced by 15%. For the same change in conditions, the inert transfer rate would be reduced by only 2%. Thus, it is especially important to design a membrane lung in such a way that blood at low oxygen tension is brought near the membrane interface, as is done by both periodic remixing and induced secondary flow (continuous mixing).

Since the blood film resistance in artificial membrane lungs appears to have been effectively reduced by the above-mentioned techniques, it seems that attention must once again be directed towards the problem of reducing the membrane resistance. The most direct and effective approach is to remove the membrane entirely. Although this is currently done in devices which contact the blood directly with the gas stream, the damage done to the blood by this process prohibits long term use. An alternate possibility is the use of a "liquid membrane" - that is, a stable, inert, immiscible liquid which can be oxygenated separately by direct exposure to the gas phase, and then contacted with the blood. Since the liquid-liquid interface cannot support shear forces, as the solid membrane does, a stagnant blood film will no longer tend to form adjacent to the interface. Thus both the membrane resistance and the blood film resistance can, in principle, be dramatically reduced. Of course other problems may be substituted, such as separation of the two phases following contact and interaction of the blood with the liquid-liquid interface, but barring the discovery of membranes with significantly improved permeabilities, liquid-liquid extraction remains an interesting concept worthy of further study.

SUMMARY AND CONCLUSIONS

Three types of theoretical models have been used to describe the convective diffusion of gases in blood. Of these the simplest, the Newtonian "macroscopic" local equilibrium model, appears to be adequate for design purposes.

Diffusion in blood can be described using heterogeneous media theory to evaluate effective diffusion coefficients. The diffusivity of oxygen in whole blood at 37°C appears to be about 1.3×10^{-5} cm^2/sec, although a wide range of values has been reported. The diffusivity of carbon dioxide in blood has not been adequately measured.

Facilitation of oxygen diffusion is not an important mechanism in whole blood. Facilitation of carbon dioxide diffusion does appear to be important. Augmentation of mass transfer by microscopic movements of the red cells in a macroscopically smooth flow does not appear to be an important effect at moderate shear rates.

Oxygenation of blood occurs faster than deoxygenation for the same net partial pressure driving force, because of the nonlinearity of the oxygen dissociation curve and the nature of the blood flow adjacent to the membrane.

For oxygen and carbon dioxide exchange between pure oxygen gas and venous blood in a silicone rubber membrane lung at 37°C, oxygen transfer is favored in a membrane-limited device and carbon dioxide transfer is favored in a blood film-limited device. To obtain physiologically balanced gas exchange in a membrane limited device, the device must be designed with an excess capacity for oxygenation, in order for adequate carbon dioxide removal to occur. To obtain balanced gas exchange in a blood film-limited device either carbon dioxide must be added to gas stream to selectively reduce the net carbon dioxide driving force, or the device must be operated hyperbarically to selectively increase the net oxygen driving force.

Techniques for inducing secondary flows in blood flowing past a membrane surface have been developed to the point where silicone rubber membrane lungs are becoming membrane-limited. In view of this, a further major advance in the design of artificial lungs depends upon the reduction of the membrane resistance. Use of a liquid-liquid system in which an inert, immiscible liquid is oxygenated separately and then contacted with the blood, is a promising concept for this next generation of artificial lungs.

NOMENCLATURE

C_{Hb}	Hemoglobin concentration (gm/cm^3)
C_{rc}	Number concentration of red cells (cm^{-3})
$D_{i,b}$	Diffusivity of species i in media b (cm^2/sec)
f	Fractional hemoglobin saturation
ξ	Probability density function
H	Hematocrit
Hb	Hemoglobin
$\vec{J}_i$	Diffusional flux of species i (cc(STP)/cm^{2}sec)
k_f	Forward reaction rate constant (sec^{-1})
k_r	Reverse reaction rate constant (sec^{-1})
n_i	Concentration of species i (cc(STP)/cm^3)
N_i	Total concentration of reversibly reacted species i (cc(STP)/cm^3)
P_i	Partial pressure of species i (mmHg)
r	Radial cylindrical or spherical coordinate (cm)
R_i	Chemical reaction rate of species i (cc(STP)/cm^{3}sec)
S	Total red cell surface area per unit volume of blood (cm^{-1})
t	Time (sec)
$\vec{v}$	Velocity vector (cm/sec)
v_j	j component of $\vec{v}$ (cm/sec)
x_j	rectangular coordinate (cm)
z	Axial cylindrical or rectangular coordinate (cm)
α_i	Solubility of species i (cc(STP)/cm^{3}mmHg)
β_k	Partition coefficient of species k
γ	Red cell oxygenation rate constant (sec^{-1})
ψ	Relative permeability of a heterogeneous medium to the suspending phase

θ_p Ratio of the diffusivity plasma to that in saline

SUPERSCRIPTS

*	Equilibrium value
'	Pertaining to the red cell interior
"	Pertaining to the red cell membrane

SUBSCRIPTS

b	Blood
g	Gas
Hb	Hemoglobin
i	Species
ik	Species k containing i
P	Plasma
rc	Red blood cell
s	Serum
T	Total
o	Initial value
∞	Value at infinity

BIBLIOGRAPHY

1. Buckles, R.G., E.W. Merrill, and E.R. Gilliland. 1968 "An Analysis of Oxygen Absorption in a Tubular Membrane Oxygenator", AIChE Jour., 14, 5:703.

2. Carslaw, H.S. and J.C. Jaeger. 1959. Conduction of Heat in Solids. The Oxford University Press, London, England.

3. Colton, C.K., and R.F. Drake. 1969. "Analysis of VIVO deoxygenation of human blood: A feasibility study for an implantable biological fuel cell". Trans, Amer Soc. Antif. Int. Organs. 15:187

4. Colton, C.K., K.A. Smith, E.W. Merrill. 1969. "Diffusion of Organic Solutes in Plasma and Whole Blood," Proceeding 8th ICMBE, Chicago, Illinois. Session 23-7.

5. Colton, C.K. 1969. Permeability and Transport Studies in Batch and Flow Dialyzers with Applications to Hemodialysis, Ph.D. Thesis, M.I.T.

6. Crank, J., and G.S. Park. 1968 Diffusion in Polymers. Academic Press, New York.

7 DeGroot, S.R., and P. Mazur. 1962. Non-equilibrium Thermodynamics. North-Holland Publishing Company, Amsterdam, The Netherlands.

8 Dobell, A. C., M. Mitri, R. Galva, E. Sarkozy, and D. R. Murphy. 1965. "Biologic Evaluation of Blood After Prolonged Recirculation Through Film and Membrane Oxygenators." Ann. Surg. 161:617

9 Drinker, P.A., R.H. Bartlett, R.M. Bialer, and B.S. Noyes,Jr. "Augmentation of Membrane Gas Transfer by induced secondary flows". (in press) Surgery

10 Enns, T., 1967. "Facilitation by Carbonic Anhydrase of Carbon Dioxide Transport". Science, 155, 3758:44

11 Enns, T., 1969. "Transport of Gases Across the Red Cell Membrane", Red Cross Symposium on the Red Cell Membrane, Washington, D. C.

12 Forster, R. E., 1964. "Rate of Gas Uptake by Red Cells", Handbook of Physiology-Respiration, 32:827

13 Fricke, H. 1924. "A Mathematical Treatment of the Electric Conductivity and Capacity of Disperse Systems I. The Electric Conductivity of a Suspension of Homogeneous Spheroids", Physical Rev., 24:575.

14 Friedlander, S.K., and K.H. Keller. 1965. "Mass Transfer in Reacting Systems Near Equilibrium", Chem. Eng. Sci. 20:121

15 Gerbode, F., J. J. Osborn and M. L. Bramson. 1967. "Experiences in the Development of a Membrane Heart-Lung Machine" American J. of Surgery 114:16

16 Gertz, K. H., and Loeschcke. 1954. "Bestimmung der Diffusions - Koeffizienten von H_2, O_2, N_2, and He in Wasser und Blutserum bei Konstant Gehaltener Konvektion," Z. Naturforschg 96:1.

17 Goldsmith, H. L. 1968. "The Microrheology of Red Blood Cell Suspensions". Biological Interfaces: Flows and Exchanges Proceedings of a Symposium sponsored by the New York Heart Association; Little, Brown and Co., Boston.

18 Goldstick, T. K. 1966. Diffusion of Oxygen in Protein Solutions, Ph.D. Thesis, University of Calif., Berkeley.

19 Handbook of Physiology-Respiration, Vol. I, II. 1964. American Physiological Society, Washington, D.C.

20 Hershey, D., and J. Hesselberth. 1967. "Oxygenation of Whole Blood by Diffusion plus Homogeneous Reversible Chemical Reaction in a wetted wall column". 60th Ann. Mtg., AIChE, New York.

21 Hershey, D., and T. Karhan. 1968. "Diffusion Coefficients for Oxygen Transport in Whole Blood", AIChE Jour., 14, 6:969.

22 Hochmuth, R. M. Unpublished data.

23 Illickal, M.M., G. E. Brown, Joseph M. Van de Water, W. H. Lee, D. B. Pall, and J. V. Maloney, Jr. 1967. "Boundary Layer Phenomenon in Membrane Oxygenators" Surgical Forum, 18:134

24 Keller, K.H., and S.K. Friedlander. 1966. "The Steady State Transport of Oxygen through Hemoglobin Solutions", J. Gen. Physiol. 49, 4:663

25 Kolobow, T., W. Zapol, and J. Pierce. 1969. "High Survival and Minimal Blood Damage in lambs exposed to long term (1 week) veno-venous pumping with a polyurethane chamber roller pump with and without a membrane blood oxygenator". Trans. Am. Soc. Artif. Int. Organs. 15:172

26 Kreuzer, F., 1953. "Modellversuche zum Problem der Sauerstoff-diffusion in der Lungen", Helv. Physiol. et Pharm. Acta., Suppl. IX

27 Kreuzer, F., and W. Z. Yahr. 1960. "Influence of the Red Cell Membrane on Diffusion of Oxygen", J. Appl. Physiol., 15, 6:1117.

28 Lande, A.J., S. J. Fillmore, V. Subramanian, R. N. Tiedemann, R.G. Carlson, J.A. Bloch, and C.W. Lillehei. 1969. "24 hour venous-arterial perfusions of awake dogs with a sample membrane oxygenator", Trans. Amer. Soc. Artif. Int. Organs, 15:181.

29 Lee, W.H., Jr., D. Krumhaar, E.W. Fonkalsrud, D.A. Schjeide, and J. V. Maloney, Jr. 1961. "Denaturation of Plasma Proteins as a Cause of Morbidity and Death after Intracardiac Operations". Surgery 50:29.

30 Martisovits, V., and S. Veis. 1967. "Quantitative description of Blood Oxygenation process in oxygenator for extra corporal circulation", Bulletin of Mathematical Biophysics. 29:485.

31 Marx, T. I., W. E. Snyder, A. D. St. John, and Calvin E. Moeller. 1960. "Diffusion of oxygen into a film of whole blood", J. Appl. Physiol. 15:1123

32 Merrill, E.W., E. R. Gilliland, G. Cokelet, H. Shin, A. Britten, and R. E. Wells. 1963. "Rheology of Human Blood Near and at Zero Flow". Biophys J. 3:199.

33 Merrill, E. W., A.M. Benis, E. R. Gilliland, T. K. Sherwood, E. W. Salzman. 1965. "Pressure-Flow Relations of Human Blood in Hollow Fibers at Low Flow Rates," J. Appl. Physiol. 20:954.

34 Merrill, E. W., and Pelletier, G. A., 1967. "Viscosity of Human Blood: Transition from Newtonian to Non-Newtonian". J. Appl. Physiol., 23:178.

35 Moll, W. 1966. "The Diffusion Coefficient of Hemoglobin", Resp. Physiol., 1:357.

36 Peirce II, E.C., J.J. Corrigan, B.B. Kent, C.L. Cate, R.L. Johnson, and A.E. Dreves. 1969. "Comparitive trauma to blood in one disc oxygenator and membrane lung". Trans. Amer. Soc. Artif. Int. Organs. 15:33.

37 Pircher, L. 1952. Helv. Physiol. Pharm. Acta., 10:110.

38 Ratan, R.S., G.F. Bennett, P.L. Bollin, W.A. McAlpine and M.W. Selman. 1967. "Experimental evaluation of a rotating membrane oxygenator", Jour. of Thoracic and Cardiovascular Surgery, 53:519.

39 Roughton, F. J. W. 1963. "Kinetics of Gas Transport in Blood", Brit. Med. Bull. 19:80.

40 Roughton, F.J.W. 1964. "Transport of Oxygen and Carbon Dioxide", Handbook of Physiology - Respiration, 31:767.

41 Satterfield, C.N., and T.K. Sherwood. 1963. The Role of Diffusion in Catalysis. Addison-Wesley, Inc. Massachusetts.

42 Shirley, Y.E., 1968. Facilitated Transport of Carbon Dioxide, M.S. Thesis, University of Minnesota.

43 Smith, J. M. 1956. Chemical Engineering Kinetics. McGraw-Hill, Inc. New York.

44 Snell, F.M. 1965. "Facilitated Transport of Oxygen through Solutions of Hemoglobin", J. Theoret. Biol. 8:469.

45 Spaeth, E.E. and S.K. Friedlander. 1967. "The Diffusion of Oxygen, Carbon Dioxide, and Inert Gas in Flowing Blood", Biophys. J. 7:827.

46 Spinak, R. B. 1968. Transport of Carbon Dioxide in Erythrocytes, M. S. Thesis, University of Minnesota.

47 Stein, T. R. 1968. Augmented Diffusion of Oxygen. Ph.D. Thesis, University of Minnesota.

48 Thews, G. 1968. "The theory of Oxygen Transport and its Application to Gaseous Exchange in the Lung", in Oxygen Transport in Blood and Tissue, Georg Thieme Verlag, Stuttgart, Germany.

49 Ward, W.J.,and Robb, W.L. 1967. "Carbon Dioxide-Oxygen Separation: Facilitated Transport of Carbon Dioxide across a Liquid Film", Science, 156:1481.

50 Weissman, M. H., and L.F. Mockros. 1969. "Oxygen and Carbon Dioxide transfer in Membrane oxygenators" Med & biol. Engng. 7:169

51 Wittenberg, J.B. 1966. "The Molecular Mechanism of Hemoglobin-Facilitated Oxygen Diffusion", J. Biol. Chem., 241:104.

52 Yoshida, F. and N. Oshima, 1966. "Diffusivity of Oxygen in Blood serum," J. Appl. Physiol., 21;915.

53 Zingg, W. 1967. "Membrane oxygenator for infants", Trans, Amer. Soc. Artif. Int. Organs. 13:334.

DESIGN CONSIDERATIONS AND LONG-TERM IN VIVO STUDIES WITH THE DISPOSABLE SPIRAL MEMBRANE LUNG

Theodor Kolobow, Warren M. Zapol, and Robert L. Sigman

Lab. of Technical Development, National Heart and Lung Institute, National Institutes of Health, Bethesda, Maryland

INTRODUCTION

Clinical blood oxygenation has now been routinely practiced for nearly two decades. Pathologic findings after extracorporeal perfusion by bubble and disc blood oxygenation techniques suggest that significant morbidity and mortality is due to damage to blood elements (1). Such perfusion methods cannot be used for long periods of time. Long term perfusion is needed in various clinical situations; for instance, a patient after a heart operation, who is unable to tolerate being taken off bypass after corrective surgery. Such patient would benefit from continued extracorporeal cardiopulmonary assistance.

Interposing a thin membrane between blood and gas phases prevents the deleterious changes noted above (2). Membrane oxygenation is therefore indispensable to prolonged extracorporeal respiratory and cardiopulmonary support. Many persons with severe acute illnesses, fatal with current therapy, may be saved by prolonged extracorporeal pump-oxygenation. Presently available membrane oxygenators are complex, expensive, difficult to operate and often unsafe. Prolonged human respiratory or circulatory assistance with these membrane lungs has only been attempted in a few instances (3). Our simple, low cost, disposable and safe membrane lung with high gas transfer efficiency is designed for prolonged human support.

In this paper we describe in detail the considerations which have entered into the design of the spiral membrane blood oxygenator briefly reported previously (4). This simple, compact, disposable unit has been used on lambs successfully for periods up to a week in high flow perfusion experiments which are also described here.

A. General Design Considerations

1. Basic Design. This lung contains a sealed membrane envelope enclosing a spacer screen and provided with oxygen inlet and outlet ports at either end. Oxygen is sucked instead of blown through the membrane envelope. Blood flows between the outer surfaces of the spiral membrane envelope (Figure 1). The membrane envelope is rolled

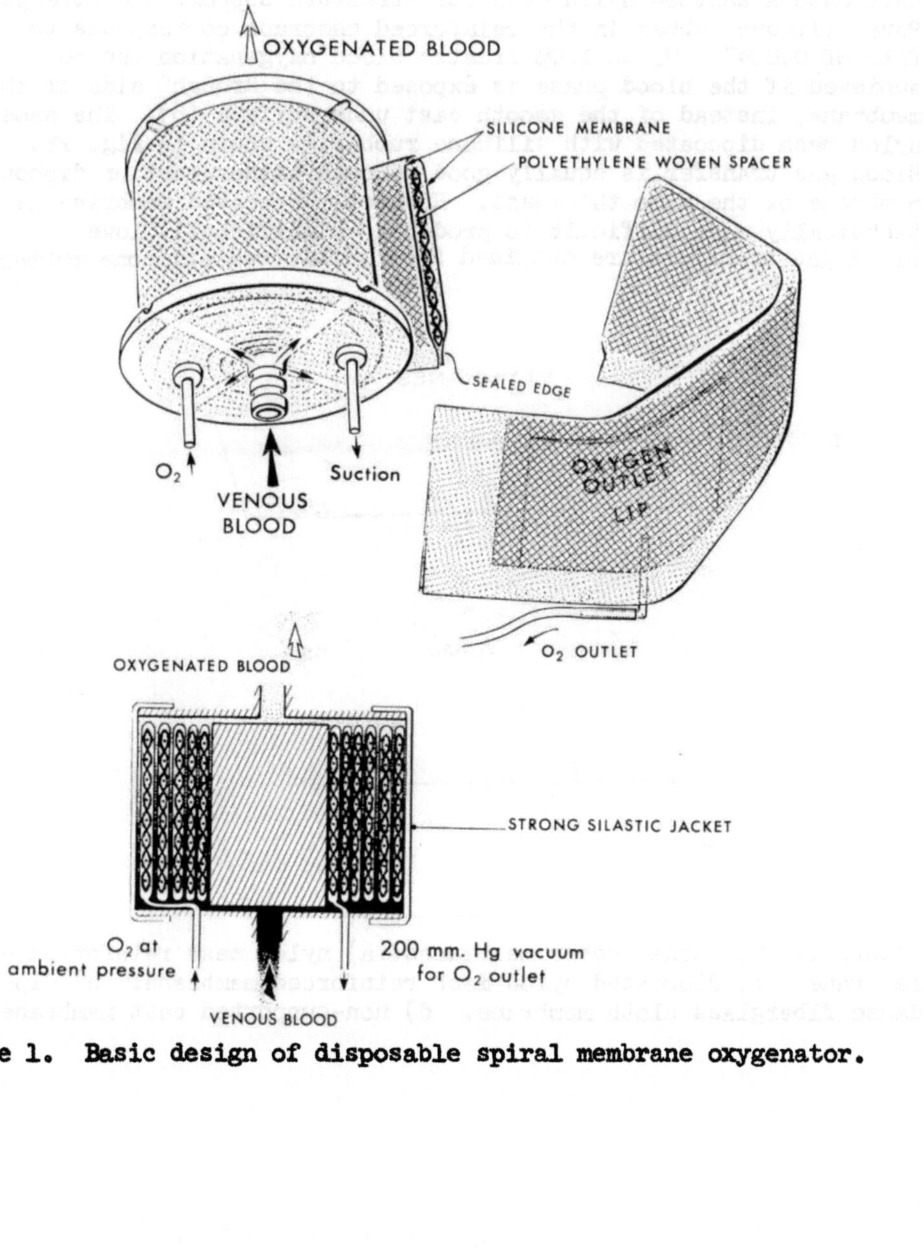

Figure 1. Basic design of disposable spiral membrane oxygenator.

around a spool, and both envelope and spool are sealed within a tight fitting silicone rubber jacket.

2. The Membrane. Using 0.0008" polyethylene film, a membrane lung with a surface area of 7.2 m^2 was built and tests showed it had an oxygenating capacity of only 1 liter/minute at a normal A-V gradient of 4 vol. %. This low oxygen transfer is explained by low permeability of polyethylene membrane (59 cc STP/$(cm)^2$(mm) (sec)(cmHg) x 10^{10} (5). Much better results were obtained with reinforced Silastic[R] membrane (Dow Corning Corporation, Midland, Michigan) with a much greater permeability (6000 cc STP/$(cm)^2$(mm) (sec)(cmHg) x 10^{10}. Figure 2A is a sketch of the membrane when cast onto a knitted nylon mesh for structure support and strength. Pure silicone rubber in the reinforced membrane corresponds to a film of 0.004". Up to 100% greater blood oxygenation can be achieved if the blood phase is exposed to the "rough" side of the membrane, instead of the smooth cast undersurface (6). The same nylon mesh dipcoated with silicone rubber is shown in Fig. 2b. Blood gas transfer is equally good through either cast or dipcoated membrane of the same thickness. However, dipcoated membrane is technically more difficult to produce. Significantly lower blood gas transfers are obtained from reinforced silicone rubber

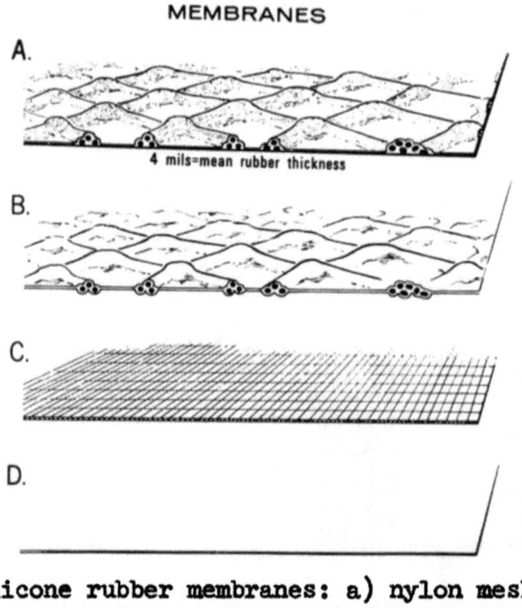

Figure 2. Silicone rubber membranes: a) nylon mesh reinforced cast membrane. b) dipcoated nylon mesh reinforced membrane. c) dipcoated dense fiberglass cloth membrane. d) non-supported cast membrane.

membrane with a smooth surface on both sides, such as a dense weave fiberglass reinforced dipcoated membrane (Figure 2c) or non-reinforced silicone membrane (Figure 2d), because of poor blood mixing encountered between smooth membrane films.

3. Spacer. The spacer screen provides within the membrane envelope a passageway for oxygen flow. The present screen is a continuous tape of woven 16 x 18 mesh of 16 mil monofilament polyethylene or Saran screen (Figure 3a) or a tape of slit screen with edges pressed smooth (Figure 3b). Smooth screen edges are important to prevent tears in the silicone membrane. To prevent undue rise in CO_2 tension within the envelope from rapid CO_2 diffusion out of the blood we have found that oxygen flow must be over 2 liters/m^2 of membrane.

If the mesh is too large, the two inner membrane surfaces will come into contact, resulting in loss of membrane diffusion area, a marked rise in blood priming volume, and diminished gas exchange efficiency from increased blood film thickness (Figures 4a and 4b). Too small a mesh will result in an increase in blood perfusion pressure, decrease in gas flow through the envelope resulting in a rise in gas phase CO_2 concentration, and decrease in CO_2 transfer. Too much contact between membrane and screen subtracts from the effective membrane area and decreases oxygen transfer.

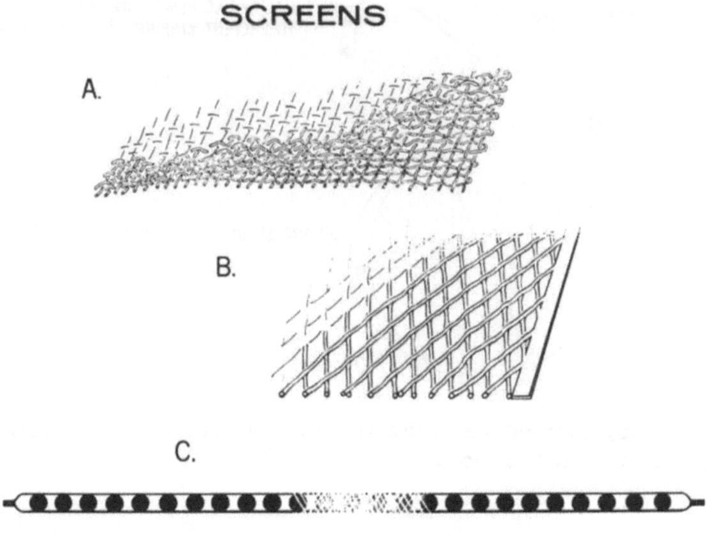

Figure 3. a) woven screen with finished selvage. b) woven screen with secondarily finished selvage. c) position of screen within the membrane envelope.

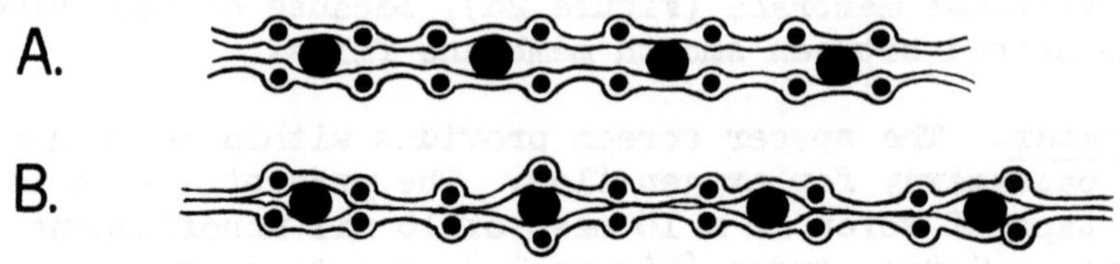

Figure 4. a) correct screen mesh. b) too large screen mesh.

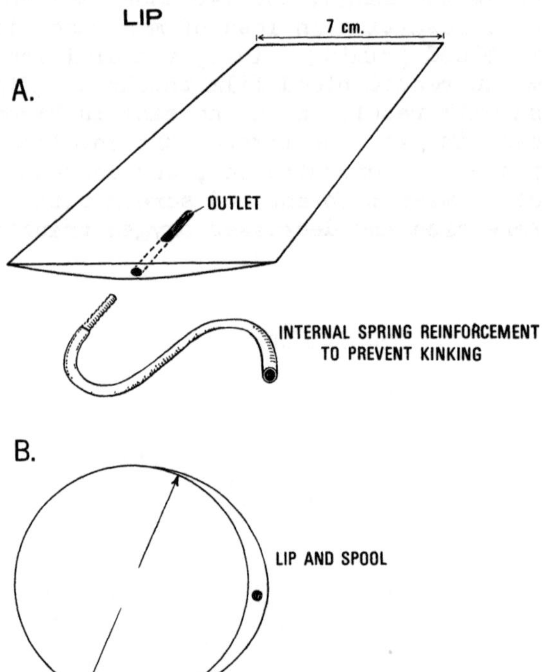

Figure 5. a) oxygen or suction port ("lip"). b) smooth taper of "lip" encircles spool.

4. Oxygen and Suction Ports. Oxygen and suction are applied through molded silicone rubber sections having a smooth taper (Figures 5a, 5b) and bonded as an integral part of the membrane envelope (Figure 6a). The smooth taper of this section ("lip") is extremely important as it avoids blood shunting in the assembled lung. An 8" length of silicone rubber tubing is pushed into the oxygen inlet and outlet ports of the molded lips (Figure 5a). A coiled stainless steel spring wire is placed within the silicone rubber tubing to prevent collapse or accidental kinking of this gas conduit.

5. Oxygen Flow. The membrane lung requires a minimum of one oxygen inlet port, and one oxygen outlet port. The non-linear relationship of gas flow to suction across a 10 cm wide 2 m. long membrane envelope is illustrated in Figure 7. Gas flow is inversely proportional to the length of the membrane envelope. Practical considerations should blood leak into the screen compartment, blood priming volume, and membrane lung efficiency require a screen with some resistance to oxygen flow. Gas flow can be increased simply by adding outlet ports as various distances along the membrane. Several such "lips" are used to divide a single long membrane envelope to avoid CO_2 build up within the gas compartment (Figure 6b). Only two ports are needed if one is interested in but oxygen transfer, to the exclusion of CO_2.

Oxygen is sucked intermittently through the membrane envelope at a gradient from minus 50 to minus 200 mm. Hg. Suction provides several advantages: 1. During oxygenator priming keeps the blood compartment passageways fully open, reducing the possibility of air entrapment. 2. Intermittent suction causes membrane motion and increases the membrane oxygen transfer due to agitation of the blood boundary layer (6). This is difficult with pressurized oxygen, as it would increase resistance to blood flow, and requires a proportional increase in perfusion pressure. 3. Safety, however, is the most important reason for choosing suction over pressurized oxygen. All membranes we have ever used had occasional pinholes. It is extremely dangerous for gas bubbles to enter the circulation and occlude capillaries. Small bubbles can then embolize into the vasculature even with use of a bubble trap. Use of hypobaric oxygen will prevent bubbles from entering the blood stream. Moreover, blood may leak into the oxygen compartment, but due to the markedly greater viscosity of blood, only minute quantities leak out.

The effective oxygen gradient across the membrane will diminish as suction in the oxygen compartment is increased. This loss of oxygen diffusion gradient averages 15%, and is compensated for partly by a decrease in silicone rubber thickness due to stretching of the membrane when suction is applied.

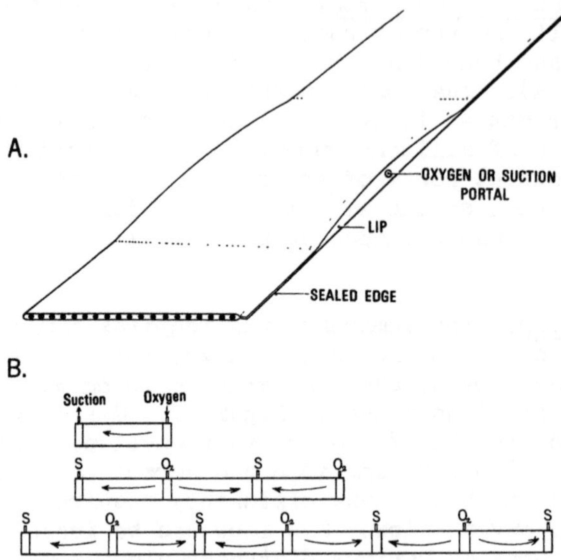

Figure 6. a) "lip" becomes an integral part of the membrane envelope. b) a long membrane envelope can be segmented by inserting several "lips" to increase CO_2 washout.

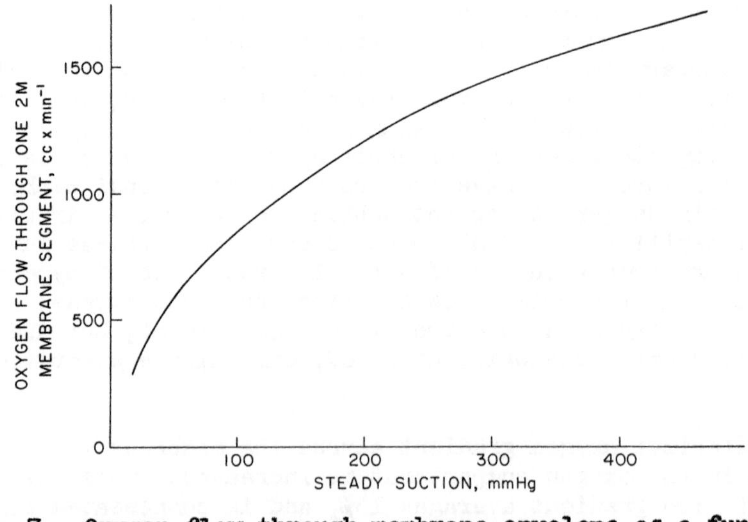

Figure 7. Oxygen flow through membrane envelope as a function of suction.

6. Blood Priming Volume. Changing suction levels within the membrane envelope alters lung prime volume by approximately 5 cc/m^2. Total blood prime volume averages 80 cc/m^2 of membrane (Figure 8). The two blood headers increase total membrane oxygenator prime to 100 cc/m^2.

7. Testing. Silicone membrane as supplied by the manufacturer is pre-tested for pinholes using Ozalid paper. Briefly, the method consists of placing the membrane to be examined in contact with a sheet of Ozalid paper and passing the pair slowly through a second (ammonia) stage of a conventional Ozalid machine (7). This test has many advantages over the laborious process of leak detection by sponging the membrane with acetone and looking for telltale marks of solvent penetration on the paper backing.

After fabrication of the sealed membrane envelope it is checked for leaks by evacuating it to minus 500 mmHg and timing the rate of vacuum drop. With a pinhole free membrane envelope the ensuing vacuum drop is entirely due to diffusion of atmospheric gases across the membrane envelope. Any shortening of this vacuum drop time is due to membrane pinholes or fabrication imperfections. To perform a final check, the membrane envelope is evacuated to minus 300 mmHg and passed through a water bath. A membrane envelope found defective by the vaccum drop test can have its defective area pinpointed and it can then be repaired. Vacuum drop tests are repeated on repaired dry membrane envelopes. The time recorded on the vacuum drop test can be permanently recorded and referred to during subsequent fabrication and sterilization steps.

8. Assembly. The membrane envelope is wrapped onto a molded polycarbonate spool coated with silicone fluid* under a tension of 1.5 kg/10 cm width of membrane envelope (Figure 9a). Gas tubes with spring wire reinforcement are drawn through the appropriate orifices in the spool forming a leak proof joint (Figure 9b).

Placing a strong outer silicone rubber jacket around the spool and membrane envelope completes the assembly process. A tubular extruded silicone rubber jacket is first expanded by vacuum assist and then released over the spool containing the spirally wound membrane envelope (Figures 10 a-f). We can thus obtain a totally enclosed blood compartment without the use of tensioning or clamping devices.

9. Blood flow resistance. The use of the above screen, membrane and winding tension results in a lung requiring a very

*Dow Corning 200 fluid. Dow Corning Corp., Midland, Michigan.

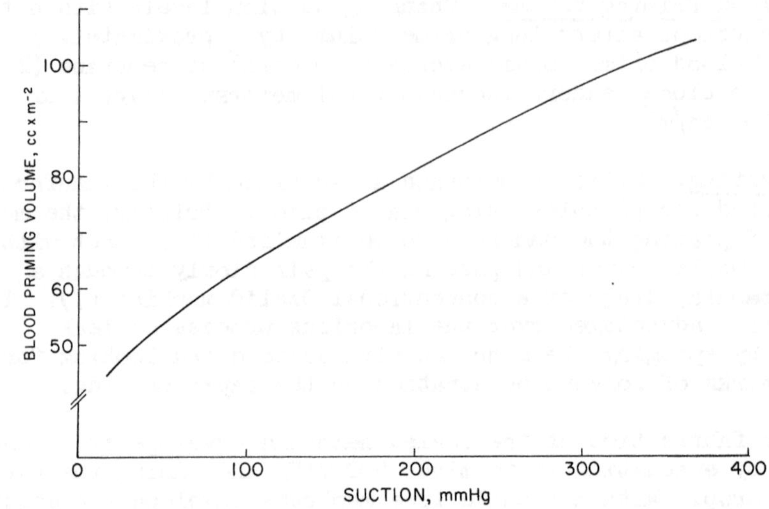

Figure 8. Membrane blood priming volume as a function of suction.

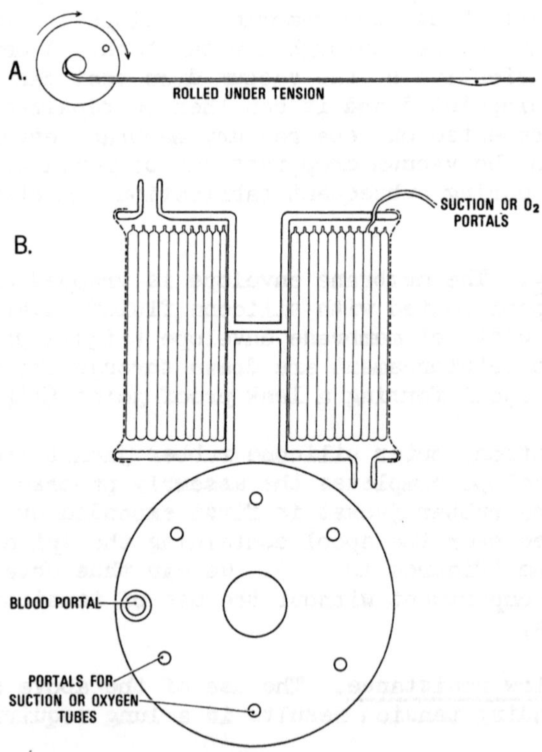

Figure 9. a) winding of membrane envelope. b) suction and oxygen ports within the spool.

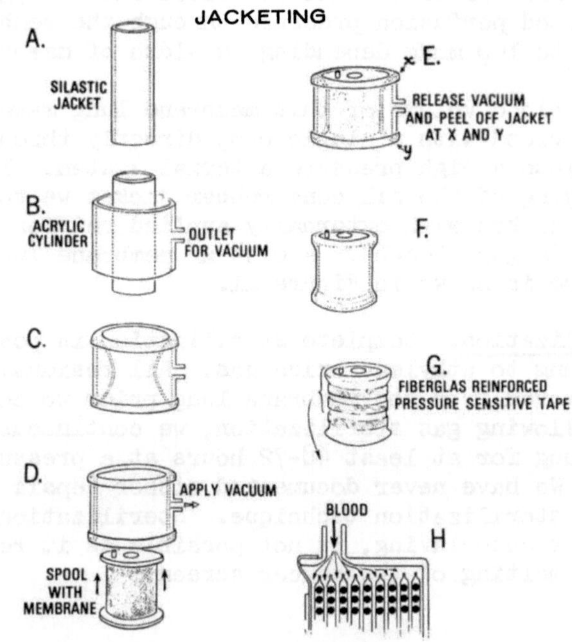

Figure 10. Placing of outer jacket around spool and membrane envelope.

Figure 11. 0.44 m^2 membrane lung with reinforcing tapes.

low blood perfusion pressure. At its rated blood oxygenating capacity the blood perfusion pressure through the membrane lung ranges from 30 to 100 mmHg depending on width of membrane envelope.

The fixed blood volume of this membrane lung makes it possible to pump venous blood with a single pump directly through the membrane lung into a high pressure arterial system. In that case, to prevent bulging of the silicone rubber jacket we then have to strengthen the jacket with externally applied reinforced pressure sensitive tape (Figure 10g-h). A 0.44 m^2 membrane lung with reinforcing tape is shown in Figure 11.

10. Sterilization. Complete sterilization is possible by exposing the lung to ethylene oxide gas. All residual sterilizing gas must be removed from the membrane lung prior to contact with blood (8). Following gas sterilization, we continuously evacuate the membrane lung for at least 48-72 hours at a pressure of less than 30 mmHg. We have never documented either sepsis or hemolysis following such sterilization technique. Sterilization at elevated temperatures, or autoclaving, is not possible as it results in deformation or melting of the spacer screen.

B. Design Considerations for Special Capacity Membrane Lungs

In the majority of potential human clinical applications the patient's arterial and venous blood gas parameters will be seriously deranged, with venous oxygen saturation uniformly under 60% despite exhaustive conventional therapy. Consequently, hypoxia may well decrease cardiac, cerebral, renal and hepatic performance, especially in the face of preexisting vascular disease. Failure of additional organ systems may require earlier initiation of extracorporeal respiratory or cardiorespiratory support.

Blood from the central venous system must be pumped through the membrane oxygenator and have its oxygen saturation raised to above 90% and CO_2 levels lowered to normal physiologic values. We advise against attempts to raise oxygenator efficiency by rapid extracorporeal recirculation of blood by shunting because it may increase blood injury from pumping. Other observers report perfusion and post-perfusion anemia syndromes (9, 10). In our experience we have not encountered progressive anemia during membrane lung blood oxygenation (11). Figure 12 and 13 show spiral membrane lung sizes from 0.2 m^2 to 9.1 m^2. Silicone rubber membrane (0.004") used in the spiral membrane lung configuration can provide the required blood oxygen transfer and CO_2 exchange by choosing the proper capacity membrane lung. Although the immediate clinical need to supplement respiratory gas exchange requires but a small capacity membrane lung, it is wise to provide reserve capacity

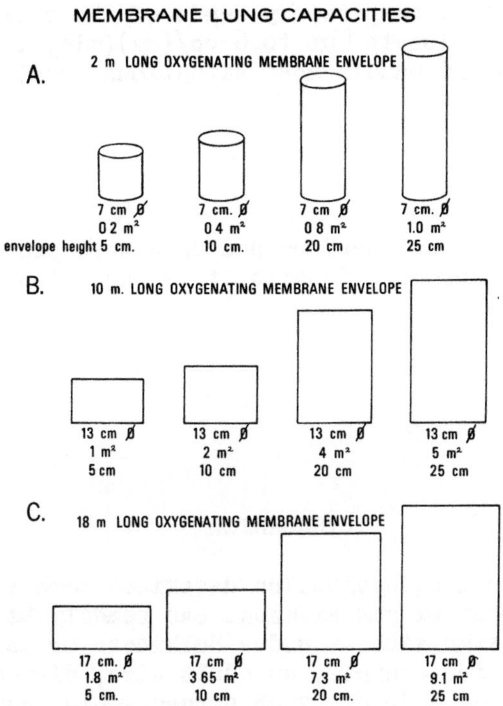

Figure 12. The spiral membrane lung of different widths and surface areas.

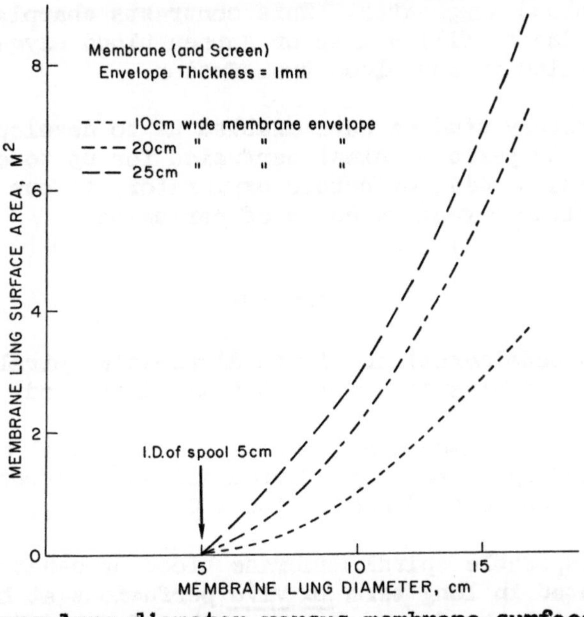

Figure 13. Membrane lung diameter versus membrane surface area.

to take over most of respiratory needs of the sick patient should his condition deteriorate [up to 6 cc/(kg)(min)]. Probably 3 or 4 of the units would handle most situations involving human use.

Editors Note: Clinical performance data were presented in the original paper. For more details,it is suggested that the authors be contacted.

DISCUSSION

The membrane lung evaluation described here suggests that the required amount of gas exchange can readily be accommodated in the spiral design without undue bulkiness or unpredictability of performance. For example, an adult size, disposable 5 m^2 membrane lung using a 30 cm. wide screen would require an 8.2 m. long membrane envelope with 5 lips (outlet ports). The time needed for manufacturing and testing is under 2 hours. Less than 500 cc. of prime volume are needed to fill an adult size disposable spiral membrane blood oxygenator. This contrasts sharply with the many liters needed to fill a disc or screen blood oxygenator, with the risks entailed in any blood transfusion.

Laboratory studies have enabled us to develop safe operating techniques to perform animal perfusion for up to one week's duration. In contrast, no disc or bubble oxygenator, to our knowledge, has had a survivor beyond 24 hours of perfusion.

SUMMARY

1. Design considerations of the disposable spiral membrane blood oxygenator have been presented and discussed.

2. The blood oxygenating capacity of the spiral membrane lung with a 0.004" nylon knit reinforced Silastic[R] membrane readily attains 60 cc/(m^2)(min), with a CO_2 exchange of 45 cc/(m^2)(min).

3. The disposable spiral membrane blood oxygenator has been evaluated in long term <u>in vivo</u> perfusions at high blood flows in alert lambs for up to one week and was found not to cause any detectable injury or impairment in the perfused animals.

REFERENCES

1. Veith, F.J., Deysine, M., Nehlsen, S.L., Panossian, A., Hagstrom, J.W.C. Pulmonary Changes Common to Isolated Lung Perfusion, Venovenous Bypass, and Total Cardiopulmonary Bypass. Surg. Gyn. Obst. 125:1047, 1967.

2. Zotti, E.F., Ikeda, S., Lesage, A.M., Sealy, W.C., Young, W.G. Prolonged Veno-venous Perfusion with a Membrane Oxygenator for Respiratory Failure. J. Thoracic and Cardiovasc. Surg. 51: 383, 1966.

3. Hill, J.D., Bramson, M.L., Hackel, A., Deal, C.W., Sanchez, P.A., Osborn, J.J., Gerbode, F., Laboratory and Clinical Studies during Prolonged Partial Extracorporeal Circulation Using the Bramson Membrane Lung. Circulation, Supplement II, 27: 139, 1968.

4. Kolobow, T., Bowman, R.L. Construction and Evaluation of an Alveolar Membrane Artificial Heart-lung. Trans. Amer. Soc. Artif. Int. Organs, 9: 238, 1963.

5. Lebovits, A. Permeability of Polymers to Gases, Vapors, and Liquids. Modern Plastics, March 1966.

6. Kolobow, T., Zapol, W., Marchus, J. Development of a Disposable Membrane Lung for Organ Perfusion. Chapter in Organ Perfusion and Preservation (Ed. J.C. Norman) Appleton Century Crofts, New York, 1968.

7. Blok, J., Humphrey, J.S., Jr., Nichold G.E. Hole Detection in Polymer Films and in Plastic Track Detectors. Rev. Scient. Inst., 40: 509, 1969.

8. Matsumoto, T., Hardaway, R.M., Pani, K.C., Sater, C.M., Bartak, D.E., Margetic, P.M. Safe Standard of Aeration for Ethylene Oxide Sterilized Supplies. Arch. Surg. 96: 464, 1968.

9. Galletti, P.M. Laboratory Experience with 24 Hour Partial Heart-Lung Bypass. J. Surg. Res. 5: 97, 1964.

10. Kusserow, B., Clapp, J.F. Red Blood Cell Survival after Prolonged Perfusion with a Blood Pump. Trans. Amer. Soc. Artif. Int. Organs 12: 122, 1965.

11. Kolobow, T., Zapol, W., and Pierce, J. High Survival and Minimal Blood Damage in Lambs Exposed to Long Term (1 week) Veno-venous Pumping with a Polyurethane Chamber Roller Pump with and without a Membrane blood oxygenator. Trans. Amer. Soc. Artif. Int. Organs, 15: 172, 1969.

12. Noble, F.W. Dual Frequency Ultrasonic Fluid Flowmeter. Rev. Sci. Inst. 39: 1327, 1968.

13. Kolobow, T., Zapol, W., Pierce, J.E., Kelley, A.F., Replogle, R.L., and Haller, A. Partial Extracorporeal Gas Exchange in Alert Newborn Lambs with a Membrane Artificial Lung Perfused via an A-V Shunt for Periods up to 96 Hours. Trans. Amer. Soc. Artif. Int. Organs 14, 328, 1968.

14. Kolobow, T., Zapol, W. A New Thin Walled Non-Kinking Catheter for Peripheral Vascular Cannulation. Surgery, in press.

A MATHEMATICAL MODEL FOR THE PREDICTION OF OXYGEN, CARBON DIOXIDE, AND pH PROFILES WITH AUGMENTED DIFFUSION IN CAPILLARY BLOOD OXYGENATORS

Fernando Villarroel, Clifford E. Lanham, Kenneth B.Bischoff*
Thomas M.Regan*, Jerry M. Calkins, Harry Diamond Laboratory,
Connecticut Ave. and Van Ness St. Washington, D.C.

1. INTRODUCTION

The mathematical analysis and modeling of the steady flow capillary tube blood gas exchanger has been attempted by earlier investigators with varying degrees of success. Analysis of the data and theoretical prediction of earlier researchers led us to conclude that a phenomenon other than ordinary diffusion was also occuring in the gas exchanger. Figure 1 shows the data collected by Buckles (1, 2) using fresh human blood and his theoretical prediction. It is evident in this figure that at the lowest flows the theoretical curve falls well below the data points. Rotation induced diffusion has been suggested previously in the literature (3) although its existence in flowing blood has not been proven conclusively.

This paper presents a mathematical analysis of the gas transport phenomena including the diffusion augmentation due to the mixing effect of the rotating erythrocytes. Blood is considered a fluid continuum with reversible gas sinks in instantaneous local equilibrium with the plasma surrounding them. The equilibrium between the gas contained in the sinks and that dissolved in the plasma is given by the well known blood gas dissociation curves for oxygen (figure 2) as well as for carbon dioxide. The resistance of the wall to the transport of the gases is also considered, although the axial transport in the wall is neglected.

---

*University of Maryland.

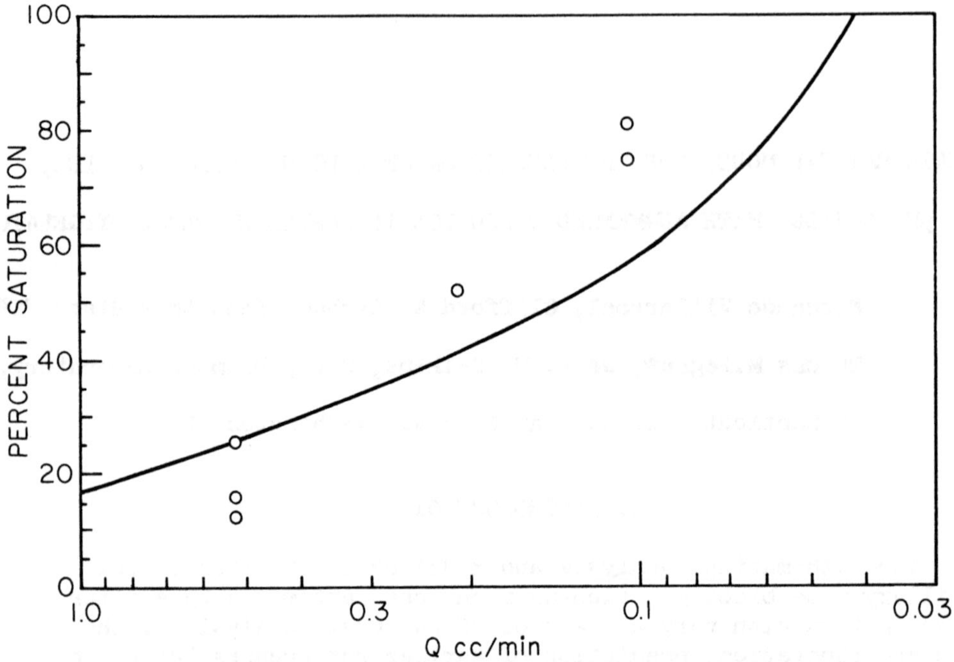

Figure 1. Buckle's theoretical prediction and data (Ref. 1, 2)

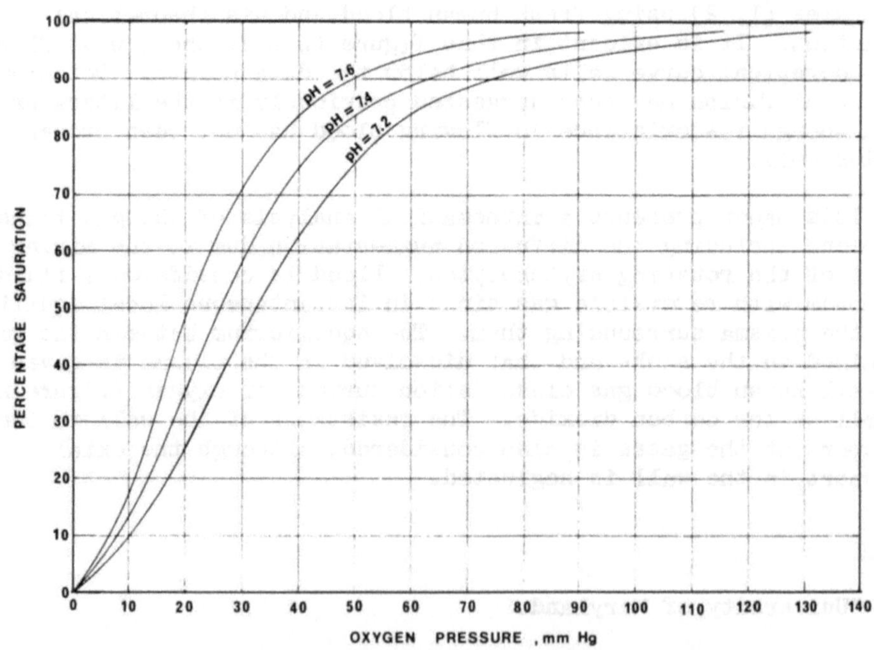

Figure 2. Oxygen Dissociation Curve for Human Blood (Ref. 11)

2. MATHEMATICAL MODEL

The model presented in this paper for the steady state transport of gases involves the following assumptions:

a. The axial diffusion through the wall is negligible,
b. The system is isothermal,
c. The blood is a homogeneous, incompressible, non-Newtonian

fluid where the velocity profile is described by an integrated form of the Casson (4) equation,

d. Each geometrical point in the fluid is in equilibrium with a sink,
e. Axial gas transport by diffusion in the blood is negligible compared with the gas transport due to the blood flow,
f. The oxygen consumed by metabolic processes is negligible, and
g. The effective gas diffusivity in blood is not a constant but rather a function of the rotation of the erythrocytes (mixing effect).

The equation of continuity in cylindrical coordinates for a reacting system neglecting the axial diffusion transport is

$$\frac{1}{r}\frac{\partial}{\partial r}\left(rD\frac{\partial C}{\partial r}\right) = V_z\frac{\partial C}{\partial z} + S \qquad (1)$$

where the reaction rate S for a diffusion limited reaction is given by the substantial time derivative (i.e., the derivative following the motion of the liquid) of the concentration of gas stored in the sinks.

$$S = \frac{D}{Dt}\text{ (stored gas)} = V_z\frac{\partial}{\partial z}\text{ (stored gas)} \qquad (2)$$

The concentration of stored oxygen is given by $\frac{T}{100}$ P and the concentration of stored carbon dioxide is given approximately by the following relation (5)

$$\text{stored } CO_2 = K_1 + K_2C \qquad (3)$$

Then for O_2

$$S_{O_2} = \frac{T}{100} V_z \frac{dP}{dC} \frac{\partial C}{\partial z} \tag{4}$$

and for CO_2

$$S_{CO_2} = K_2 V_z \frac{\partial C}{\partial z} \tag{5}$$

The velocity profile is described by an integrated form of the Casson's equation

$$V_z = \frac{2Q}{\pi R^2} \frac{(1-r^{*2}) - \frac{8}{3}\gamma^{1/2}(1-r^{*3/2}) + 2\gamma(1-r^*)}{1 - \frac{16}{7}\gamma^{1/2} + \frac{4}{3}\gamma - \frac{1}{21}\gamma^4} \tag{6}$$

and the diffusivity, D, incorporates the augmentation due to the rotating erythrocytes

$$D = D_o - \beta \frac{dV_z}{dr} \tag{7}$$

where β is the blood mixing coefficient. This coefficient can be related to the square of the diameter of the red cell as predicted by Keller and Leonard (3).

The ordinary diffusivity for oxygen, D_o, was obtained using Buckles (1) theoretical approach based on Fricke's (6) equation and the data obtained by Klug et al (7) for solutions for hemoglobin in plasma.

$$D_o = D_p \frac{1 + \phi X}{1 - \phi} \tag{8}$$

where

$$\phi = \frac{H}{100}\left(\frac{D_{RBC}/D_p - 1}{D_{RBC}/D_p + X}\right)$$

$$X = X\left(\frac{D_{RBC}}{D_p}, \frac{h}{\delta}\right)$$

For human blood, Buckles (1) showed that

$$D_o = 1.89 \times 10^{-5} \left(\frac{100 - 0.648H}{100 + 0.531H} \right) \tag{9}$$

The value of the ordinary diffusivity of carbon dioxide in blood can be estimated (8) assuming that the ratio between the diffusivity of carbon dioxide and the diffusivity of oxygen in whole blood is approximately the same as this ratio for pure water

$$\frac{D_o\ (CO_2)}{D_o\ (O_2)} = 0.9833 \tag{10}$$

Substituting the values of D, V, and S into equation (1) the following differential equation is obtained

$$\frac{F_2 + N_F F_1^{\,2}}{r^*} \left[\left(F_2 + \frac{N_F F_1 r^{*\,1/2}}{1 + N_F F_1^{\,2}} \right) \frac{\partial C}{\partial r^*} + r^* \frac{\partial^2 C}{\partial r^{*2}} \right] = F F_3 \frac{\partial C}{\partial z^*} \tag{11}$$

The boundary conditions used to solve this equation are the following for $C = C\ (r^*, Z^*)$

1) $C\ (r^*, 0) = C_o$

2) $\frac{\partial}{\partial r^*} \left[C(0, Z^*) \right] = 0$

3) $D \frac{\partial}{\partial r^*} \left[C(1, Z^*) \right] = F_4 \frac{D_W}{\ln OR^*} \left[C_\infty - C(1, Z^*) \right]$ (12)

The first boundary condition defines the initial value of the concentration at C_o. The second one establishes that the concentration gradient at the center of the tube is zero. The last boundary condition determines the gas transport through the wall. The coefficient F_4, expresses the interface resistance between the blood and the wall.

The carbon dioxide solution of equation (11) permits the computation of the pH at any point of the tube using the Henderson-

Hasselbach equation assuming that the concentration of bicarbonate remains constant through the tube. This assumption is valid in the range of pH changes observed during laboratory experimental testing.

The numerical solution of equation (11) was developed for a simultaneous solution for oxygen and carbon dioxide. The equation was approximated by a finite difference equation and solved using a digital computer.

3. Results and Discussion

A comparison of the mathematical model with the data obtained by Buckles using fresh human blood is shown in figure 3. The diffusivity of oxygen in medical grade silicone rubber was obtained from reference (9). The surprising result is that the best agreement between the theory and the data is obtained for $\beta = 0$; i.e., the micromixing phenomenon is insignificant. Another comparison was performed using data obtained by Weissman and Mockros (10) for cattle blood assuming a hemoglobin concentration of 15 g percent (which was not measured in all their experiments). The data points and the theoretical prediction are shown in figure 4.

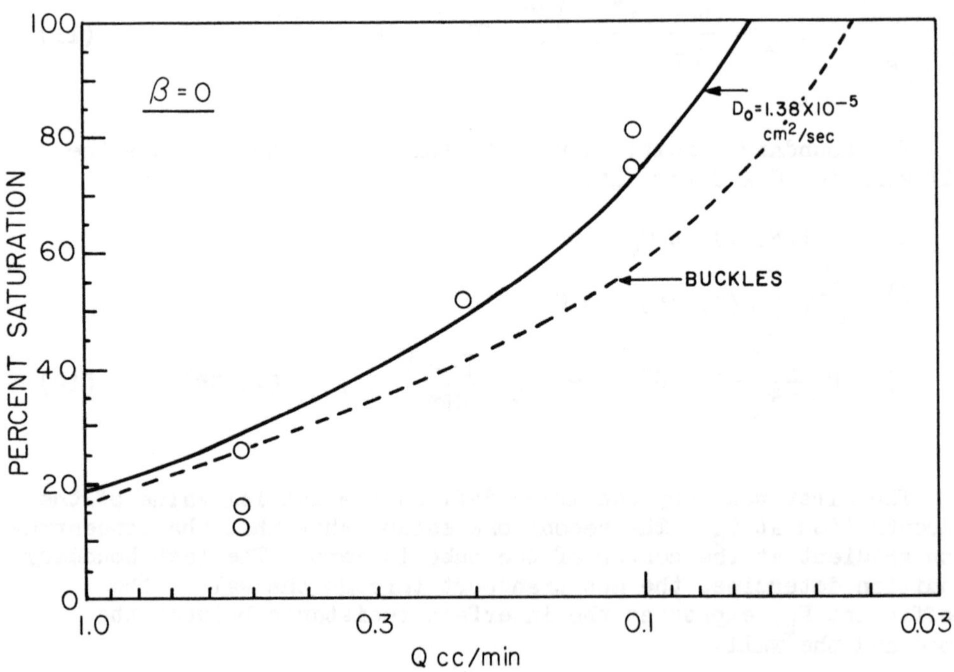

Figure 3. Theoretical prediction and data obtained by Buckles (Ref. 1, 2)

The theoretical prediction for $\beta = 0$ is in excellent agreement with the data published by these investigators. The excellent agreement between the theory and the data can be observed in this figure. Some dispersion of the data with respect to the theoretical curve may be expected due to the unreported hemoglobin concentration which, in some cases, could have been considerably different from the assumed average 15 g percent.

Preliminary data obtained by the authors using rejuvenated 21 day old human blood confirm the validity of the theory presented in this paper. Still some questions have been raised about the possibility of significant augmented diffusion during the oxygenation of non-rejuvenated aged blood due to the uneven surface of its red cells. Theoretical and experimental research using aged human blood as well as <u>in vivo</u> animal tests will continue.

The computer results for a typical case, assuming $\gamma = \beta = 0$, are given in figure 5. The physical parameters are the following

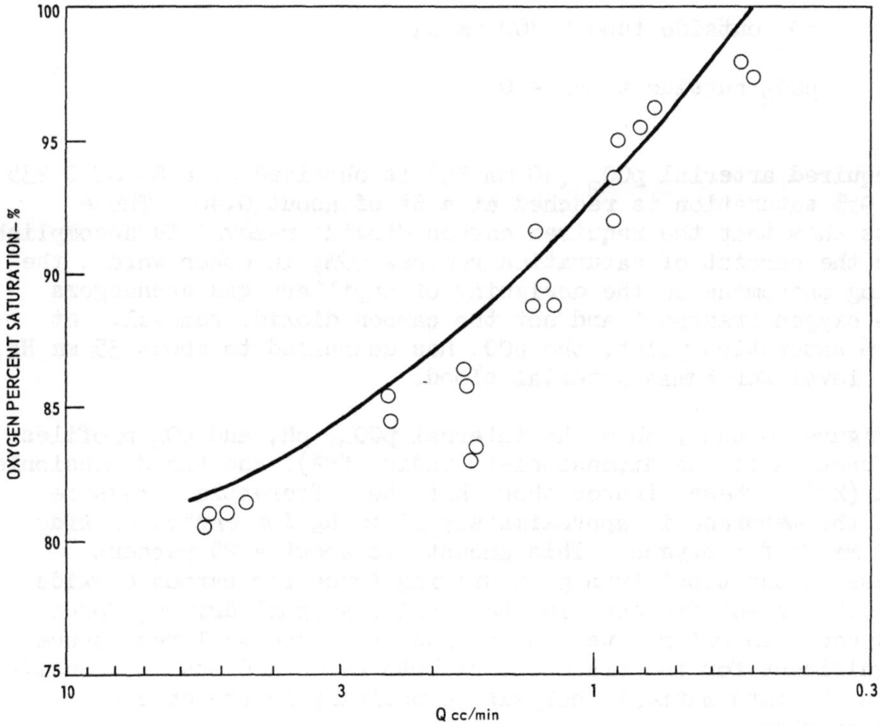

Figure 4. Comparison of theoretical prediction with data obtained by Weissman and Mockros (Ref. 10)

hematocrit = 42%

hemoglobin = 15 g%

D_o = 1.125 x 10^{-5} cm^2/sec (equation 9)

Wall O_2 permeability = 1210 $\frac{cc(STP)\ mil}{min\ m^2\ atm}$ } silicone rubber medical grade

Wall CO_2 permeability = 6310 $\frac{cc(STP)\ mil}{min\ m^2\ atm}$

R = 0.0060 in

OR = 0.0125 in

pO_2 initial = 31 mm Hg (60% oxygen saturation)

pCO_2 initial = 50 mm Hg

pH initial = 7.4

pO_2 outside tube = 760 mm Hg

pCO_2 outside tubes = 0

The required arterial pCO_2 (40 mm Hg) is obtained at a Z* of 0.235 while 95% saturation is reached at a Z* of about 0.40. These results show that the required carbon dioxide removal is accomplished before the percent of saturation reaches 95%; in other words, the limiting phenomena in the designing of capillary gas exchangers is the oxygen transport and not the carbon dioxide removal. At the 95% saturation point, the pCO_2 has decreased to about 35 mm Hg, a safe level for human arterial blood.

Figures 6 and 7 show the internal pCO_2, pH, and pO_2 profiles as a function of the dimensionless radius (r*), and the dimensionless length (Z*). These figures show that the differential pressure across the membrane is approximately 12 mm Hg for carbon dioxide and 40 mm Hg for oxygen. This amounts to about a 25 percent decrease in the total transport driving force for carbon dioxide and a 5.5 percent decrease in the total transport driving force for oxygen. Therefore, we can conclude that the wall resistance is significant for the transport of both gases and must be considered in the mathematical analysis, especially in the case of carbon dioxide.

The internal pH curves show that the value of pH at the wall

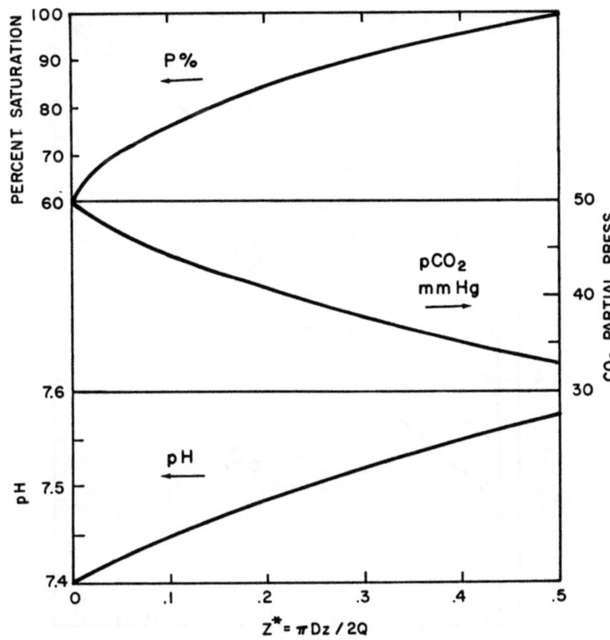

Figure 5. Average percent saturation, pCO_2 and pH function of the dimensionless length.

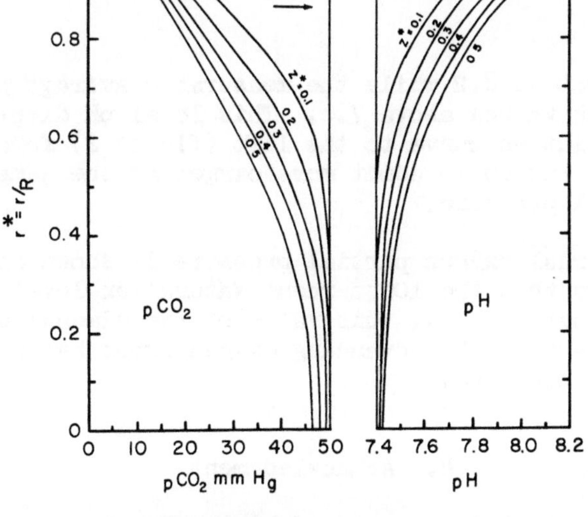

Figure 6. pCO_2 and pH internal profiles.

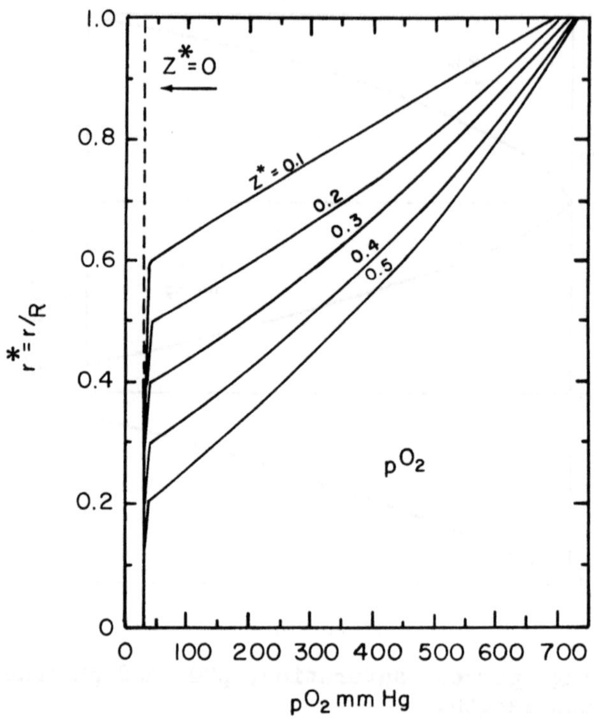

Figure 7. pO_2 internal profile.

may be as great as 8.2 while the measurable average pH (figure 4) does not reach values above 7.6. This local pH displaces the oxygen dissociation curve to the left (figure 2) forcing the red cells in that region to admit more oxygen at the same plasma oxygen partial pressure.

The internal oxygen partial pressure is shown in figure 6. Figure 4 shows that the 100 percent saturation level is reached at a Z* of about 0.5. At this value of the dimensionless length, figure 6 shows that the advancing oxygen front has not reached the center of the tube.

4. Acknowledgment

This work is being sponsored by the Development Directorate

of the U.S. Army Medical Research and Development Command.

SYMBOLS

C concentration - g-mole/l

C equivalent concentration outside the wall, corresponding to a plasma partial pressure equal to the outside gas partial pressure

C_o initial concentration - g-mole/l

D effective diffusivity - cm²/sec

D_o ordinary diffusivity in whole blood - cm²/sec

D_p ordinary diffusivity in plasma - cm²/sec

D_{RBC} ordinary diffusivity in the erythrocytes - cm²/sec

D_w ordinary diffusivity in the wall - cm²/sec

$F = 1 - \gamma^2 - 8/3\,\gamma^{1/2}(1 - r^{*3/2}) + 2\gamma(1 - r^*)$, for $r^* > \gamma$

$\quad = 1 + 2\gamma - \dfrac{\gamma^2}{3} - \dfrac{8}{3}\gamma^{1/2}$ for $r^* < \gamma$

$F_1 = r^{*1/2} - \gamma^{1/2}$, for $r^* > \gamma$

$\quad = 0$, for $r^* < \gamma$

$F_2 = 1.0$, for oxygen

$\quad = 0.9833$, for carbon dioxide

$F_3 = 1 + \dfrac{T}{100}\dfrac{dp}{dc}$ for oxygen

$\quad = 10$, for carbon dioxide

$F_4 = (1 + Dw/kR \ln OR^*)^{-1}$

h = thickness of a red cell

H = hematocrit

$K_1 = 1.14 \times 10^{-2}$

$K_2 = 9$

N_F = $\tau_o \beta / \gamma \eta D_o$

OR outside radius of the tube - cm

OR* = OR/R

P oxygen percent of saturation (given by the oxygen dissociation curve)

pCO_2 partial pressure of CO_2 - mm Hg

pH -Log $[H_3O^+]$

pO_2 partial pressure of O_2 - mm Hg

Q blood flow - cc/sec

r radius - cm

r* = r/R

r_o radius of the plug flow region - cm

R inside radius of the tube - cm

S reaction rate

t time - sec

T concentration in the red cells at 100 percent saturation - g-mole/liter

V_z axial velocity - cm/sec

Z axial distance - cm

Z* = $2\gamma\eta D_c Z / \tau_o R$, or for $\gamma = 0$ $Z^* = \frac{\pi DZ}{2Q}$

Greek Symbols

β blood mixing coefficient - cm^2

γ = r_o/R

δ diameter of the red cell

π = 3.1416

τ_0 = 0.07 dynes/cm^2

η = 0.0283 poises

REFERENCES

1. Buckles, R.G., Ph.D. Thesis, M.I.T., Cambridge, Mass (1966).

2. Buckles, R.G., Merrill, E.W., & Gilliland, E.R. AIChE J. 703, Vol 14, No. 5.

3. Keller, K.H. & Leonard, E.F. 12th Advanced Seminar AIChE, Part I.

4. Casson, N., Rheology of Disperse Systems, Mill, C.C. (Editor), Pargaman press, Oxford.

5. Physiology Study Notes, Staff of the Physiology Department, Northwestern University Medical School, 1964 edition.

6. Fricke, H., Physical Rev., 23, 575 (1924)

7. Klug, A., Krenzer, F., Roighton, F.J.W., Helvet Physiol. et Pharmacol Acts, 14, 121 (1956).

8. Perry, R.H., Chemical Engineers Handbook, 4th Edition, McGraw Hill, New York, 1963.

9. Galletti, P.M., Snider, M.T., and Gilbert-Aiden, D., Med. Res. Eng., 20, Vol 5, No 2 (1966).

10. Weissman, M.H., and Mockros, L.F., J. of Eng. Mech. Div., Proc. ASCE, 225, Dec 1967.

11. Handbook of Respiratory Data in Aviation, National Research Council. Committee on Aviation Medicine, Subcommittee on Oxygen Anoxia, Office of Scientific Research and Development, Washington, D.C. 1944.

DEVELOPMENT OF CAPILLARY MEMBRANE BLOOD OXYGENATORS

G.W. Harris, F.C. Tompkins, R.P. deFilippi
and J.H. Porter, Abcor, Incorporated
Cambridge, Massachusetts

M.J. Buckley, Massachusetts General Hospital
Boston, Massachusetts

Figure 1 presents a schematic diagram of an assembled oxygenator unit. In operation, oxygenating gas is introduced into the central distributor from both ends, passing out radially over the capillary bundle through a series of perforations in the distributor, and exiting through the two central ports to the atmosphere. Venous blood is introduced through one of the two outer ports, entering a small mixing chamber from which it passes through a header into the capillary bundle. Gas exchange is accomplished as the blood proceeds through the capillaries. The oxygenated blood then exits via the header on the other end.

Figure 2 presents a disassembled view of the oxygenator. The capillary bundle, including central distributor and epoxy headers, is removable from the cylindrical shell. "O" rings provide seals between headers and shell, end caps and shell, and distributor and end caps. This arrangement has proved compact and easy to handle.

ANALYSIS

A. Introduction

In the analysis, the quantities to be fixed are the capillary diameter and wall thickness; the capillary membrane material; the total blood-flow resistance through the device (pressure drop); and the oxygenating gas pressure and composition. For purposes of evaluating experimental data, the analysis will yield the amount of O_2 and CO_2 transferred, or arterial PO_2 and PCO_2, for a device

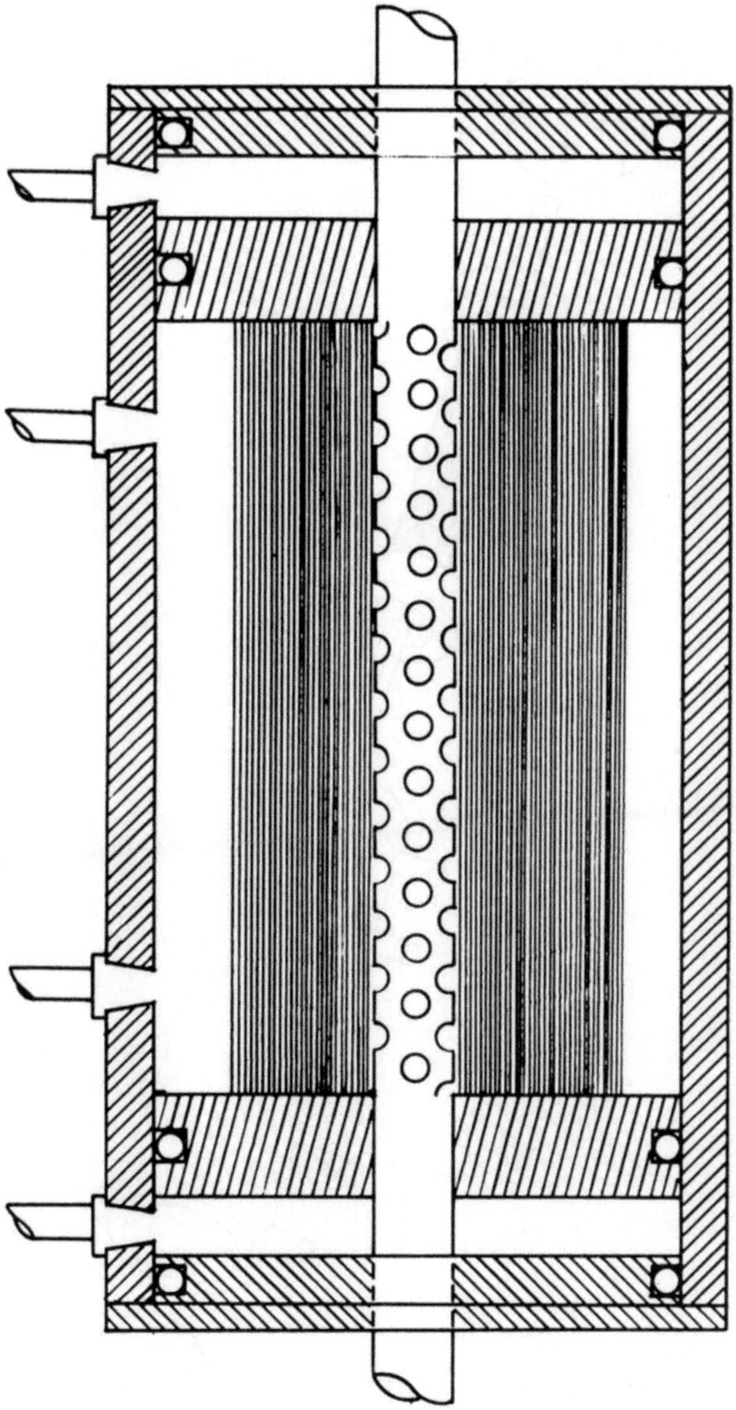

FIGURE 1 ASSEMBLED OXYGENATOR UNIT.

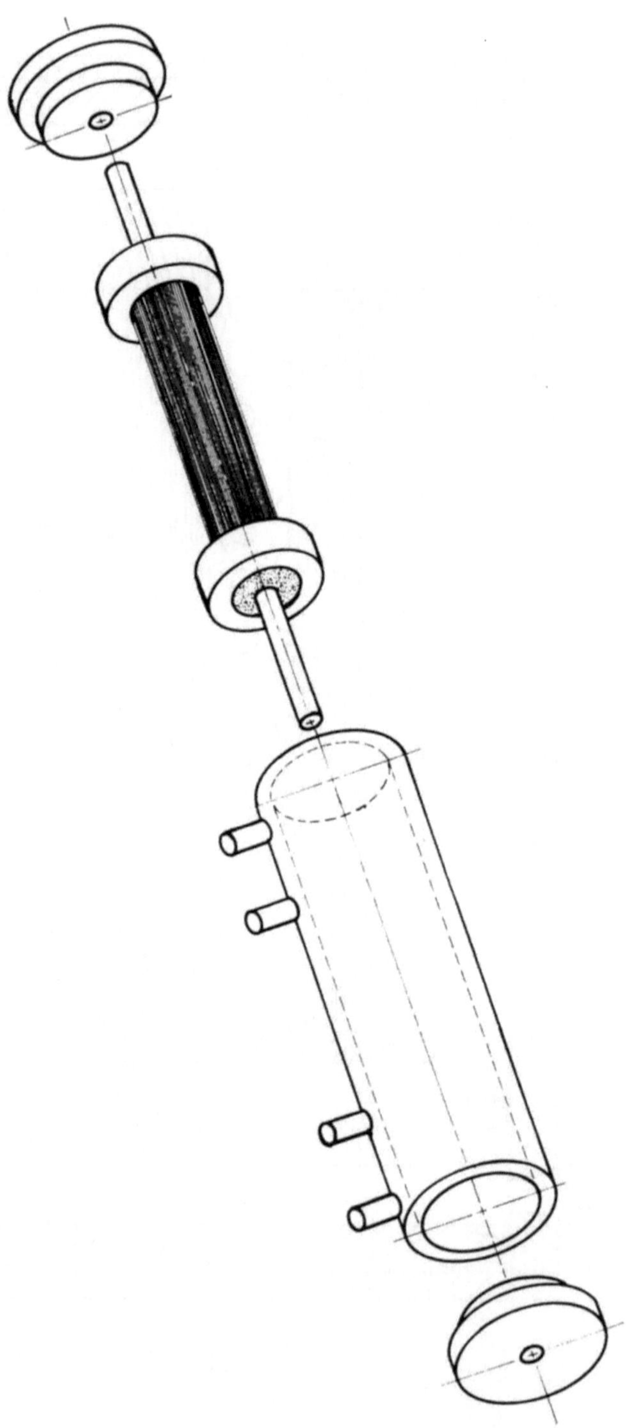

FIGURE 2 CAPILLARY MEMBRANE OXYGENATOR EXPERIMENTAL DEVICE

with a given capillary length and number of capillary tubes. For design purposes, the results of the analysis will fix the length and number of capillary membranes for given arterial and venous oxygen and carbon dioxide partial pressures.

It is important to note that the design calculation specifies an oxygenator with the minimum number of parallel capillaries. Capillary bundles with more parallel channels of shorter length would result if a smaller pressure drop were fixed. Because the difficulty of fabrication, and hence the cost, would probably increase with increasing number of capillaries, the optimum design appears to be that which would have the minimum number of parallel capillaries, and thus the highest pressure drop tolerable. In some potential applications, such as with infants flow resistance should be very low, and a large number of shorter capillaries indeed would be used.

The principle assumptions behind the analysis are the following:

1. The mass transfer resistance is distributed between the membrane and blood alone, with negligible resistance in the gas phase;

2. The distribution of blood flow among all capillary tubes is uniform.

3. Equilibrium reactions with blood gases are rapid compared to diffusion.

The analytical expressions consist of differential mass balances for both O_2 and CO_2 in the blood, plus a momentum balance (Poiseuille's law) on the overall blood flow to introduce the pressure drop term. A schematic diagram of the system analyzed is shown in Figure 3.

B. Material Balance

A material balance on oxygen results in the following relationship for the cylindrical geometry:

$$\frac{\alpha_B D_B}{r}\frac{\partial}{\partial r}\left(r\frac{\partial P}{\partial r}\right) = V(r)\left[\alpha_B \frac{\partial P}{\partial x} + NH\frac{\partial y}{\partial x}\right] \quad (1)$$

where

α_B = oxygen solubility in blood $\left(\frac{\text{g moles}}{\text{mmHg-cc}}\right)$

D_B = whole blood oxygen diffusivity (cm^2/sec)

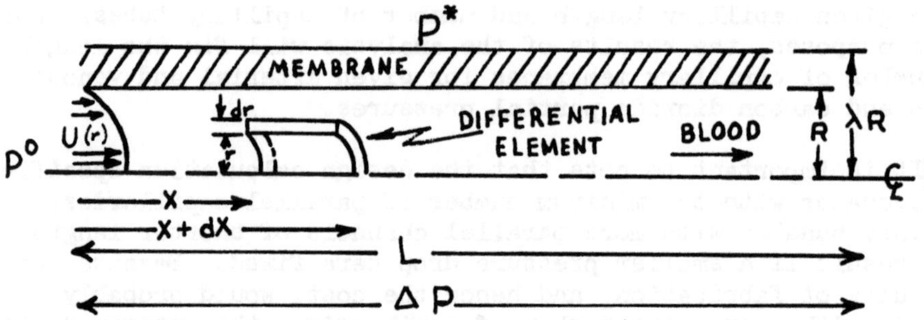

Where pO = entering gas tension (O_2 or CO_2)

p* = gas phase partial pressure (O_2 or CO_2)

pF = mixed outlet gas tension (O_2 or CO_2)

U(R) = velocity profile

ΔP = hydraulic pressure drop

FIGURE 3 DIFFERENTIAL ANALYSIS OF OXYGENATOR

r = radial coordinate (cm)

p = oxygen tension (mmHg)

V(r) = velocity profile (cm/sec)

N = moles oxygen/mole hemoglobin at saturation

H = hemoglobin concentration (moles/cc)

y = fractional oxygen saturation

x = axial coordinate (cm)

This equation may be manipulated into a more convenient form for our purposes. Noting that y, the fractional oxygen hemoglobin saturation, is a function of oxygen partial pressure, the partial derivative $\frac{\partial y}{\partial x}$ may be expanded by the chain rule of differential calculus to become $\frac{\partial y}{\partial P} \cdot \frac{\partial P}{\partial x}$. Dividing both sides of equation (1) by α_B and grouping terms yields:

$$\frac{D_B}{r}\frac{\partial}{\partial r}\left(r\frac{\partial P}{\partial r}\right) = V(r)\left[1+\frac{NH}{\alpha_B}\frac{\partial y}{\partial P}\right]\frac{\partial P}{\partial x} \qquad (2)$$

The group $\frac{NH}{\alpha_B}\frac{\partial y}{\partial P}$ represents the ratio of moles of hemoglobin bound oxygen per mole of physically dissolved oxygen in the blood on a unit volume basis, and is arbitrarily given the symbol $S(x)$. Now dividing both sides by $1+S(x)$ yields:

$$\frac{D_B}{[1+S(x)]}\frac{\partial}{\partial r}\left(r\frac{\partial P}{\partial r}\right) = r\,V(r)\frac{\partial P}{\partial x} \qquad (3)$$

Written in this form, the equation is exactly analogous to the classical Graetz problem in heat transfer solutions of which are well characterized. The above equation suggests the use of a modified diffusivity

$$\overline{D}' = \frac{D_B}{[1+S(x)]}$$ where $S(x)$ is related to the instantaneous slope of the oxygen saturation function. Solutions to the Graetz problem for mass transfer are expressed in terms of two dimensionless parameters, the Sherwood group and the modified Graetz group.

$$N_{SH} = \frac{kd}{D_B} \qquad (4)$$

$$N_{GR} = \frac{4Q\,(1+S(x))}{D_B x} \qquad (5)$$

where

k = mass transfer coefficient (cm/sec)

d = capillary inside diameter (cm)

D_B = whole blood oxygen diffusivity (cm^2/sec)

Q = blood flow rate/tube (cm^3/sec)

x = distance from tube entrance (cm)

Note that, although S(x) represents an instantaneous value, for purposes of representing experimental data, the chord slope of the saturation function is used. This is very accurate in the mid-range of the curve, but does lose accuracy at both low and high saturation levels where the curve bends sharply.

The solution holds for a uniform oxygen tension (P_{O_2}) at the inside wall of the capillary, which is related to the gas phase composition through a material balance across the tube wall. This relationship is incorporated into a boundary condition for the original equation and has the effect of accounting for a finite wall resistance to mass transfer not considered in the original Graetz case. The equation to be solved along with accompanying initial and boundary conditions is shown in Figure 4.

Exactly analogous relationships may be written for carbon dioxide, the principle difference being that the influence of carbon dioxide tension (through pH) on the oxygen saturation curve must be accounted for. Implicit in the development thus far are mathematical descriptions of the following:

1. Blood velocity profile in the capillary
2. oxygen dissociation curve for hemoglobin
3. influence of carbon dioxide partial pressure on blood pH
4. carbon dioxide dissociation curve for blood

C. Velocity Profile

Blood flows through the capillary under a pressure gradient established by an external pump or the heart. Blood is slightly non-Newtonian in nature and the velocity profiles are best fitted by Casson's equations as developed by Benis, et.al.(1). The velocity profile is described by the following equation:

$$V(\eta) = \frac{\Delta P R^2}{4 \mu L}\left[(1-\eta^2) - \frac{8}{3}\xi^{1/2}(1-\eta^{3/2}) + 2\xi(1-\eta)\right] \qquad (6)$$

$$\xi = \frac{2 \tau_y L}{\Delta P R} \qquad (7)$$

$$\eta = r/R \qquad (8)$$

where

ΔP = pressure gradient (dynes/cm^2)

$$\bar{D}_B' \frac{\partial}{\partial r}\left(r\frac{\partial P}{\partial r}\right) = r\,V(r)\frac{\partial P}{\partial x}$$ material balance on oxygen

Boundary Conditions

$$P_B\big|_R + R\left(\frac{D_B\alpha_B}{D_M\alpha_M}\right)\ln \lambda R \frac{dP_B}{dr} = P^*$$

Equation relates to oxygen tension on blood side of capillary wall ($P_B|_R$) to gas phase partial pressure and blood and membrane properties.

where

D_M = oxygen diffusivity in membrane (cm^2/sec)

α_M = oxygen solubility in membrane $\frac{\text{moles}}{\text{cc-mmHg}}$

$\frac{\partial P}{\partial r}$ = 0 at r = 0, all x

This relationship follows symmetry considerations

Initial Conditions

P = P* at x = 0, all r

Uniform entering oxygen tension

FIGURE 4 MATERIAL BALANCE RELATIONSHIP.

R = capillary radius (cm)

L = capillary length (cm)

μ = blood viscosity (g/cm-sec)

τ_y = yield stress of blood $\frac{\text{dynes}}{\text{cm}^2}$

$V(\eta)$ = velocity as function of radial position (cm/sec)

r = radial coordinate (cm)

Physical constraints restrict the value of ξ to less than or equal to one. For small values of ξ, blood flow approaches Newtonian behavior. In general, at flow rates typically encountered, ξ is very small and essentially Newtonian behavior is obtained.

D. Oxygen Dissociation

The oxygen dissociation curve for hemoglobin is described by the Margaria equation (2);

$$Y_{O_2} = \frac{(m-1)k^4 P_{O_2}^4 + kP_{O_2}(kP_{O_2}+1)^3}{(m-1)k^4 P_{O_2}^4 + (kP_{O_2}+1)^4} \tag{9}$$

where

Y_{O_2} = fractional oxygen saturation

P_{O_2} = oxygen partial pressure (mm Hg)

m = empirical constant

k = empirical constant dependent upon pH and temperature

Analysis of available data in the literature (3) yielded the following expressions for the adjustable constants:

$$m = 124 \tag{10}$$

$$k = \left\{0.796\left(\frac{PH-7.0}{7.0}\right) + 0.152\left(\frac{PH-7.0}{7.0}\right) + 0.0211\right\} \exp\left\{-17.72\left(\frac{T-293}{T}\right)\right\} \tag{11}$$

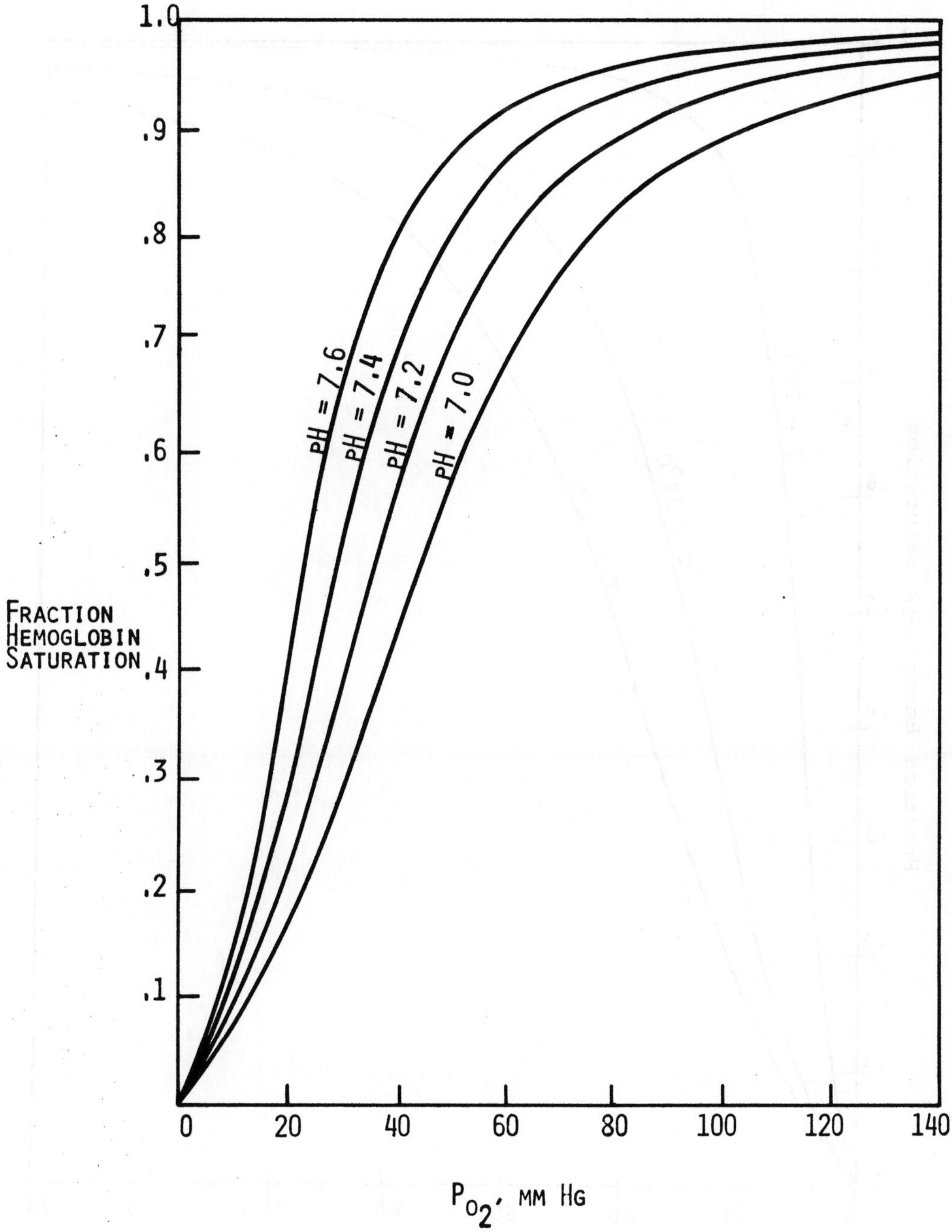

FIGURE 5 EFFECT OF pH ON HEMOGLOBIN SATURATION AT 38°C.

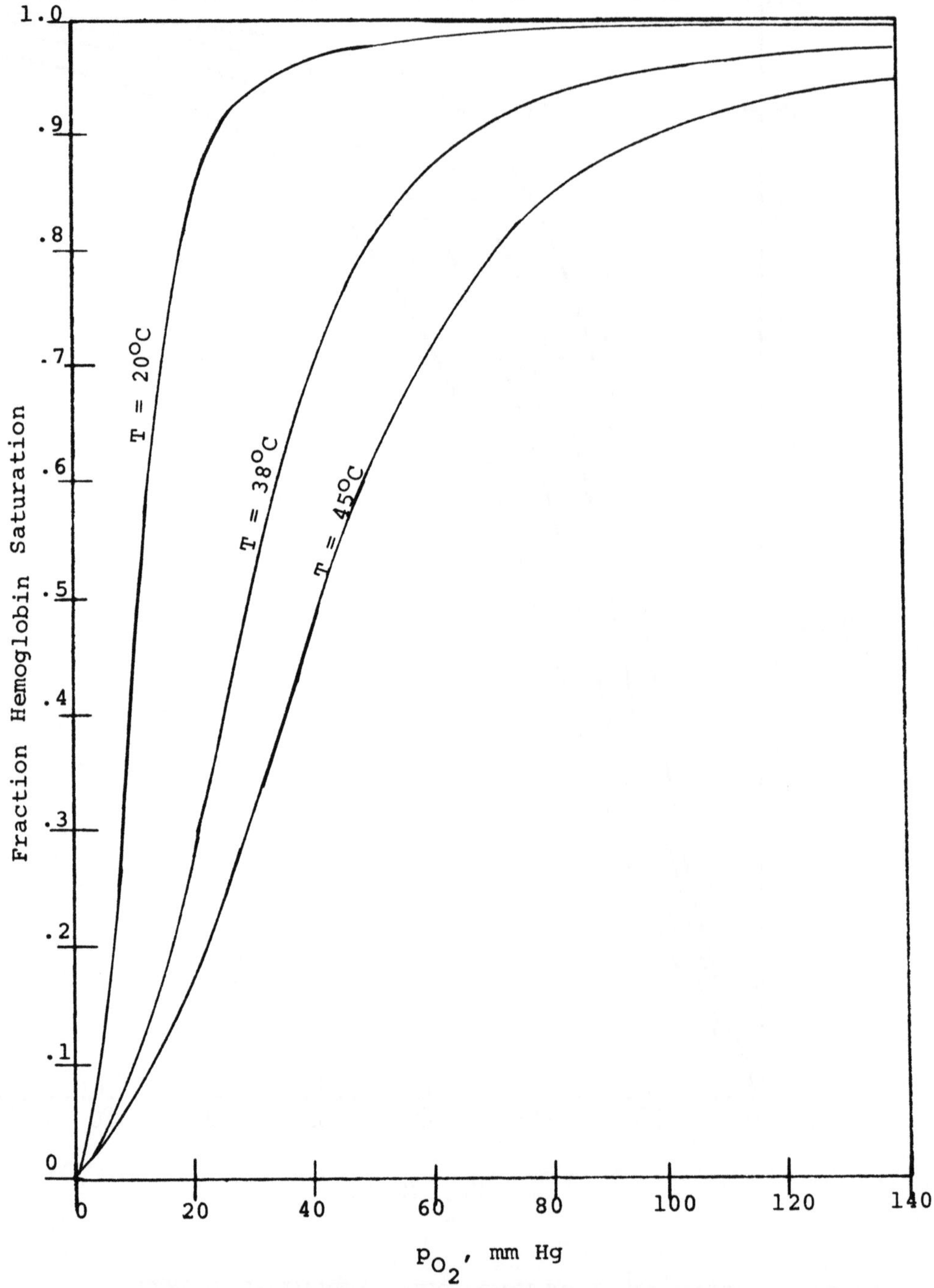

FIGURE 6 EFFECT OF TEMPERATURE ON HEMOGLOBIN SATURATION AT pH = 7.4.

These relationship give good results over the following range:

$$6.8 \leq pH \leq 7.8 \quad (12)$$

$$293^oK \leq T \leq 318^oK \quad (13)$$

Figures 5 and 6 illustrate the considerable effects that both pH and temperature have on both the slope and the absolute values of the saturation curve for hemoglobin. The pH effect is especially important because pH varies along the length of the capillary due to CO_2 transport.

E. CO_2 - pH Relationship

For normal blood (protein content 7%), the relationship of pH to P_{CO_2} is:

$$pH = -0.3782 \ln P_{CO_2} + 8.806 \quad (14)$$

Again, this represents an empirical relationship derived from available data in the literature (3).

F. Carbon Dioxide Dissociation

The carbon dioxide dissociation curve for normal blood was obtained by curve fitting to data reported by Guyton (4). The resulting functional relationship is given below:

$$Y_{CO_2} = \frac{k' P_{CO_2}}{1+k' P_{CO_2}} \quad (15)$$

where

Y_{CO_2} = fractional CO_2 saturation

P_{CO_2} = CO_2 tension (mmHg)

k' = empirical constant

A value of $k' = 0.02386$ was found to best fit data over the range $0 \leq P_{CO_2} \leq 140$. Figure 7 shows the comparison of the calculated

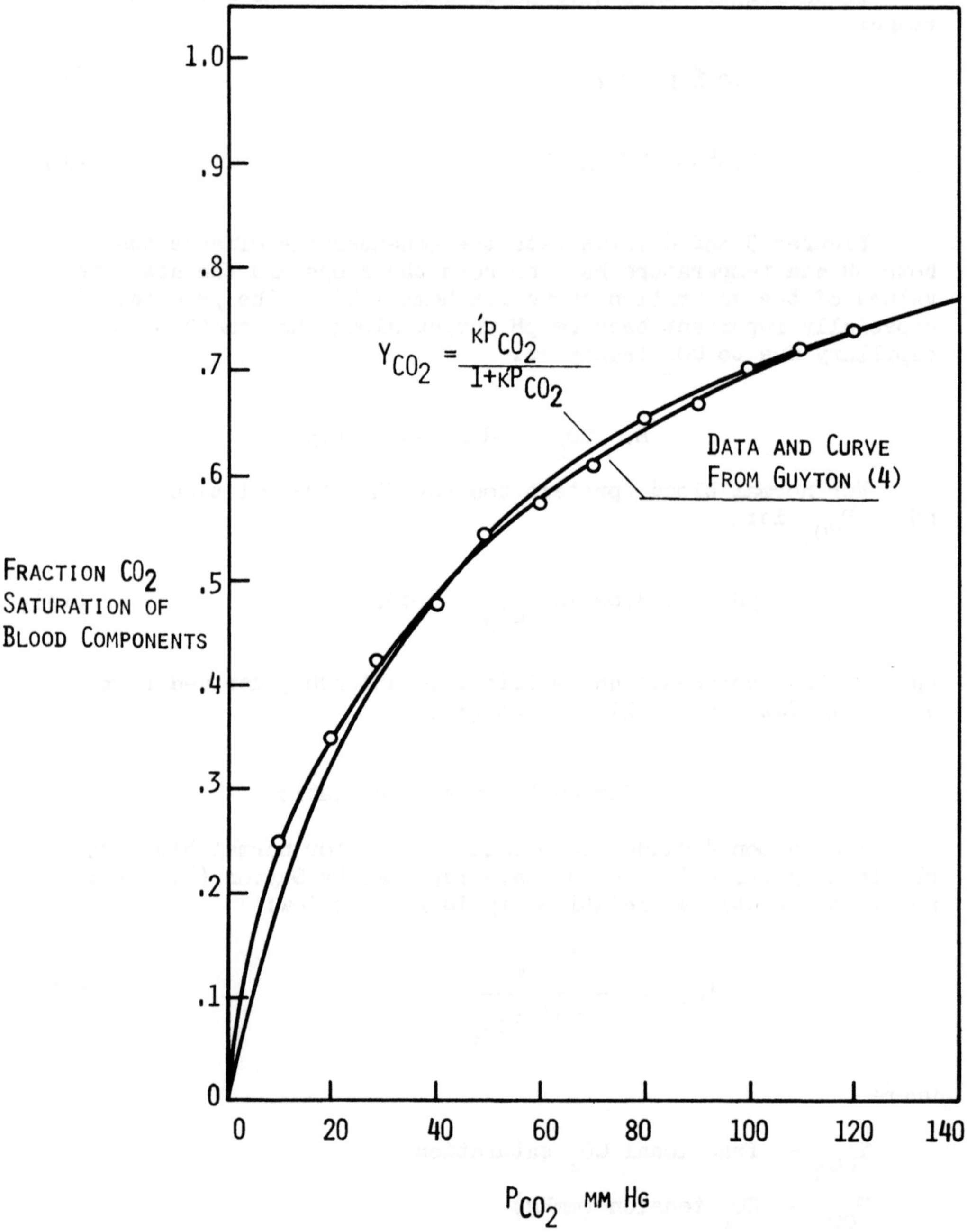

FIGURE 7 CARBON DIOXIDE SATURATION CURVE.

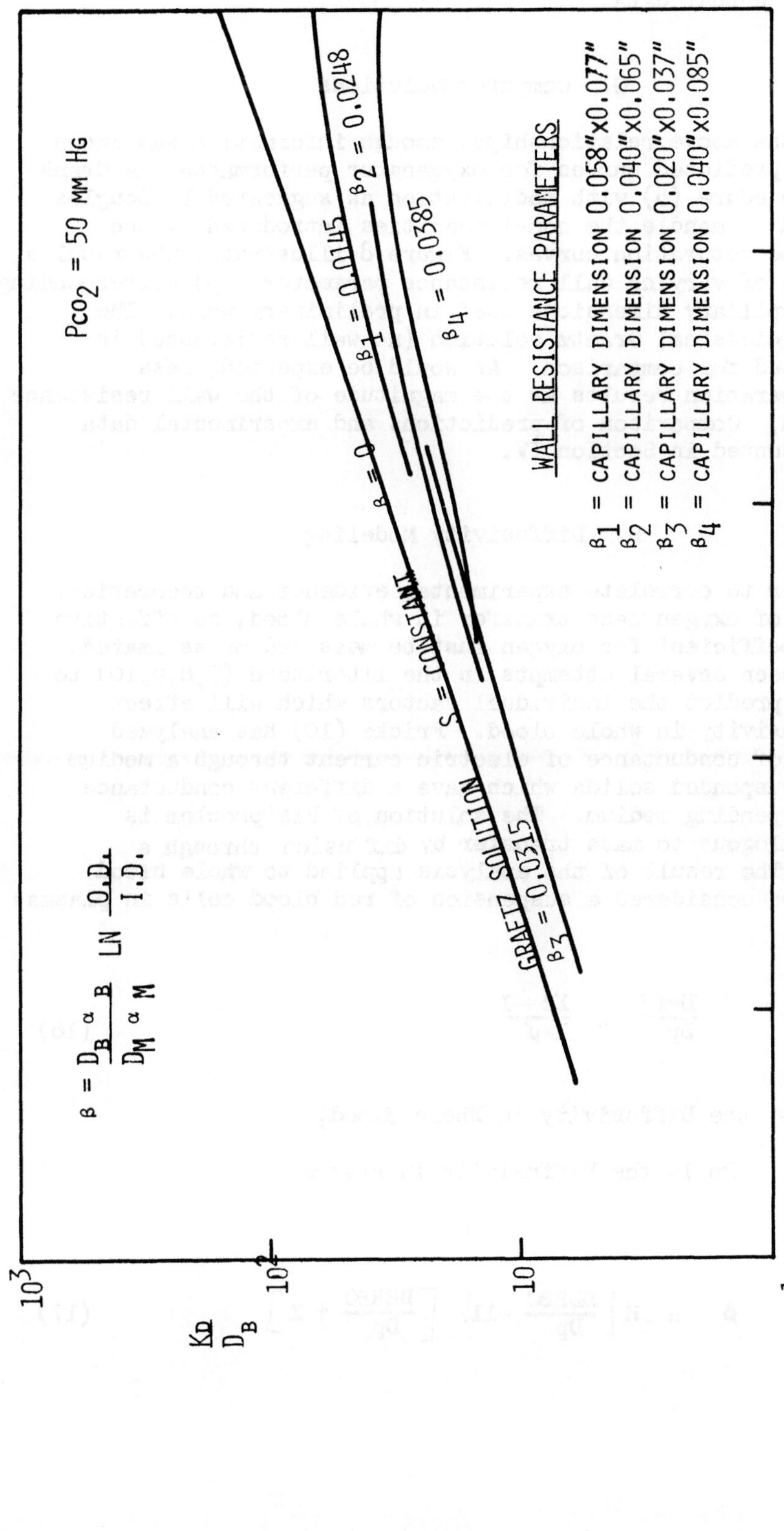

FIGURE 8 PREDICTED BUNDLE PERFORMANCE.

curve to the actual data.

G. Computer Solutions

Given the above relationships, enough information was known to generate predicted curves for oxygenator performance. A Crank-Nicolson procedure (5) with modification as suggested by Douglas (6) was used to handle the non-linearities introduced by the nature of the saturation curves. Figure 8 illustrates the results for the case of varying wall resistance parameters (β) corresponding to actual capillary dimensions used in preliminary work. The case of the classical Graetz solution (no wall resistance) is also presented for comparison. As would be expected, less efficient operation results as the magnitude of the wall resistance is increased. Comparison of predictions and experimental data will be presented in Section IV.

H. Diffusivity Modeling

In order to correlate experimental evidence and theoretical predictions of oxygen mass transfer in whole blood, an effective diffusion coefficient for oxygen must be measured or estimated. There have been several attempts in the literature (7,8,9,10) to measure and predict the individual factors which will effect oxygen diffusivity in whole blood. Fricke (10) has analyzed the problem of conductance of electric current through a medium containing suspended solids which have a different conductance than the suspending medium. The solution of his problem is directly analogous to mass transfer by diffusion through a suspension. The result of the analysis applied to whole blood (which can be considered a suspension of red blood cells in plasma) is given by:

$$\frac{Deff}{Dp} = \frac{X\phi + 1}{1-\phi} \qquad (16)$$

where Deff = the Diffusivity in Whole Blood,

Dp is the Diffusivity in Plasma

and

$$\phi = H\left[\frac{DERBC}{Dp} - 1\right] / \left[\frac{DERBC}{Dp} + X\right] \qquad (17)$$

where DERBC = the effective diffusivity in a red blood cell,

H = the hematocrit (volume fraction red cells in whole blood)

and

X = a characteristic parameter which depends on the ratio DERBC/Dp and the geometry of the suspended particles.

X = calculated by the rather cumbersome relation;

$$X = \frac{[(DERBC/Dp-1) - (DERBC/Dp)\, B]}{(DERBC/Dp-1-B)} \tag{18}$$

$$B = \frac{1}{3}\left[\frac{1}{1+(DERBC/Dp-1)\frac{M}{2}} + \frac{1}{1+(DERBC/Dp-1)(1-M)}\right](DERBC/Dp-1) \tag{19}$$

$$M = \frac{\theta - .5 \sin 2\theta}{\sin 3\theta} \quad \text{where } \theta = \cos^{-1} a/b \tag{20}$$

where a and b are the dimensions of the red blood cell

red blood cell

Fricke (10) applied his equation to the electrical conductivity in sand, water suspensions and to whole blood and found very good agreement between theoretical and experiment measurements. Based upon this analysis the Fricke model is used as the starting point for predicting oxygen diffusion coefficients in whole blood.

A major adaptation must be made to this analysis, however, since it assumed that the effective diffusivity in the red blood cell remains constant throughout the suspension.

Work by Keller and Friedlander (7, 8) in measuring oxygen diffusion coefficients in hemoglobin-saline solutions has shown that diffusion of the oxy-hemoglobin molecule enhances the overall rate of oxygen transport. Furthermore, they show the enhancing effect depends on the fraction of hemoglobin which is oxy-hemoglobin. This in turn, depends on the local oxygen concentration which is assumed in equilibrium with hemoglobin. If a red blood cell is considered as a hemoglobin solution container, the results of their analysis can be applied directly to give the following expression for oxygen diffusivity in a red blood cell.

$$DERBC = DO2RBC + DHBO \frac{4N}{\alpha} \frac{dy}{dp} \tag{21}$$

This equation assumes that the red cell membrane offers a minimal resistance to oxygen transport and oxygen is in equilibrium with oxyhemoglobin.

DO2RBC is the diffusion coefficient of oxygen in oxy-hemoglobin

DHBO is the diffusion coefficient of oxyhemoglobin in red cells

N is the concentration of total hemoglobin in a red cell

α is the solubility coefficient of oxygen in red blood cells

p is the local oxygen partial pressure in equilibrium with hemoglobin

y is the fraction of total hemoglobin existing as oxy-hemoglobin

Using this relationship, we find that the effective diffusion coefficient depends on the slope of the oxygen-hemoglobin saturation curve (dy/dp). An extensive search through the literature (11) yielded the best estimates of the above parameters necessary to characterize whole blood oxygen diffusion. The results of the Keller-Friedlander and Fricke analyses are given in Figure 9. This shows the ratio of the whole blood to plasma diffusion coefficients as a function of local PO_2 and pH (pH effects the saturation curve (2)). An interesting feature is the occurrence of a maximum in the effective whole blood diffusivity corresponding to the maximum slope in the oxygen saturation curve.

The next figure shows the same results for a pH of 7.4 (normal) presented in a more useful form for evaluating experimental data. The integral average diffusivity is presented as a function of inlet and outlet oxygen tensions. Therefore, given the experimental ΔP_{O_2}, an average value for D_B may be directly obtained from Figure 10 for use in the Graetz and Sherwood parameters. Note that the enhancement effect is not very significant in the clinical range (ΔP_{O_2} = 50 to 100 mmHg and pH = 7.4). The critical parameters in the model are presented below:

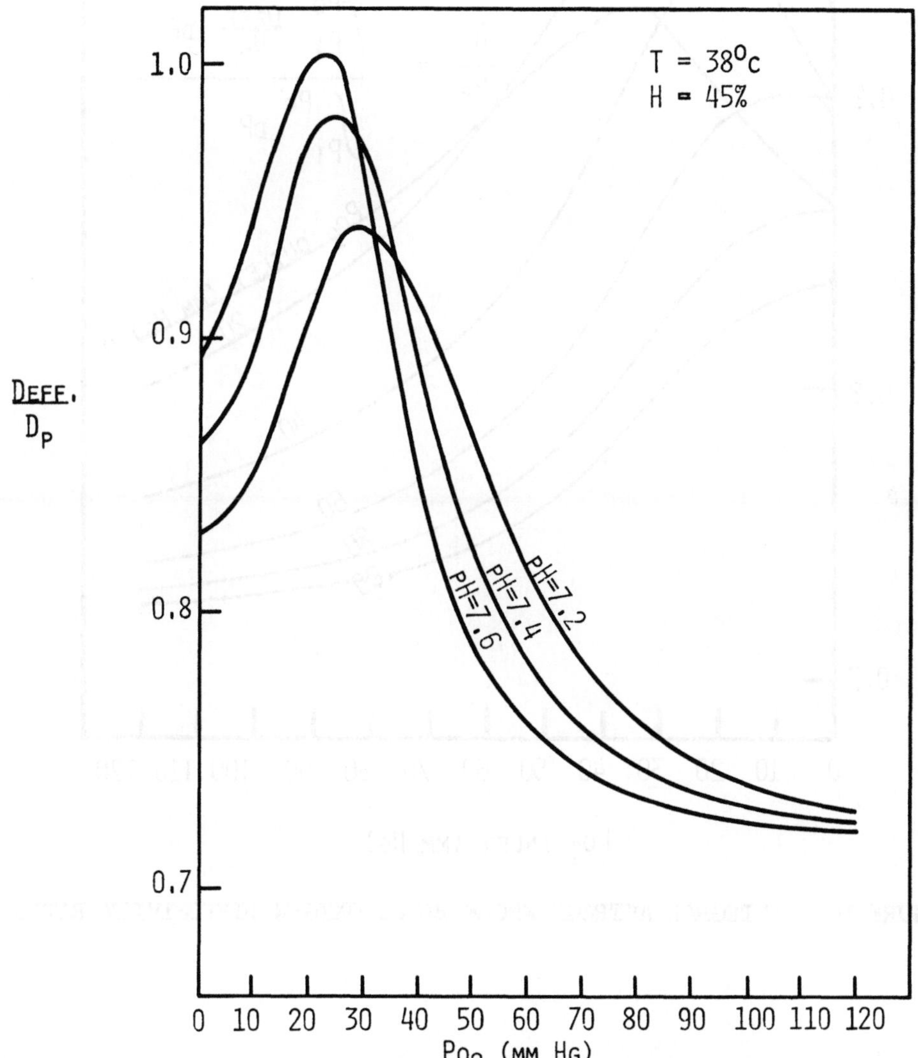

FIGURE 9 EFFECTIVE WHOLE BLOOD OXYGEN DIFFUSIVITY RATIO.

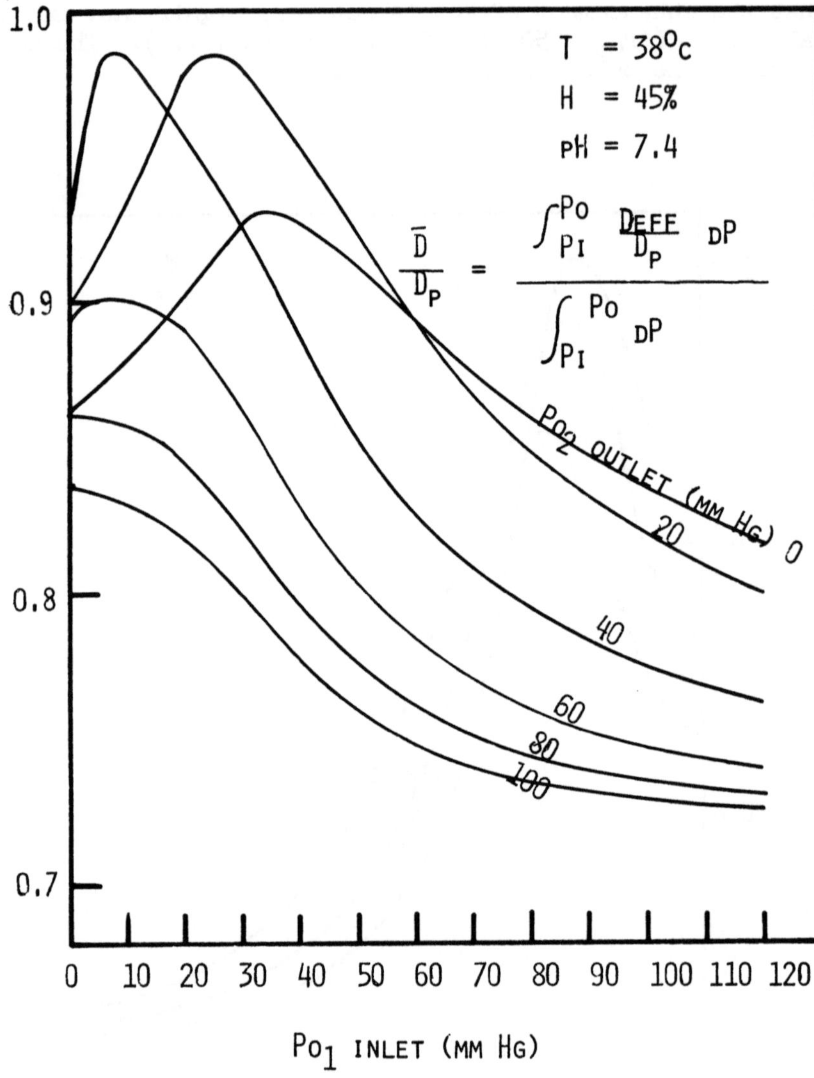

FIGURE 10 INTEGRAL AVERAGE WHOLE BLOOD OXYGEN DIFFUSIVITY RATIO.

D_{HBO} = 2.5×10^{-8} cm^2/sec at 38°C

D_{O_2RBC} = 7.5×10^{-6} cm^2/sec at 38°C

D_p = 1.67×10^{-5} cm^2/sec at 38°C

H = 0.45 volume RBC/volume WB

a = 1.7 μ

b = 8 μ

N = 5.32×10^{-6} g moles Hb/cc RBC

α = 1.40×10^{-9} g moles O_2/cc-mmHg

Editors Note: Clinical data on the Abcor Oxygenator are available from the authors.

REFERENCES

1. Benis, A.M., Sc.D. Thesis, M.I.T., 1964.

2. Margaria, R., "A Mathematical Treatment of the Blood Dissociation Curve for Oxygen, "Clinical Chemistry, 9, 745 (1963).

3. DeFilippi, R.P., Tompkins, F.C., Porter, J.H., and Harris, G.W., "Annual Report - Development of a Capillary Membrane Blood Oxygenator", prepared for Artificial Heart Program under Contract pH 43-67-1405, February, 1969.

4. Guyton, A.C., Textbook of Medical Physiology, 3rd. ed., W.B. Saunders, Philadelphia. 1966, p. 584.

5. Crank, J., and Nicolson, Proc. Cambridge Phil. Soc., 43, 50 (1947).

6. Douglas, J., Trans. Am. Math. Sec., 89, 484 (1958).

7. Keller, K.H., Friedlander, S.K., "The Steady State Transport of Oxygen Through Hemoglobin Solutions," Journal of General Physiology, 49, 663 (1966).

8. Keller, K.H., Friedlander, S.K., "Diffusivity Measurements of Methemoglobin," Journal of General Physiology, 49, 681 (1966).

9. Buckles, R.G., Merrill, E.W., and Gilliland, E.P., "An Analysis of Oxygen Absorption in a Tubular Membrane Osygenator," Accepted for publication, A.I.Ch.E. Journal.

10. Fricke, H., "A Mathematical Treatment of the Electric Conductivity and Capacity of Disperse Systems, "Physical Review, April, 1924, pp. 575-587.

11. Tompkins, Frederick C., Jr., "Program Review - Capillary Membrane Oxygenator", Abcor Internal Report, June, 1969.

ACKNOWLEDGEMENT

This research was supported by the Artificial Heart Program, National Heart Institute, under Contract No. ph 43-67-1405.

OXYGENATION OF BLOOD FOR CLINICAL APPLICATIONS

William J. Dorson, Jr.,

Chemical Engineering Faculty, Arizona State University

Tempe, Arizona

The analytical description of blood oxygenation is of both practical and theoretical importance. The design of clinical oxygenators in the past has been based on experimental trials, whereas the newer generation of such devices rely upon theoretical analysis or implications. Initial clinical use of oxygenators was for short-term open heart surgery, and direct contact of blood with the gas phase was allowed (1). Longer term lung prosthesis or assistance would be possible if the trauma associated with this direct contact could be avoided (2). For this reason, membrane oxygenators have been proposed where the flowing blood and gas phases are separated by a permeable membrane. The interposing of such a membrane results in several problems associated with the additional resistance to the transfer of O_2 and CO_2 compared with the direct contact oxygenators. Not only is there a resistance associated with the membrane itself, but the blood adjacent to the membrane approaches stagnant conditions. The zero flow membrane region results in diffusion controlled mass transfer which is generally more significant in dictating oxygenator size (and complexity) than is the membrane resistance. A recent proliferation of membrane oxygenator design innovations reflect attempts to augment diffusion with convection in the direction perpendicular to the flow (3,4,5). Whereas diffusion is an extremely efficient method of transfer over short distances (10^{-4} cm.) (6), it is ineffective in practical channel dimensions (0.1 cm.). Therefore, an understanding of diffusional processes in a flowing blood stream is of immediate practical importance.

ANALYSIS

There are several approaches which may be taken to analyze the transfer of chemical species in a flowing blood stream. Most involve the use of continuum concepts applied within channel dimensions many times larger than an individual red blood cell. The first, and more rigorous method, involves the application of a generalized convective-diffusion balance separately to the plasma and red blood cells. The general equation which is the basis for analysis is given as (see Nomenclature)

$$\frac{DC_i}{Dt} = -\nabla \cdot J_i + g_i \tag{1}$$

or, with expansion of the substantial derivative and definition of a diffusive flux

$$\frac{\partial C_i}{\partial t} + \text{div}(\vec{v}\, C_i) = \nabla \cdot D_i \nabla C_i + g_i \tag{2}$$

When applied to oxygen transfer with constant diffusivity in the continuous incompressible plasma phase, Equation 2 becomes

$$\frac{\partial C_p}{\partial t} + \vec{v} \cdot \nabla C_p = D_p \nabla^2 C_p \tag{3}$$

where it is assumed that there is no direct association of O_2 with plasma constituents such as proteins. Plasma therefore contains only physically dissolved oxygen. The application of Equation 2 to the stroma of the red blood cell can take several forms depending upon the mental construct one has of the membrane boundary and internal structure. If membrane permeability is considered independent of internal diffusion (constant diffusivity) and convection within the stroma is negligible, the following set of equations are obtained:

$$\frac{\partial C_1}{\partial t} = D_1 \nabla^2 C_1 + g_1 \tag{4}$$

$$\frac{\partial C_2}{\partial t} = D_2 \nabla^2 C_2 + g_2 \tag{5}$$

A single dissolved O_2 component has been assumed along with a single oxyhemoglobin specie. A more rigorous development is possible by including different forms of hemoglobin-oxygen combination as well as associated oxygen. However, there is not sufficient data at the present time to warrant this additional complexity. In the above formulation, boundary conditions would be required between the plasma and red blood cell for each independent variable differential order.

Spatial boundary conditions should include allowance for different solubilities in the plasma, cell membrane, and stroma. By assuming a continuous partial pressure gradient, concentrations could be stipulated as well as flux equality through the membrane. The membrane transfer might then be considered as pure diffusion of a dissolved species. Because of the small intracellular distances, equilibrium should not be assumed a priori, rather the kinetics of oxyhemoglobin formation and release must be included during rapid transfer (7). For time variable analysis, an initial concentration distribution is required. Several underlying assumptions in the analytical description may not be strictly valid although the more important known mechanisms are probably encompassed. If the hemoglobin within the red blood cell is highly organized (8), then the well defined separation of simple membrane resistance from internal diffusion with reaction is not valid. In addition, an isotropic homogeneous media is tacitly assumed. Dintenfass (9) has demonstrated possible internal motion within the red blood cell when exposed to high shear gradients. This observation may invalidate the above neglection of internal convection. Other similar problems will be discussed later. The above analysis has as its advantage the use of physically defined quantities such as the various diffusion coefficients, membrane permeability, and solubility coefficients in separate phases. However, not all of these values are readily available and the mathematical difficulties are severe.

A second method of analysis is based upon the inclusion of both red blood cells and plasma within a differential balance. Physical constants lose their separable identity in favor of "lumped" or "effective" values. However, the resultant model more closely coincides with the usual macroscopic experimental observations made within channels many times larger than the individual red blood cells. Blood is modelled as a single phase with two components, dissolved and combined O_2. The differential balance becomes:

$$\frac{\partial C_3}{\partial t} + \vec{V}\cdot\nabla C_3 = \nabla\cdot D_3 \nabla C_3 + g_3 \qquad (6)$$

and

$$\frac{\partial C_4}{\partial t} + \vec{V} \cdot \nabla C_4 = \nabla \cdot D_4 \nabla C_4 + g_4 \tag{7}$$

D_3 and D_4 would commonly be considered constant, and C_4 would be a known function of C_3 through either a lumped kinetic relation or the Bohr equilibrium when applicable. Since the diffusivities and reaction terms in Equations 6 and 7 are, at best, interpreted as effective values, a balance can be made to include all oxygen species (in our example, two) as

$$\frac{\partial}{\partial t}(C_3 + C_4) + \vec{V} \cdot \nabla (C_3 + C_4) = -\nabla \cdot (J_3 + J_4) \tag{8}$$

where, by conservation of oxygen, the reaction terms sum to zero. There are several excellent examples in recent literature of the application of Equation 8 to flowing blood systems. Computer solutions are generally required as well as assumed equilibrium (or near equilibrium) between C_3 (dissolved O_2) and C_4 (oxyhemoglobin). Mathematical descriptions of the Bohr equilibrium are available (10, 11, 12).

Weissman and Mockros solved Equation 8 for straight (13) and coiled (14) permeable tube flow using an empirical (exponential) equilibrium relation and a single effective diffusion coefficient defined by

$$-\nabla \cdot (J_3 + J_4) = D_e \nabla^2 C_3 \tag{9}$$

A similar approach, with consideration given to the effects of a non-Newtonian velocity profile, was taken by Buckles (15). In both cases, the steady state dimensional expression of Equation 8 reduces to

$$V_z \left(1 + \frac{\partial C_4}{\partial C_3}\right) \frac{\partial C_3}{\partial z} = \frac{D_e}{r} \frac{\partial}{\partial r}\left(r \frac{\partial C_3}{\partial r}\right) \tag{10}$$

Whereas Weissman utilized a constant wall concentration (zero wall resistance), Buckles included the more general wall flux matching condition. Another difference involved the treatment of D_e. Weissman, working with constant inlet saturation of 75%, used D_e to accomplish a match between theory and data for straight tubes.

The adequacy of a single value of D_e in this match as flow rate, tube diameter, and tube length were varied served as verification of their analytical approach. Buckles estimated the required physical parameters from available and measured data and then compared theory directly with data.

An analysis of tubular flow compared with experiment requires the integration of a variable transfer process over the axial distance. Therefore, the data represents average results possibly representative of multiple changing mechanisms within the lumen. For this reason, the work of Spaeth and Friedlander (16) is significant in that local phenomena could be better isolated for analysis and testing in a rotating disk transfer experiment. Sufficient estimation of physical parameters were performed to allow direct comparison of theory with data. As in the work of Weissman and Buckles, Spaeth exhibited a reasonable agreement of theory with data. In addition, partly due to the "local" nature of their work, Spaeth and Friedlander were able to test the concepts of local equilibrium and facilitated (variable D_e) transport. Their conclusions were that local equilibrium was valid in gross geometries common to clinical devices which are diffusion controlled and that facilitated transport in the O_2 transport system is probable. Facilitated transport had been previously reported by Wittenberg (17) and Scholander (18) in myoglobin and hemoglobin solutions under appropriate oxygen partial pressure conditions (partial saturation).

The following concept of O_2 transfer in blood is commonly employed and partly supported by the work above. Oxygen is dissolved within the plasma phase starting from either the human or artificial lung exposure to sufficient partial pressure. The solubility of O_2 in plasma is reduced compared to pure water by electrolyte-water association (salting effect) and protein-water binding, both passive processes. Absorption potential is not considered significant in terms of total O_2 content. Oxygen diffuses through the continuous plasma phase while simultaneously encountering the red blood cell membrane. Here, whether the cell membrane exhibits a significant resistance to the transfer of O_2 is questionable. Roughton (19) concluded that the membrane is a significant hindrance to oxygen reaching the stroma while Kreuzer (20) reaches the opposite conclusion. Kreuzer cites the crenation of individual red blood cells as an explanation of Roughton"s results. We tend to favor Kreuzer's analysis since oxygen is a relatively small molecule and the membrane resistance to transfer of water is negligible compared to other postulated O_2 transfer mechanisms. Once inside the RBC (red blood cell) stroma, O_2 undergoes simultaneous diffusion and reversible reaction. The diffusion includes both dissolved and bound oxygen. The solubility of oxygen is known to be enhanced in hemoglobin solution independent of the oxyhemoglobin reaction (21). If the overall processes within the RBC are more rapid than the external (tortuous) transport of O_2 by diffusion within the plasma, augmented transfer will occur. In particular, if the reaction kinetics of oxyhemoglobin formation are faster than diffusive transfer, and if there is sufficient communication between hemoglobin molecules, then significant increases in oxygen transfer can occur within partially oxygenated blood regions due to facilitated transfer. If membrane resistance was much greater than the total of internal (stroma) transfer hindrance, no effect of partial saturation would be noticeable. That is, if membrane resistance predominates over internal stroma processes, then there should be no effect of PO_2 on the observed transfer rate. The foregoing discussion is based upon purely physical reasoning, independent of synergistic mechanisms.

The additional variable of RBC orientation and/or movement (i.e., tumbling) in a flow field must be considered in convective systems. While it is implied from rheological studies that there are certain limited conditions which should result in significant mass transfer effects due to RBC motion and position (22), two reports from one laboratory refute this contention (15,23). The importance of RBC orientation may be difficult to determine from experimental transfer data. To demonstrate, Spaeth (24) calculated the effective O_2 blood diffusivity from the heterogeneous model of Fricke (25) with a diffusivity ratio (stroma:plasma) of 0.62 and compared his results to the "all parallel" and "all series" orientation from Fatt and La Force (26), shown here as Figure 1. The heterogeneous theory falls between the series and parallel calculations, which have a maximum separation of 6%. This difference is entirely too small to be measureable. The RBC stroma to plasma diffusivity ratio used in constructing Figure 1 is thought to represent inert transfer typical of diffusion through completely oxygenated blood, and compares favorably with measurements made with saturated blood (15, 27). If facilitated transfer becomes significant, the slope on Figure 1 would be progressively more positive as the saturation level of blood is decreased.

The studies which led to the present construct of blood oxygen transfer are useful in the design of practical artificial lung devices. However, because there are still unanswered questions, measurements and correlation with geometries analogous to the contemplated design are desirable. None of the cited references have considered the effects of simultaneous O_2 and CO_2 transfer (Bohr and Haldane shifts). Simplified analytical approaches would be of value in investigating the importance of simultaneous O_2 and CO_2 transfer. The constancy of the effective diffusion coefficient (13) and the observation that O_2 transfer rate was relatively unaffected by shifts in the Bohr curve (15) leads to the possibility of analytical simplification. Diffusion measurements reported in the literature are generally the result of either physical or geometric simplifications (i.e., hemolyzed versus whole blood and plasma, steady state transfer through stagnant thin films, etc.). Of particular interest are the reports of advancing front concepts (28, 29, 30) and analytical solutions pertaining to particular geometries (27, 31, 32). The degree of simplification and its basis is important, since the essential character of the transfer process must be maintained. A model based on significant axial streaming in permeable tube flow of red blood cells would have, as an assumption, the limit of radial plasma diffusion near the wall (33). Such a model is too gross to match blood oxygenation data over a wide range of variables. On the other hand, the

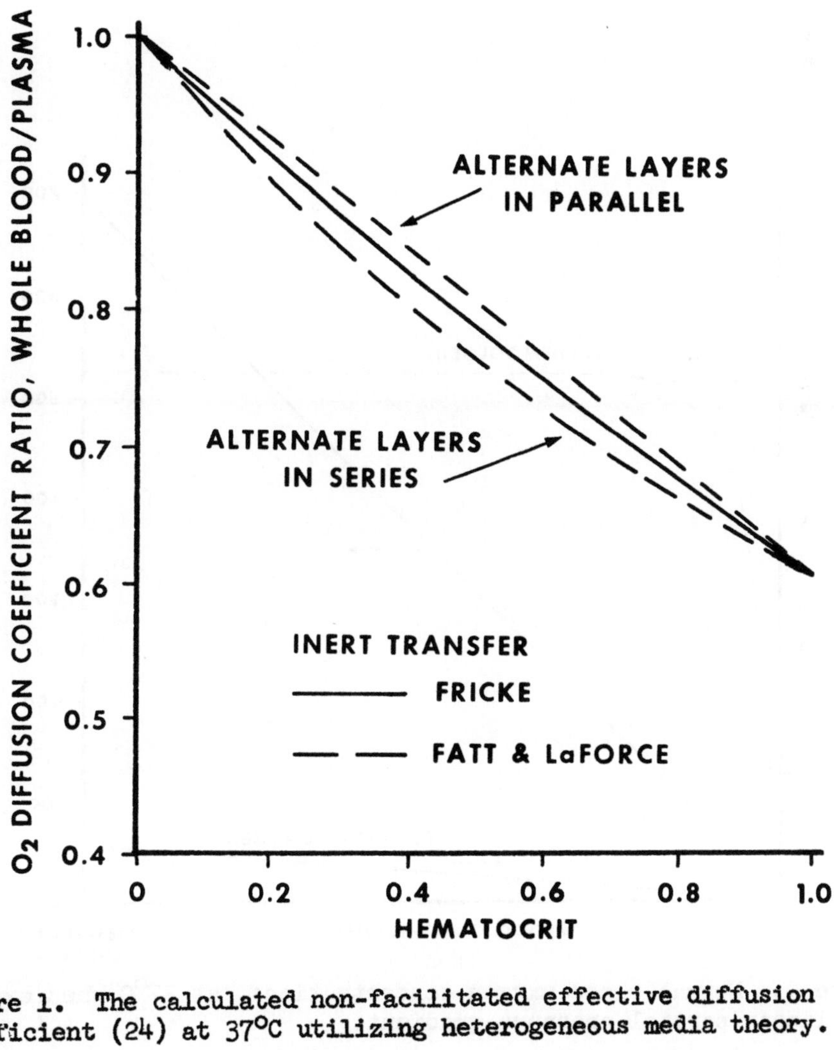

Figure 1. The calculated non-facilitated effective diffusion coefficient (24) at 37°C utilizing heterogeneous media theory.

advancing front simplifications should be well founded for blood oxygenation in artificial devices. An oxygen partial pressure of ~700 mm. Hg is common at the membrane surface exposed to blood. The oxygen saturation is nearly complete for all partial pressures greater than 100 mm. Hg, while the steep slope of the dissociation curve occurs between 10 and 60 mm. Hg. A typical equilibrium distribution of dissolved oxygen and oxyhemoglobin in a blood film is shown in Figure 2 when exposed to a linear O_2 partial pressure varying from zero to 700 mm. Hg.

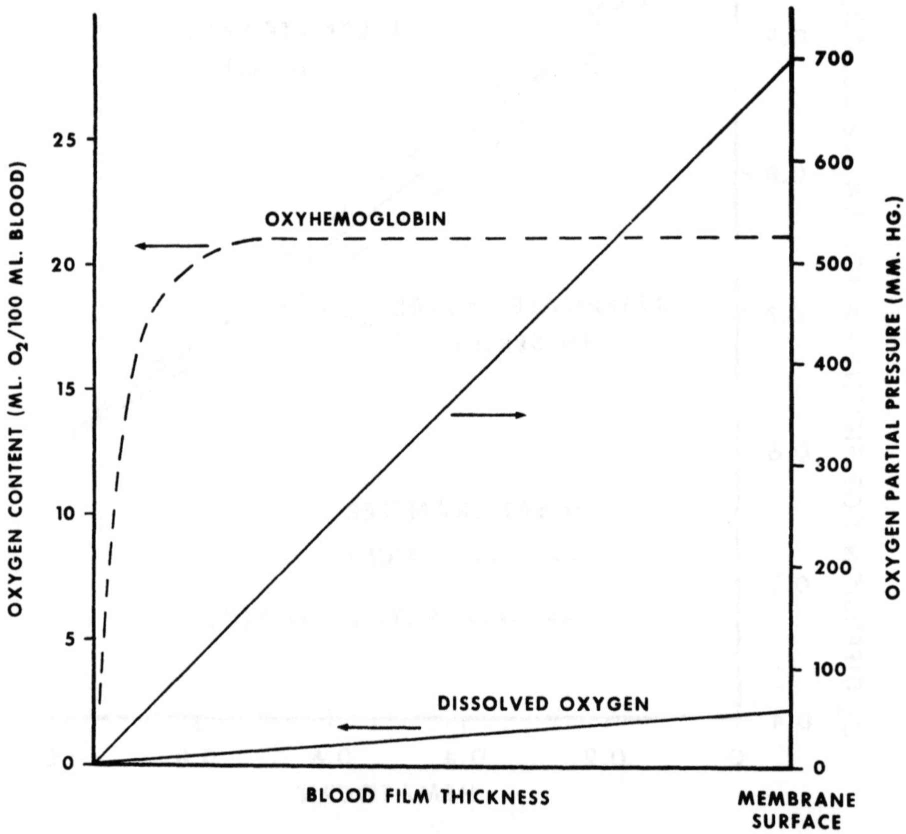

Figure 2. Normal blood oxygen concentrations, at 37°C when exposed to a linear partial pressure gradient.

Under normal arterial conditions, the oxyhemoglobin content is ~65 times the dissolved O_2 content. When exposed to large PO_2 gradients, blood behaves as if it were completely oxygenated at all high PO_2 values (behind the advancing front) while below the critical partial pressure blood oxygen level would remain at initial conditions until this hypothetical front moves inward. This movement is fed by diffusion inward through the fully oxygenated zone.

Advancing front theory has been utilized successfully in other fields of endeavor (34) with similar characteristics to blood oxygenation. However, little practical exploration of this concept has evolved in this latter application. The success of this approximation can be measured by the comparison of theoretical with experimental results. Although the idea behind advancing fronts can be applied to any geometry and flow field, the development here will be restricted to oxygenation of blood flowing within straight permeable tubes, and will not consider the simultaneous problem of CO_2 transfer. Following the note of Lightfoot (29), the blood channel in laminar flow is divided into a fully oxygenated region near the wall ($p < r < a_i$) and a core of unoxygenated (inlet condition) blood ($o < r < p$). Within the oxygenated zone heterogeneous diffusion predominates represented by

$$\frac{d}{dr}(rJ_r) = 0 \tag{11}$$

where

$$J_r = -D_5 \frac{dc}{dr} \tag{12}$$

Here, we deviate from Lightfoot in stipulating a wall matching flux condition (30) as

$$D_5 \frac{\partial c}{\partial r}\bigg|_{r=a_i} = D_m \frac{\partial c_m}{\partial r}\bigg|_{r=a_i} \tag{13}$$

along with the condition that $C = C_o = \alpha_b P_o$ @ $r = p$. Diffusion of dissolved oxygen in the permeable wall is described by equations similar to Equations 11 and 12. If the intermediate variable wall concentration C_w, is stipulated at $r = a_i$ then the oxygenated zone is described by

$$-J_r = \frac{D_5 (C_w - C_o)}{a_i \ln a_i / p} \tag{14}$$

A similar solution is obtained to describe the transfer of O_2 inside the membrane wall (from which C_w can be eliminated in favor of the constant gas concentration, C_g). The flow of unoxygenated hemoglobin within the central core is given as

$$\dot{m} = 2\pi C_{bo} \int_0^p V_z \, r \, dr \tag{15}$$

which can be integrated for a parabolic velocity profile to yield

$$\dot{m} = 4\pi C_{bo} \bar{V} \left[\frac{p^2}{2} - \frac{p^4}{4a_i^2} \right] \tag{16}$$

Equation 16 provides the cross-sectional flow of unaffected blood as a function of the venous core dimension, p. A local mass balance can be made between the reduction in the venous core and the oxygen flux evaluated at the wall.

$$-2\pi r J_r \Big|_{r=a_i} = -\frac{d\dot{m}}{dz} \tag{17}$$

Equation 17 can be utilized to determine p as a function of z once C_w is eliminated in Equation 14. Equation 13 is utilized along with the condition of partial pressure equality expressed as

$$\left.\frac{C_w}{\alpha_b}\right|_{\text{blood}} = \left.\frac{C_w}{\alpha_m}\right|_{\text{membrane}} \quad @\ r = a_i \tag{18}$$

Combining Equations 18, 17, 16, 14 and 13 yields the differential equation,

$$\left(\frac{P_g - P_o}{C_{bo}}\right)\frac{\alpha_b\, 2\, D_5}{\bar{V} d^2} = \left(\ln \eta - \gamma\right)\left(\eta - \eta^3\right)\frac{d\eta}{dz} \tag{19}$$

where

$$\eta = \frac{r}{a_i} \tag{20}$$

and

$$\gamma = \frac{D_5 \alpha_b \ln a_o/a_i}{D_m \alpha_m} = \frac{D_5 \alpha_b \ln a_o/a_i}{\mu} \tag{21}$$

Equation 19 can be integrated between the limits $z = 0$ and $\eta = 1$ at the extrance to $z = z$ and $\eta = \eta$ at an arbitrary position.

$$\left(\frac{P_g - P_o}{C_{bo}}\right)\frac{\alpha_b D_5 z}{\bar{V} d^2} = \frac{3}{32} + \frac{1}{4}\ln\eta\left(\eta^2 - \frac{\eta^4}{2}\right) - \frac{1}{8}\left(\eta^2 - \frac{\eta^4}{4}\right) - \frac{\gamma}{4}\left(\eta^2 - \frac{\eta^4}{2} - \frac{1}{2}\right) \tag{22}$$

where

$$\left(\frac{P_g - P_o}{C_{bo}}\right)\frac{\alpha_b D_5 z}{\bar{V} d^2} = \left(\frac{P_g - P_o}{C_{bo}}\right)\frac{\alpha_b z}{Re\, Sc\, d} \tag{23}$$

The fractional saturation of inlet hemoglobin can be expressed as

$$f = \frac{\dot{m}_o - \dot{m}}{\dot{m}_o} = 1 - 2\eta^2 + \eta^4 \tag{24}$$

where m_o is the entry rate of unoxygenated hemoglobin. The fractional saturation, f, is related to the absolute saturation, s, by

$$f = \frac{s(\eta) - s_o}{1 - s_o} \tag{25}$$

Comparison of the advancing front theory with data has been presented (30) for limited test conditions. Equation 22 can be solved for the diffusion coefficient, D_5, as

$$D_5 = \frac{\zeta}{\left(\frac{P_g - P_o}{C_{bo}}\right)\frac{\alpha_b \pi NL}{4Q} - \left(\frac{\alpha_b \ln a_o/a_1}{4\mu}\right)\beta} \tag{26}$$

where

$$\zeta = \frac{3}{32} + \frac{1}{4}\ln \eta\left(\eta^2 - \frac{\eta^4}{2}\right) - \frac{1}{8}\left(\eta^2 - \frac{\eta^4}{4}\right) \tag{27}$$

and

$$\beta = \frac{1}{2} - \eta^2 + \frac{\eta^4}{2}$$

Equation 26 is presented for the general case of N parallel tubes of length L with a total flow rate, Q. In the solution of the diffusion coefficient, f, ζ and β are all functions of η . The experimental value of f is therefore used to calculate both ζ and β . When C_{bo} and α_b were not reported, the following approximations were utilized

$$C_{bo} = 1.34(35H)(1 - s_o) \tag{28}$$

and (21)

$$\alpha_b = \alpha_{rc}H + \alpha_p(1 - H) \tag{29}$$

EDITORS NOTE: Clinical results were contained in the original paper. Also in the original paper computer solutions of the governing equations were compared with the experimental data.

SUMMARY OF RESULTS

Theoretical approaches to the analysis of O_2 and CO_2 transfer to blood flowing in a straight channel are presented and discussed. Computer solution of a detailed macroscopic balance which assumes chemical equilibrium has been shown to be successful under limited conditions. However, the required effective diffusivity can only be approximated a priori. A simplified advancing front theory shows promise as a method of analyzing and interpreting experimental data obtained from finite channel lengths. Again, the effective diffusion coefficient cannot be accurately predicted. Part of the problem is assigned to experimental difficulties, and recommendations are made for future studies. In general, unless significant increases can be made in experimental accuracy, investigations made in finite channels have limited ability to produce fundamental data. The concept of facilitated transport and its quantification within a blood channel is yet to be definitely proven, although evidence indicates its existence.

The application of continuum mechanics seems valid for the design of clinical blood oxygenators. However, until a better understanding of the basic transfer processes in blood is obtained, we are analytically restricted to simple geometries. In order to achieve clinical efficiencies complex designs are being explored. Due to lack of theoretical support, these oxygenator systems are being developed on an empirical basis. The common denominator in modern membrane devices is the gentle production of convective currents perpendicular to the main stream flow which augments diffusion. Applicable theory would greatly enhance clinical progress by assuring maximum transfer efficiency. Increased efficiency appears to be important in long term perfusion since exposed area and prime volume is thereby minimized. Even empirical designs show promise for prolonged lung assistance.

NOMENCLATURE

a_i = inside tube radius

a_o = outside tube radius

C_{bo} = concentration of unsaturated hemoglobin at the tube inlet

C = concentration

C_i = concentration of component i

C_p = concentration of dissolved O_2 in plasma

C_1 = concentration of dissolved O_2 in RBC stroma

C_2 = concentration of oxyhemoglobin in RBC stroma

C_3 = concentration of dissolved O_2 in whole blood

C_4 = concentration of oxyhemoglobin in whole blood

C_g = concentration O_2 in the membrane at $r = a_o$

C_o = inlet dissolved O_2 concentration

C_m = concentration of O_2 in the membrane material

C_w = concentration of dissolved O_2 at the wall (a_i)

d = tube inside diameter

D_i = diffusion coefficient of component i

D_p = diffusion coefficient of O_2 in plasma

D_1 = diffusion coefficient of dissolved O_2 in RBC stroma

D_2 = diffusion coefficient of oxyhemoglobin in RBC stroma

D_3 = diffusion coefficient of dissolved O_2 in whole blood

D_4 = diffusion coefficient of oxyhemoglobin in whole blood

D_5 = diffusion coefficient of O_2 in the radial direction

D_e = effective diffusion coefficient in whole blood

D_m = diffusion coefficient of O_2 in the membrane material

f = fractional saturation compared to inlet conditions

H = hematocrit

J_i = flux of component i

J_3 = flux of dissolved O_2 in whole blood

J_4 = flux of oxyhemoglobin in whole blood

J_r = flux in the radial direction

L = tube length

$\dot{m}$ = cross sectional rate of flow of unsaturated hemoglobin

N = number of tubes in parallel

p = radial position at which inlet conditions cease

P_o = partial pressure of inlet blood

P_g = partial pressure in the gas phase

PO_2 = partial pressure of O_2

PCO_2 = partial pressure of CO_2

q_i = reactive production rate of component i per unit volume

q_1 = reactive production rate of dissolved O_2 in RBC stroma

q_2 = reactive production rate of oxyhemoglobin in RBC stroma

q_3 = reactive production of dissolved O_2 in whole blood

q_4 = reactive production of oxyhemoglobin in whole blood

Q = total flow rate

r = radius

Re = Reynolds number

Sc = Schmidt number

s_o = inlet saturation

$s(\eta)$ = saturation corresponding to η

s_e = exit saturation

t = time

v = velocity

v_z = velocity in the axial direction

$\bar{v}$ = average velocity

z = axial distance

α_b = solubility of O_2 in blood (expressed as ml. O_2 (STP) per 100 ml. blood per atm.)

α_m = solubility of O_2 in the membrane material

α_p = solubility of O_2 in plasma

α_{rc} = solubility of O_2 in RBC stroma

β = defined function of η

γ = dimensionless tube wall material resistance

μ = permeability of tube wall material

η = p/a_i, fraction of tube radius occupied by inlet condition

ζ = defined function of η

π = 3.14159

LITERATURE CITED

1. Galletti, P.M., and Brecher, G.A., "heart-Lung Bypass," Grune and Stratton, New York (1962).
2. Dorson, W. Jr., Baker, E., and Hull, H., "Trans. Amer. Soc Artif. Int. Organs, 14, 242 (1968).
3. Rush, B.F. Jr., Boone, R., and Malette, W., Trans. Amer. Soc. Artif. Int. Organs, 15, 178 (1969).
4. Lande, A.J., Parker, B., Subramanian, V., Carlson, R.G., and Lillehei, C.W., Trans. Amer. Soc. Artif. Int. Organs, 14, 227 (1968).
5. Kolobow, T., and Bowman, R.L., Trans. Amer. Soc. Artif. In. Organs, 9, 238 (1963).
6. Riggs, D.S., "The Mathematical Approach to Physiological Problems," p. 179, Williams and Wilkins Co., Baltimore (1963).
7. Friedlander, S.K., and Keller, K.H., Chem. Eng. Sci., 20, 121 (1965).
8. Whittam, R., "Transport and Diffusion in Red Blood Cells," Williams and Wilkins Co., Baltimore (1964).
9. Dintenfass, L., Acta haemat., 32, 299 (1964).

10. Adair, G.S., J. Biol. Chem., 63, 529 (1925).

11. Margaria, R., Clin. Chem., 9, 745 (1963).

12. Hill, A.V., J. Physiol. (London), 40, 4p (1910).

13. Weissman, M.H., and Mockros, L.F., J. Eng. Mech. Div., A.S.C.E., 93, 225 (1967).

14. Weissman, M.H., and Mockros, L.F., J. Eng. Mech. Div., A.S.C.E., 94, 857 (1968).

15. Buckles, R.G., Merrill, E.W., and Gilliland, E.R., A.I.Ch.E. Journal, 14, 703 (1968).

16. Spaeth, E.E., and Friedlander, S.K., Biophys. J., 7, 827 (1967).

17. Wittenberg, J.B., J. Gen. Physiol., 49, 57 (1965).

18. Scholander, P.F., Science, 131, 585 (1960).

19. Roughton, F.J.W., Progr. Biophys., Biophys. Chem., 9, 55 (1959).

20. Kreuzer, F., and Yahr, W.Z., J. Appl. Physiol., 15, 1117 (1960).

21. Sendroy, J. Jr., Dillon, R.T., and Van Slyke, D.D., J. Biol. Chem., 105, 597 (1935).

22. Dorson, W.J. Jr., and Hershey,D., Dig. 7th. Inter. Conf. Med. Biol. Eng., 374 (1967).

23. Colton, C.K., Ph.D. thesis, Mass. Inst. Tech., Cambridge (1969).

24. Spaeth, E.E., Ph.D. thesis, Calif. Inst. Tech., Pasadena (1967).

25. Fricke, H., Physical Rev., 24, 575 (1924).

26. La Force, R.C., and Fatt, I., Trans. Far. Soc., 58, 1451 (1962).

27. Hershey, D., and Karhan, T., A.I.Ch.E. Journal, 14, 969 (1968).

28. Marx, T.I., Snyder, W.E., St. John, A.D., and Moeller, C.E., J. Appl. Physiol., 15, 1123 (1960).

29. Lightfoot, E.N., A.I.Ch.E. Journal, 14, 669 (1968).

30. Dorson, W. Jr., Baker, E., Cohen, M.L., Meyer, B., Molthan, M., Trump, D., and Elgas, R., Trans. Amer. Soc. Artif. In. Organs, 15, 155 (1969).

31. Miller, C.J., M.S. thesis, Univ. Cincinnati, Cincinnati (1965).

32. Landino, E., McCreary, J.G., Thompson, W.A., and Powers, J.E., A.I.Ch.E. Journal, 12, 117 (1966).

33. Dorson, W.J., Jr., Proc. Conf. Eng. Med. Biol., 20, 29.1(1967).

34. Bird, R.B., Steward, W.E., and Lightfoot, E.N., "Transport Phenomena," John Wiley, New York (1960).

35. Buckles, R.G., Ph.D. thesis, Mass. Inst. Tech., Cambridge(1966).

36. Weissman, M., Ph.D. thesis, Northwestern Univ., Evanston(1967).

37. Van Slyke, D.D., and Neill, J.M. J. Biol. Chem., 61, 523 (1924).

38. Goldstick, T.K.,Ph.D. thesis, Univ. Calif., Berkeley (1966).

39. Power, G.G., J. Appl. Physiol., 24, 468 (1968).

40. Hedley-Whyte, J., and Laver, M.B., J. Appl. Physiol., 19, 901 (1964).

41. Goldstick, T.K., Allen, B.J., and Fry, W.A., Proc. 8th. Inter. Conf. Med. Biol. Eng., 1.2 (1969).

42. Dorson, W.J., Jr., Baker, E., Hull, H., Molthan, M., Meyer, B., Fargotstein, R., and Cohen, M.L., Ann. Thorac. Surg., 8, 297 (1969).

INDEX

MIX
Papier aus verantwortungsvollen Quellen
Paper from responsible sources
FSC® C105338

If you have any concerns about our products,
you can contact us on
ProductSafety@springernature.com

In case Publisher is established outside the EU,
the EU authorized representative is:
Springer Nature Customer Service Center GmbH
Europaplatz 3, 69115 Heidelberg, Germany

Printed by Libri Plureos GmbH
in Hamburg, Germany